Series in BioEngineering

The Series in Bioengineering serves as an information source for a professional audience in science and technology as well as for advanced students. It covers all applications of the physical sciences and technology to medicine and the life sciences. Its scope ranges from bioengineering, biomedical and clinical engineering to biophysics, biomechanics, biomaterials, and bioinformatics.

More information about this series at http://www.springer.com/series/10358

Amine Nait-Ali
Editor

Biometrics under Biomedical Considerations

Springer

Editor
Amine Nait-Ali
Laboratoire Images, Signaux et Systèmes Intelligents
University Paris-Est Créteil (UPEC)
Vitry-sur-Seine, France

ISSN 2196-8861 ISSN 2196-887X (electronic)
Series in BioEngineering
ISBN 978-981-13-1143-7 ISBN 978-981-13-1144-4 (eBook)
https://doi.org/10.1007/978-981-13-1144-4

Library of Congress Control Number: 2018957673

This Springer imprint is published by the registered company Springer Nature Singapore Pte Ltd.
The registered company address is: 152 Beach Road, #21-01/04 Gateway East, Singapore 189721, Singapore

Preface

In general terms, Biometrics can be defined as a technology which recognizes individuals by their physical or behavioural characteristics. This common definition is often employed by considering Biometrics as a security solution to deal with the identification or verification of individuals. However, one can raise the following question: Is there a reason to specifically use the term "Biometrics" within this context? Probably, the answer would be a yes. Nowadays, Biometric technology is not restricted to law enforcement or government departments by identifying criminals or by controlling borders, but it is also accessible and used in almost everyday life. It allows a more secure access as well as making our life easier. Due to the fact that Biometric systems are commonly integrated into general purpose devices such as smartphones and computers, the attitude of the general public has changed towards them because it has become more acceptable. In addition, commercial Biometric industry is still growing and is continually being innovated, so it is not surprising the conventional metal key and passwords will not be used anymore as we know it today. In fact, Biometrics has been integrated into many products to make them smart, simple, and secure. Consequently, this has certainly had an impact on the way the term Biometrics is employed. Within this context, if an average person on the street is asked about their knowledge on biometrics, the answer can roughly range from a collection of keywords such as security, fingerprint, voice recognition to face recognition. However, Biometrics does not explicitly refer to security; there are many scientific communities that deal with Biometrics from a statistical point of view mainly because it considers the analysis and the study of measurable biological characteristics applied in fields such as agriculture and medicine. Furthermore, the term Biometrics once split into two halves, "Bio" and "Metrics", each half then highlights a clear and a significant meaning whereby "Bio" refers to "Biology" and "Metrics" refers to "Measurement". Therefore, if one achieves a measurement on any living being, it can be also considered as Biometrics whichever the purpose of the application field is, and this includes security, medicine and healthcare, sport, gaming, and agricultural botanical or veterinary sciences.

This book highlights an aspect of the "Philosophy" discussed above. The main purpose is to connect Biometrics to Biomedical engineering from different point of views, including providing an overview on the different applications used in this field. This book also explains the borders between both fields. It provides a clearer understanding of some grey areas by giving a clear-cut distinction between the two fields. For example, Biometrics such as in security Biometrics may require some biomedical knowledge and prior information to reach the final decision provided by biometric systems. In contrast, some biomedical engineering applications may share a scheme almost as similar as the one used for security Biometrics except that the purpose of the last block of the generic scheme is to identify specific pathologies rather than a person's identity. In some other situations, Biometrics and Biomedical engineering may complement each other leading to some specific and useful applications. In conclusion, Biometrics has significantly evolved over the last decades, so its terminology should correspond to the pace in order to prevent any confusion or misinterpretation. For instance, terminologies such as medical Biometrics, security Biometrics, sport Biometrics, and gaming Biometrics ought to be used to clearly define the purpose of each application.

The book contains thirteen chapters in which five sections are addressed, including the influence of pathologies on some security Biometric modalities like face, iris, and fingerprint identification, medical and security biometrics, and behavioural biometrics. Finally, instrumentation wearable technologies and imaging and some case studies are presented in the last chapters.

Entitled, *Biometrics under Biomedical Considerations*, the book is organized as follows:

Chapter "Influence of Skin Diseases on Fingerprints". This chapter deals with fingerprint modality from a biomedical point of view. In particular, it considers the cases where a user's fingers suffer from a dermatologic disease leading to damage of ridges. Such situations have a negative impact on both identification and verification processes.

Chapter "Iris Recognition in Cases of Eye Pathology". The purpose of this chapter is to provide an insight into iris recognition system performances for users suffering from some ocular pathologies. Such situations are analysed in terms of influences, and possible means are discussed to take this case into account during the phase of iris samples matching.

Chapter "Facial Rejuvenation Modeling". Face modelling is discussed in this chapter. It describes the process of normal facial growth trajectory within the categories of different age groups. The de-ageing study is based on a 2D generic face model highlighted by providing numerous illustrations showing digital face rejuvenation from adulthood to childhood appearance. At the end of this chapter, an extension of a 3D generic model is proposed.

Chapter "Facial Ageing Modeling". In this chapter, facial biometrics is discussed from the ageing point of view. Both geometrical and textural aspects are discussed. This challenging biometrics can be useful in numerous applications, including security, biomedical engineering, plastic surgery, and forensics.

Chapter "Lifestyle Facial Ageing Models". This chapter is an extension of the previous chapter whereby digital face ageing is explained by taking into account lifestyle factors. The objective is to be able to predict the biometric facial appearance by including some prior and behavioural information like addictions of some drugs, alcohol, and smoking.

Chapter "3D Face Analysis for Healthcare". It describes 3D face acquisition aspects and protocols. It discusses computer-aided three-dimensional face biometrics in health for surgery, prosthesis, ophthalmology, dermatology, and cosmetics purposes.

Chapter "Facial Biometrics: From Security to Biomedical Applications". The aim of this chapter is to bridge the gap between classical facial recognition approaches which are commonly considered for identification or verification, i.e. security applications with some extended or adjusted systems related to Biomedical engineering field.

Chapter "Eye Movement Analysis in Biometrics". In this chapter, eye movement in biometrics is addressed. An overview of acquisition aspects and applications are discussed, and it considers using this behavioural modality for recognition purposes.

Chapter "Biometric Legged Locomotion Fundamentals". This chapter can be classified into the category of behavioural Biometrics. The general purpose approach is to either analyse and identify pathologies oridentify individuals from their gait ground reaction force signals. Based on a classical biometric scheme, acquisition, analysis, preprocessing, and feature extraction are discussed.

Chapter "Wearable Technologies in Biomedical and Biometric Applications". The purpose of this chapter is to discuss the challenging and emerging technology, namely "wearable technology". Based on bio-signal acquisition, numerous applications in biomedical and biometrics are discussed. After highlighting some available systems in the market, the chapter focuses on a particular technology using an armband. Biometrics is then considered from different points of view.

Chapter "Biometrics from Cellular Imaging". In this chapter, cellular imaging is considered from medical Biometrics of cells. In particular, the chapter brings together different aspects of the cellular imaging from microscopy to cell biology, and from image processing to genomics.

Chapter "Medical Devices Design: Vital Sign Monitoring Case Study". It presents the proof of the concept of vital signs monitoring devices designed and conceived in laboratories. These types of designs can be used or adjusted in many applications related to Biometrics such as in medical Biometrics, clinical Biometrics, and for bio-signals authentication purpose.

Chapter "From Medical to Security Biometrics: Some Case Studies". This chapter focuses on some study cases involving medical image processing. The first case study specifically describes one kind of medical Biometrics applied to mammography images. The same concept is then extended to security Biometrics particularly on hidden Biometrics.

This book has been edited thanks to the efforts of a number of international researchers working in both Biometrics and Biomedical engineering fields. It is suitable for final-year undergraduate students, postgraduate students, engineers, and researchers in the field of computer engineering, information technology, Biometrics, Biomedical engineering and applied digital signal and image processing.

Vitry-sur-Seine, France Prof. Dr. Amine Nait-Ali

Contents

Influence of Skin Diseases on Fingerprints

Martin Drahanský and Ondřej Kanich

Abstract This chapter deals with fingerprint modality from a biomedical point of view. In particular, it considers the cases where a user's fingers suffer from a dermatologic disease leading to damage of ridges. Such situations have a negative impact on both identification and verification processes.

1 Introduction

Fingerprints are a well-known topic in the field of biometrics. Therefore it is surely not necessary to have a long introductory part here. On the other hand, some basic facts in a general scope regarded to biometrics have to be mentioned. Fingerprints were the leader in biometric technologies for a long time. The reasons are obvious—fingerprints are used from the 19th century in criminal police praxis as an evidence and can confirm unambiguously the identity of a person. Furthermore, we know that our fingerprints are unique, what was confirmed by experimental and theoretical research works in the past. The acceptance by users is pretty high, because we are using it in our daily lives (e.g. smart phones, access control systems). Other biometric technologies are tightening the market share of fingerprints in the last years. New technologies for iris recognition, 3D face, hand/finger veins recognition etc. are very popular and report very promising error rates and entropic information (i.e. suitability for use for large population). The trend of biometrics is going to solutions, which are not bothering the users, i.e. technology called "on the fly". This means that you just pass the biometric device and nearly without any action your biometric characteristic will be acquired and you will be recognized (preferred method is identification). This is possible for 2D/3D face, iris, and newly for fingerprints as well. We can find some pioneer solutions on the market, which enable to acquire the fingerprints on your hand just during passing the device. The user just only needs to

M. Drahanský (✉) · O. Kanich
Faculty of Information Technology, Centre of Excellence IT4Innovations,
Brno University of Technology, Brno, Czech Republic
e-mail: drahan@fit.vutbr.cz

A. Nait-Ali (ed.), *Biometrics under Biomedical Considerations*,
Series in BioEngineering, https://doi.org/10.1007/978-981-13-1144-4_1

move the hand in appropriate position and speed in a concrete area. Everything else is done automatically and the user can pass the gate or is rejected.

Nearly all the fingerprint technologies actually available on the market have very precise recognition rates and many of them have built-in anti-spoofing (liveness detection) mechanisms with the aim to detect a fake or a non-living sample. However nearly all of these systems count with a fingerprint with an appropriate quality. If the quality is low, the fingerprint is immediately rejected. What should the users with diseased fingers do? Their fingers attacked by any dermatologic disease have skin exposures leading to damage of ridges (papillary lines) and therefore unsuccessful attempt of identification or verification. And this is the topic which we want to deal with in this chapter.

2 Fingerprints

We will start with a very short introduction to history of fingerprints. The first evidence of the use of biometrics comes from the 19th century. At that time fingerprints began to be used in crime investigation. The preserved written materials regarded to biometrics are originating from the following authors: *William James Herschel* (1858—was the English Governor of India (Hooghly) and began to use fingerprints for track-recorders to confirm people's identity [1]); *Francis Galton* (1865 to 1892—a study of the inheritance of physical properties, eugenics, anthropometry and fingerprints [2, 3]); *Alphonse Bertillon* (1882—anthropometry called *Bertillionage*). In 1924 the Federal Bureau of Investigation (FBI) was established and started to use an automated fingerprint identification system (AFIS) with 810,000 fingerprints in 1965. In 2010 the AFIS at the FBI contained a total of 66 million ten-prints of fingerprints (fingerprints from each hand); on average, there was done 162,000 searches per day and the response to remote search in the database was approximately 1 h and 10 min (processing an electronic request in an urgent case took 10 min). In 2017 (September) the next generation identification (NGI) at FBI contained a total of almost 120 million ten-prints of fingerprints; on average there was 210,358 ten-print fingerprints processed per day; criminal response time (urgent) was on average 9 min and 58 s and civil response time was on average 36 min and 11 s [4].

Each of us has ridges (papillary lines) on the surface of our hand palms and feet soles. The structure of ridges clearly determines the physical identity of a concrete person. The only exception could be found in people with various types of illnesses and skin disorders. Ridges have their own graphic representation—this is the *fingerprint*. Ridges are formed during the embryonic development. The structure of ridges is unchanged during our life, unless they get damaged. Each finger is a unique pattern, i.e. there are no two identical fingers in the world [5].

Names associated with the history of fingerprints: *Nehemiah Grew* (1684—the study of papillary lines and the location of sweat pores on hands); *Johann Christoph Andreas Mayer* (1788—a work on the uniqueness of fingerprints for each individual); *Jan Evangelista Purkyně* (1823 – a work on 9 classes of fingerprints);

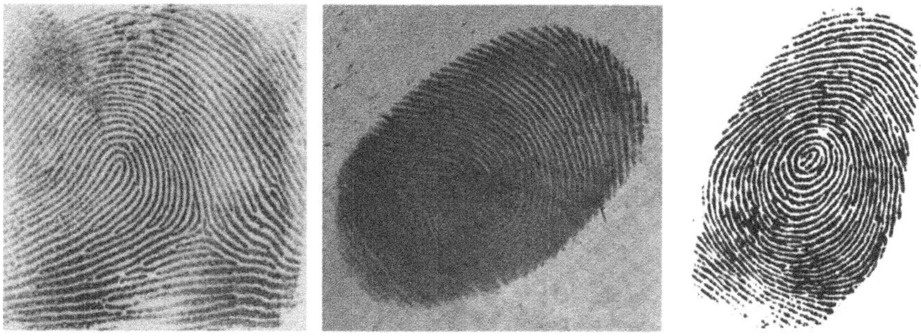

Fig. 1 Fingerprints (from left): rolled, latent, plain

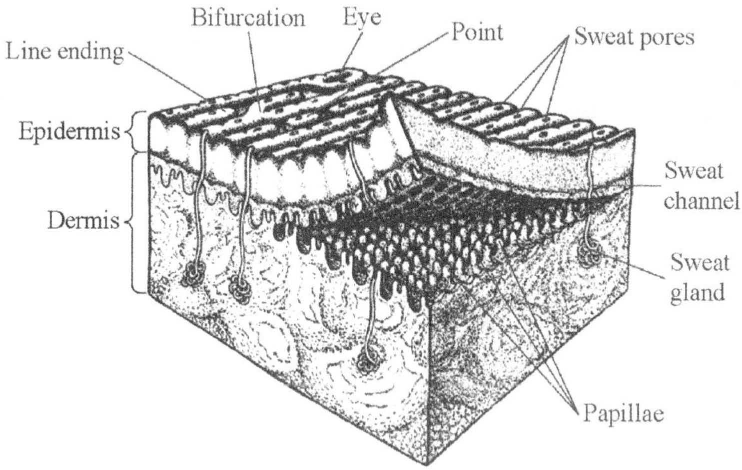

Fig. 2 Cut of the skin with papillary lines [14]

Henry Faulds (1880—a work on the usability of fingerprints for the purpose of identifying people); *Juan Vucetich* (1891—introduction of the use of fingerprints for criminalistics purposes).

In the dactyloscopic practice, we can see the following three kinds of fingerprints that differ not only in appearance and location, but also in the way of scanning (Fig. 1): *rolled*, *latent* and *plain*.

If we speak about fingerprints, we have to start with *ridges* (*papillary lines*)—(see Fig. 2), a fingerprint is a pattern formed by a structure of ridges. The height of the papillary lines lies between 0.1 and 0.4 mm and the width of the papillary lines is ranging from 0.2 to 0.5 mm.

Considering the global shape, the ridges in the fingerprint create a pattern called the *fingerprint class* [6]—we distinguish the following fingerprint classes: *arch*, *tented arch*, *whorl* (spiral), *left loop*, *right loop*. For classification of fingerprints in AFIS systems, the *Henry's classification scheme* is used. The classification is used,

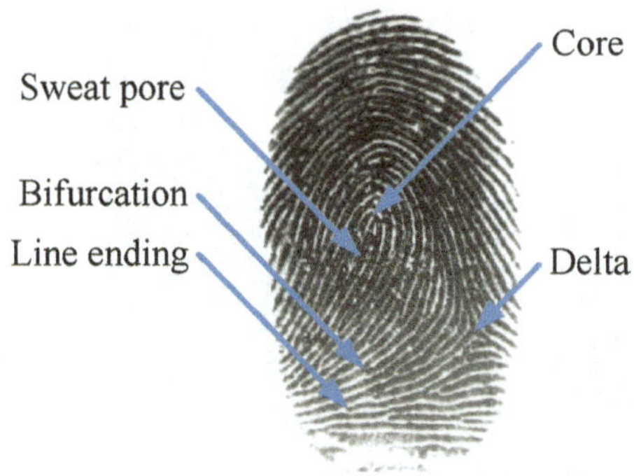

Fig. 3 Significant features of a fingerprint

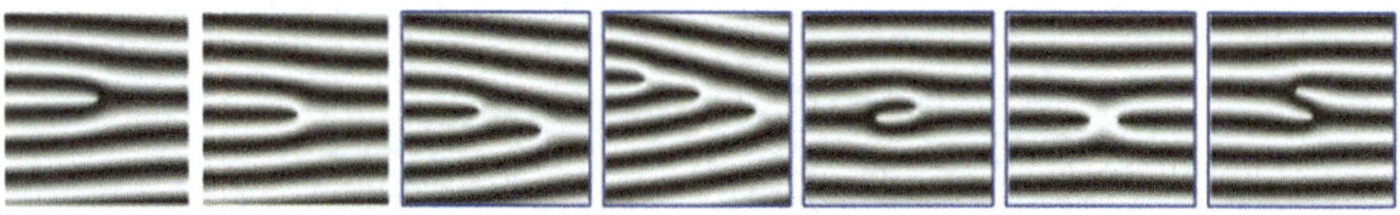

Fig. 4 Basic selected types of minutiae (left two: line ending and bifurcation)

as mentioned above, for dactyloscopic purposes, to classify fingerprints into smaller subclasses so that it would not be necessary to search the whole database to find the identity of the criminal. The fingerprint classification algorithm is based on the information contained in the fingerprint. Basic terms related to the classification (see Fig. 3) [6]:

- *Delta*—a fingerprint location where the papillary line runs in three directions.
- *Core*—the center of the fingerprint, located at the lowest arch of the run of papillary lines in the fingerprint.

Fingerprints are recognized on the basis of special patterns that form ridges—they are called *minutiae* points. The basic minutiae include (see Fig. 4, respectively): *line ending*, *single fork* (*bifurcation*), double fork, triple fork, hook, cross, side contact and many others. In dactyloscopic systems a lot of minutiae points is used, only two types are used for access systems—line ending and bifurcation.

2.1 Fingerprint Acquisition

In the past, there were used only fingerprinting methods using ink, nowadays clean fingerprinting (use of chemical solution for making fingerprints visible). These methods are known from the police praxis. Till today this method is used to have stored so called dactyloscopic cards (see Fig. 5) on nearly all police departments. The dactyloscopic card consists of the description margins and the fingerprints themselves—the

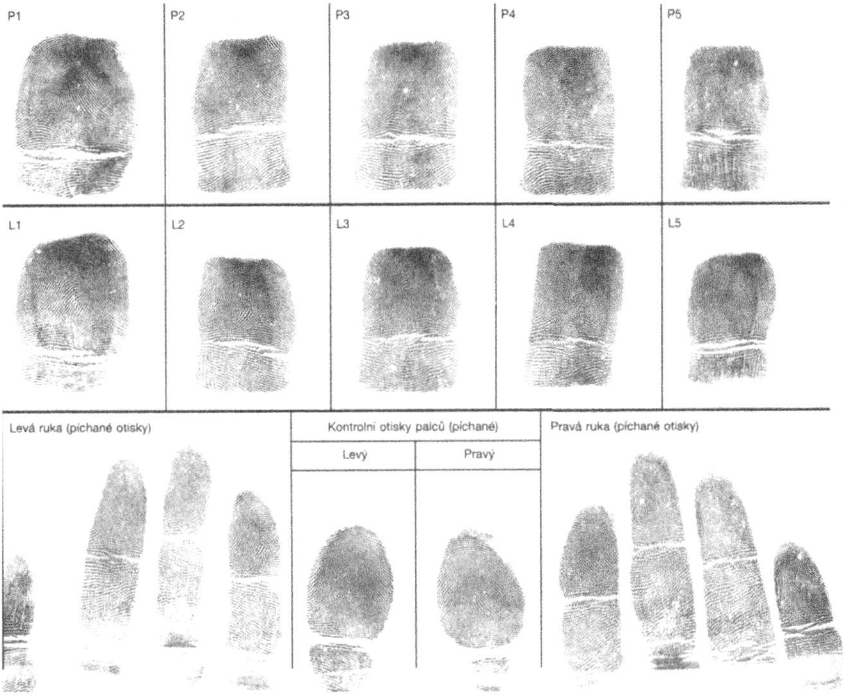

Fig. 5 Example of a Czech dactyloscopic card (without description margins)

upper part prints all 10 fingers in the form of rolled fingerprints and the control (plain) fingerprints of the fingertips are printed at the bottom. On the other side of the dactyloscopic card there are palm-prints. The only change is the use of electronic fingerprint scanners, which allow to acquire all fingerprints, incl. rolled version.

Nevertheless, the mostly used fingerprinting methods use one of the following electronic sensor technologies [17]: *optical*, *capacitive*, *RF*, *ultrasonic*, *e-field*, *electro-optical*, *pressure*, *thermal*, *MEMS*, *optical coherence tomography* and *optical TFT* (e.g. Jenetric solution). These technologies will be briefly described in the following subsections.

2.1.1 Optical Technology

The light source (often LED) illuminates the finger surface that is attached to the sensor's glass surface (there are also contactless 3D optical sensors, i.e. the finger has not to touch the surface in all cases—one example is the contactless fingerprint reader from the Swiss company Touchless Biometric Systems) and the camera scans the image. Another type of optical sensor uses optical coherence tomography (OCT)

[7]. It is very expensive but it gets the image from a deeper layer of the skin, which is harder to spoof. It can also obtain a view where sweat pores are clearly visible.

2.1.2 Capacitive Technology

The sensor is composed of a matrix of small conductive areas on which a layer of non-conductive silica dioxide is applied. The fineness of these conductive surfaces is higher than that of the papillary lines. By attaching a finger, capacitors (one electrode is the papillary line at the point of contact with the face of the sensor matrix and the other is just matrix plane) make the output of which is the value corresponding to the overlap of the surface area. To create swipe sensor one of the dimension is shortened. A related technology is so-called e-field technology.

There is one modification of the capacitive sensor that is worth mentioning. It is a combination of e-field and capacitive technology. It uses a low *radio frequency* (RF) signal and because of that it is often known as *RF technology*. This signal is sent to the skin, and because of that, between the signal references plane and the live (conductive) layer of the skin an RF electric field is created. Its equipotential contours mimic the shape of the live layer of the skin so when it is measured by an antennae array we get a fingerprint image.

2.1.3 Other Technologies

The *ultrasonic technology* is based on a rotating ultrasonic transducer, which incorporates a receiver. This rotates along the circular path and scans the fingerprint. Ultrasonic waves penetrate even under the surface of the skin. This technology can easily detect fake fingers.

The sensor based on *electro-optical technology* consists of four layers, with the finger pressure eliciting contact of the black coaxial layer of emitting light in the phosphor layer. This radiation passes through the base layer into the sensor.

The *pressure technology* sensor consists of three layers, with a non-conductive gel inserted between the electro-conductive layers. By attaching a finger to the sensor surface, the non-conductive gel is pressed at the point of contact with the papillary lines and brings both electro-conductive layers in touch.

The *thermal technology* principle is based on thermal radiation. The papillary lines have a higher heat radiance than the valleys between them. The finger is swept over a pyroelectric cell that senses this thermal radiation.

In *e-field technology*, the sensor consists of a drive ring and a matrix of antennas. The drive ring generates a sinusoidal radio frequency signal and the matrix of active antennas receives that signal modulated by the skin structure, or, more precisely, modulated by the dermis structure, because the electric field passes the upper parts of the skin (the epidermis).

The *MEMS* (Micro-Electro-Mechanical-System) *technology* uses micro parts to scan a fingerprint. One of the methods uses piezo-resistive micro beams. The user sweeps his finger along the sensor which consists of three rows of piezo-resistive gauges. Their parallel deflection will create a voltage variation which is measured and transformed into the fingerprint. The resulting image is only binary-colored which is the big disadvantage of this type of sensor technology.

2.2 Fingerprint Recognition

The biometric system is used in two modes—*registration* and *verification/identification* (generally *authentication*). In *registration* the user usually has to place the registered finger more times on the sensor surface, and the system calculates the average of these fingerprints to produce the template with a high quality. Comparison (*authentication*) works in a similar way—there is done a matching against the stored template from the database. The fingerprint recognition process is shown in Fig. 6. This process consists of the steps described in the following text.

The first step of fingerprint processing is to *capture* (*acquire*) an input image, preprocess and extract ridges. Through this process, a proper description of the course of the ridges is obtained from the fingerprint image and can be further processed. There is a large amount of noise in the input image, which requires subsequent *image enhancement*. When scanning, it is necessary to distinguish between rolled/plain and possibly latent fingerprints. In addition, it is necessary to take care of the effects of damaged fingerprints, injuries, etc. It is necessary to check the liveness of the finger (anti-spoofing), whether there is not any spoof used instead of a real finger from a genuine user.

For the computation of the *orientation field*, it is necessary to calculate the direction of every ridge from the surroundings (according to the gray color tone) at each point of the image. First, the orientation field for each pixel is calculated. In the second step, the transformation to the *block orientation field* is done. The block orientation field is then mapped on the original fingerprint image. This step is used only for fingerprint processing algorithms in criminal expertise area (dactyloscopy).

Further *enhancement* of the image and *thresholding* will provide black ridges and white background (valleys). This image enhancement includes, e.g. histogram scaling or filtering in frequency domain (after Fourier transformation) [8, 9]. It is necessary to check the quality of an input image—one of the popular algorithms is

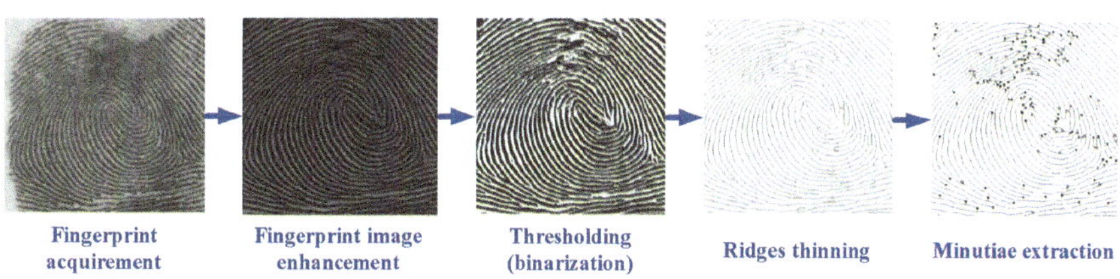

Fig. 6 Scheme of the fingerprint processing for biometric recognition

so called NFIQ (NIST Fingerprint Image Quality). For thresholding the Regional Average Thresholding (RAT) scheme is often used.

From the previous step we obtain a black-and-white image of ridges, each of which could have various widths. Now, it is necessary to *thin* the ridges to a 1 point thickness. It must be true that the papillary line must not dwindle in any direction in order to avoid a problem with the position of the minutiae.

For the *detection* and *extraction of minutiae*, the method of detecting ridges according to Hong [10] is used (among other possible methods). This method is based on the fact that the ridges run parallel to each other and reach the maximum level of gray in the middle of the line. Generally, two basic types of minutiae are detected: *line ending* and *bifurcation*, while other types of minutiae are a combination of these two basic types. The following data is stored for each minutia: the *position* of the minutia (coordinates x and y), the *type* of the minutia (line ending/bifurcation) and the *gradient* (the orientation of the papillary line).

The result of the minutiae extraction is then *compared* (*matched*) with the stored template from the database. The methods for comparing fingerprints are as follows [6, 11–15]:

- Methods based on minutiae
  - They use position, type and gradient (direction)
  - Generally, issues of pattern comparison
- Methods based on correlation
  - 2D correlation between input and template
  - Computationally demanding but implementable in hardware
- Methods based on the properties of papillary lines
  - Orientation and frequency of papillary lines, line shape, texture information, etc.
  - Low resolution, but often used as support information for the minutiae-based method.
- Methods based on high resolution or 3D properties of the finger or papillary lines
  - Usage level 3 features e.g. sweat pores placement and distribution, ridge edge features, etc.
  - Usage of 3D features such as curvature of the finger, depth values, etc.
  - Can be combined with traditional 2D methods mentioned above.

The method based on minutiae is the most commonly used method. This is practically a problem of comparing patterns—two sets of minutiae. Two basic methods for comparison using the minutiae are: *Hong method* [16] and *Ratha method* [17]. Both methods are based on two main steps: generating a global overlap (alignment) and locating a local overlap (compare)—see Fig. 7.

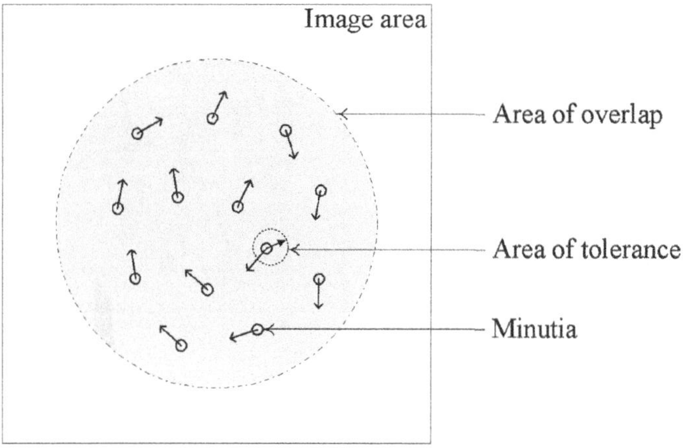

Fig. 7 Areas of overlap and tolerance for minutiae matching

3 Influences of Fingerprint Images

In the real world we cannot find an undamaged or a perfect fingerprint—this is just only a theory. The only way how we can get a "perfect" fingerprint is to synthetically generate it. The finger itself is often damaged, acquirement method is not perfect, environmental conditions are not ideal, what leads to getting the damaged fingerprint image. These damages can be basically divided into three groups: *finger and user condition*, *sensor effects* and *influence of the environment*.

A "perfect" fingerprint has nicely visible ridges, which are clearly distinguishable from a background (valleys). The majority of sensors delivers fingerprint images in grayscales, i.e. sometimes it is difficult to distinguish ridges from the background (valleys), especially if there are some distortions on finger or in the environment. If a histogram of an ideal fingerprint is computed, there should be only two peaks visible—one representing black (ridges) and one representing white (background/valleys). This is very often not the case—see Fig. 8 (left fingerprint acquired by Biolink optical reader, right one by Veridicom capacitive reader).

3.1 Finger and User Condition

The *dirt on the finger* which can be caused, e.g. by a few grains of dust, small particle or a greasy finger from some meal. Liquids or other conductive materials are one of the most problematic types of dirt. For every sensor technology the most problematic type of dirt is different. These troubles can be in conjunction with the *dry* or *moist finger*. The effect of moisture or dry finger is very significant. It is also a very frequent way of damaging the fingerprint. The *physically damaged finger* is

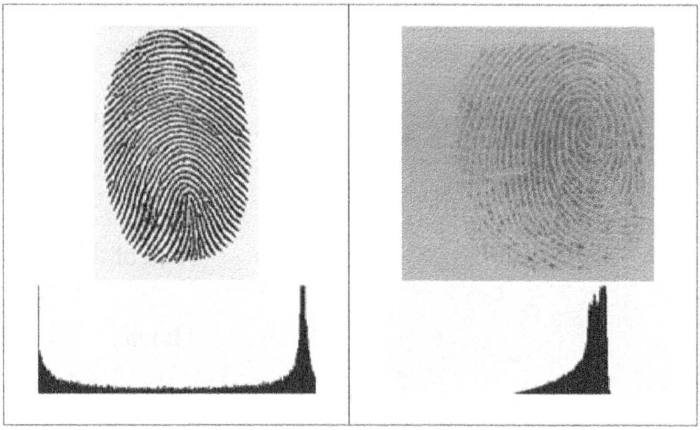

Fig. 8 Images and histograms of high (left) and low (right) quality fingerprints

common in some groups of users, namely among people who are doing a manual work, because some abrasions or cuts could be found on their fingers. If it is only a small injury or everyday wear damage, the ridges will regenerate. On the contrary, a deep wound will affect the ridges forever. Fingers can be also damaged by *skin diseases*.

The second part of this category are phenomena caused by users when acquiring the image. One of them, usually caused by inexperienced users, is small or insufficient *contact region*. That simply means that the finger was presented to the sensor in a way, which allows only a partial acquisition of the fingerprint. The small contact region can also be caused by extremely low *pressure* to the surface. The other way around, very high pressure can create a black oval instead of the fingerprint image. When the pressure is somewhere between these extremes, it can cause either very thick or very thin ridges. There is an endless number of movements that users can do when showing the *non-cooperative behavior*. It can be dynamic or static behavior, which is done on purpose, i.e. it is not an accident. The line between an unintentional wrong usage and a non-cooperative behavior is very thin. Nevertheless, there are users who resent biometric systems or just try to push to its limits. They can move the finger or change the pressure while the sensor is acquiring image. The other possibility is the static behavior like a rotation of the finger, a small contact region with just a side of the finger, etc.

3.2 *Sensor Effects*

These phenomena are dangerous, because they influence all users using the device. The first phenomenon is the *dirt on the surface* of the sensor. While the dirt on the finger is associated with grease or small dust, the surface of the sensor can be polluted much more. There can be grease from several fingers, it can be wooden, metallic or earth dust or fine sand because the sensor is used in a factory or outside. When the amount of dirt is higher, the sensor cannot even recognize that there is a finger laid on its sensing area. When the dirt on the surface comes from the ridges, it is called the *latent fingerprint*. This is fingerprint residues from previous users and not only that it can create fake ridges for the current user but sometimes it can be reactivated. Latent fingerprints are typical for crime scene, where they are left by an offender (if not wearing gloves). There is also a possibility of the *physical damage* of the sensor. Whether it is intentional or caused by an accident, the sensor can be damaged, however usually not totally destroyed. The resulting damage is different for every sensor technology. The *sensor technology* itself has a large impact on how the fingerprint will look like. Some technologies get ridges as a white color, some get ridges as a black color and some technologies get colorized image (basically photo) of the fingerprint.

3.3 Influence of the Environment

The last group influence usually both sensor and user/finger. The *temperature* usually influences the fingerprint indirectly by high moisture of the finger or on the other hand extreme dryness of the finger when it is very cold. But the temperature can also have an impact on the sensor technology. There are technologies that are more sensitive to the surrounding temperature than the others. The *vibration* can create a blurred image or damage the device internally. It is common in mobile devices when they are used in the means of transport or similar situations. The *electro-magnetic radiation* is another phenomena. In some cases it can influence the device as a whole. It can even change the information that is transported from one part of the biometric system to the other. This can then in more sensitive sensor technologies lead to a blurred image. The *surrounding light* is another sensor technology specific phenomenon. It influences only sensors, which have a light sensing unit, i.e. the optical or electro-optical technology. A problematic situation occurs when the sensor is large and the finger is small. The uncovered edges of the sensor create the fingerprint image from the surrounding light and not from the finger itself.

4 Diseases Influencing the Fingertips

Skin is constantly being regenerated. A keratinocyte ("skin cell") starts its life at the lower layer of the epidermis (the basal layer), which is nourished by blood vessels and is supplied with nerve endings from dermis. The cell migrates upward from the basal layer to the *stratum corneum* (the outermost skin layer). During four weeks the cell undergoes a series of changes, gradually flattening out and moving toward the surface. Then it dies and is shed. This physiological process can be negatively affected in many *diseases* of the skin [18].

The first group of *skin diseases* represents diseases causing *histopathological changes of the epidermis and dermis*—these diseases usually cause problems for all kinds of fingerprint scanners, because they can influence either the color or the internal structure of the skin. The most common representatives of this group are [19, 20]: *hand and fingertip eczema*, *dyshidrosis*, *tinea*, *pyoderma*, *pitted keratolysis*, *pyogenic granuloma*, *systemic sclerosis* or *Raynaud's phenomenon* [18].

The second group represent diseases causing *skin discoloration*—these diseases may cause problems for optical fingerprint sensors and also for sensors which use a fingerprint anti-spoof detection check based on the color or spectral analysis of the human skin. Typical representatives are [19, 20]: *macular drug eruptions and rashes* in infectious diseases, *pitted keratolysis*, *Raynaud's phenomenon*, *xanthomas*, *carotenosis* or *hereditary hemorrhagic teleangiectasia* [18].

The last group represent diseases causing *histopathological changes at the junction of the epidermis and dermis*—these diseases could cause structure changes underneath the skin at the junction between dermis and epidermis. Typical representatives are [19, 20]: *hand eczema*, *verruca vulgaris* (*warts*), *psoriasis* or *epidermolysis bullosa* [18].

We created a database with diseased fingerprints in cooperation with University hospital Olomouc. Each image in the database has anonymized information about the patient, severity and type of disease. There are more than two thousand of fingerprints in our database. Some examples of diseased fingerprints from our database are shown in Figs. 9, 10 and 11.

Fingertip eczema (see Fig. 12) is a very dry, inflammatory, non-infectious disease, which occurs on the palmar surface or the fingertips. The skin becomes cracked and scaly, and usually starts peeling off which results in exposition of red and tender skin surfaces [19–21].

As the number of fingerprints with fingertip eczema in the database is large, a wide range of typical features was observed. There are two groups of these fingerprints: (i) less and (ii) more severely damaged. In the first group of fingerprints, occurrence of thin lines of different directions was typical. These lines often connect or cross each other. In some cases, small round white spots were present, and in others, occasional dark areas make the papillary lines partially unreadable. In the second group, the damage is more severe. Fingerprints are usually almost completely damaged, straight lines cover the entire fingerprint area and create grids by crossing each other. The background is darker and large irregular spots can be seen. As the papillary lines cannot be seen at all, this type of damage is by no means recoverable.

Psoriasis vulgaris (see Fig. 13) is a common, chronic and inflammatory disease of the skin which is often indistinguishable from a serious form of hand eczema. It is characterized by dry and scaling plaques covered with dry scales that peel in layers [19–21].

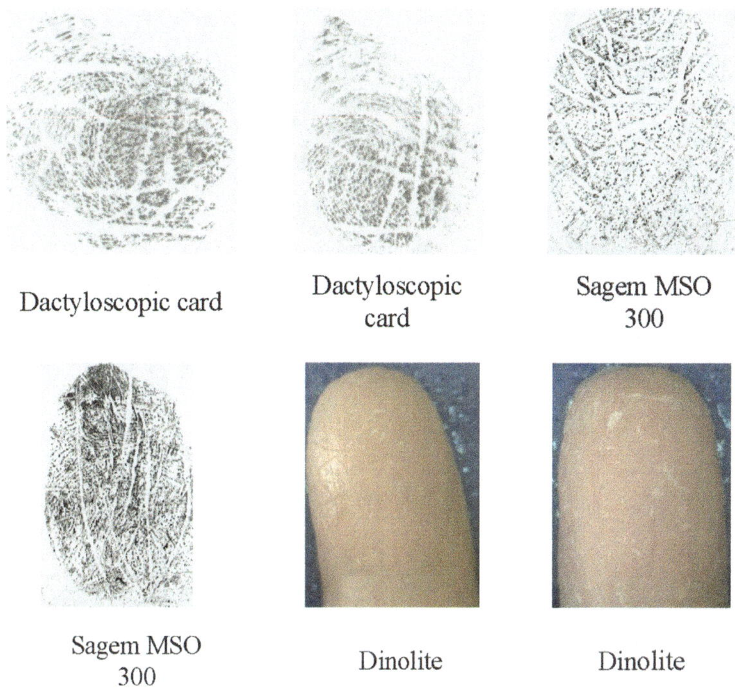

Fig. 9 Fingertip eczema—a severe form

The vast majority of fingerprints affected by psoriasis are completely damaged. Ridges are mostly unreadable. The most frequent feature is a large irregular dark spot bounded by a white border. Apart from this feature, the presence of larger dark areas or thick lines is also common, as well as round and oblong spots.

Dyshidrotic eczema also known as *pompholyx* is a variant of hand and foot dermatitis that makes skin extremely dry. Its typical features are itching vesicles and scales located on the palms and sides of fingers [19].

Fingerprint images damaged by dyshidrotic eczema are generally covered with irregular blurred shapes with no specific form. Another typical feature is a thick line. These fingerprints were divided into two groups, according to how severe the damage is. In the first group of less severely affected fingerprints, the entire area of a fingerprint is often covered, but ridges remain visible. Ridges are usually disrupted at multiple places and irregular blurred white spots may appear. Fingerprints in the second group are seriously damaged and cannot be repaired. The image area is typically covered by thicker lines in combination with large blurred white spots. Ridges are not sufficiently visible.

Hyperkeratotic eczema (see Fig. 14) is a chronic form of hand eczema characterized by the occurrence of orange and brown scales with cracks between them [19–21]. Only one third to one half of the fingerprint area is usually affected. Sometimes, only

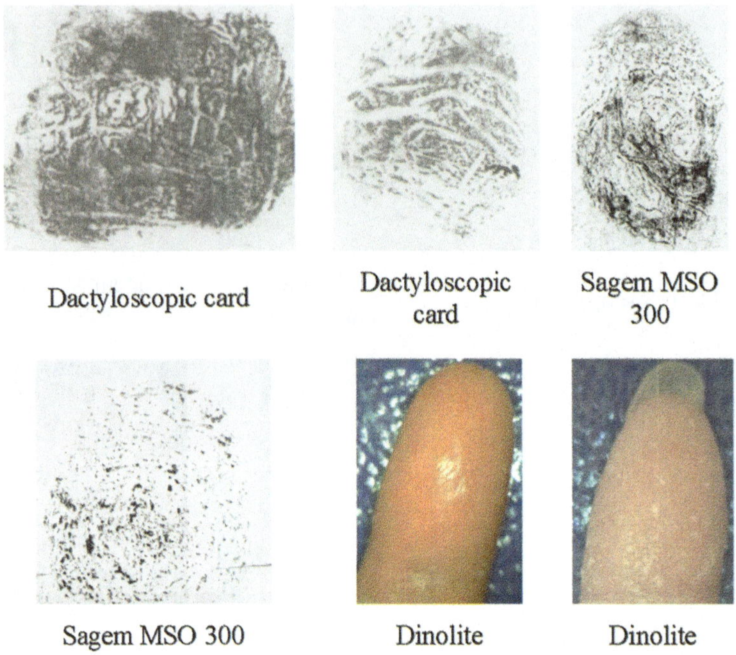

Fig. 10 Psoriasis—a full seizure

the ridges are multiply disrupted. In other cases however, ridges are distorted and their direction is difficult to determine. Small to medium round spots are likely to be present.

Verruca vulgaris (*warts*) (see Fig. 11) is a very common skin disease, characterized by the presence of stiff elevated bumps on the skin surface. They grow in size, which is in average about 5 mm but can reach up to more than 1 cm. On their surface, tiny black dots may appear [19–21]. The influence of this disease on the fingerprint images is minor and easily removable. Typically, 1–4 round white spots occur, sometimes with black dots in their center.

Systemic scleroderma (see Fig. 15) is characterized by the appearance of hard, smooth and ivory-colored areas. In the early stage, affected areas are red and swollen; later, they become completely immobile and lose their natural peaked contour [19–21]. The fingerprints in the database did not show any signs of damage. It can be therefore concluded that the number of acquired fingerprints was not sufficient to describe the disease's influence on fingerprint images.

Acrodermatitis continua or *dermatitis repens* (see Fig. 16) is a chronic inflammatory disease of the hands and feet, and one of the less frequent types of psoriasis vulgaris. The outbreak of the disease is accompanied by asymmetric formation of pustules of the fingertips, and continues with eruption of fresh pustules with hyperkeratotis and crusting. As the disease progresses, nails can even float away [19–21].

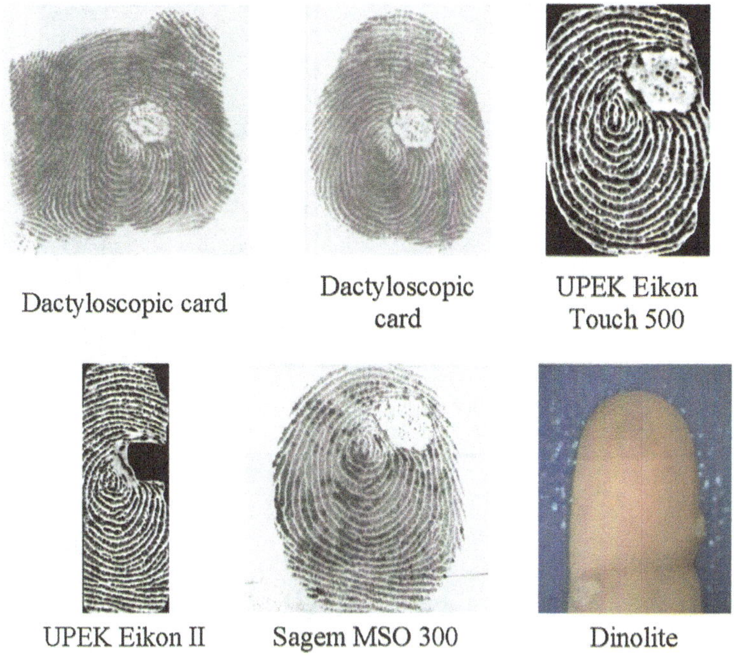

Fig. 11 Verruca vulgaris (warts)

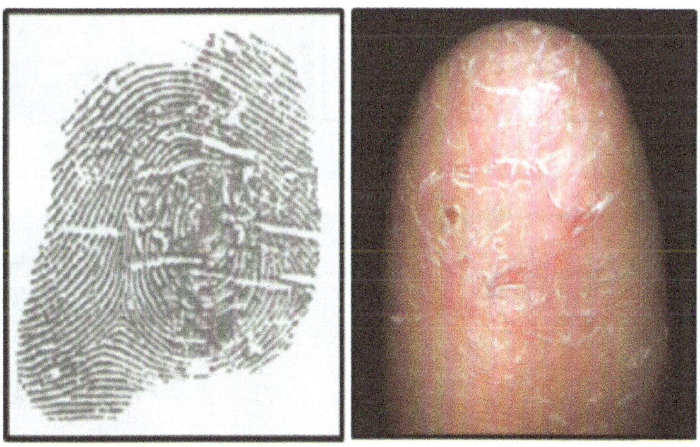

Fig. 12 Fingertip eczema. *Source* Our database and [19]

Fingerprint images are typical for the occurrence of small round spots that look like a cheetah skin and cover usually the whole fingerprint area. Larger oblong or round spots occur as well and straight lines or cracks are also not uncommon. Ridges cannot be recognized at all, and the original structure of the fingerprint is completely covered. Larger dark areas are often present and the spots can be blurred together.

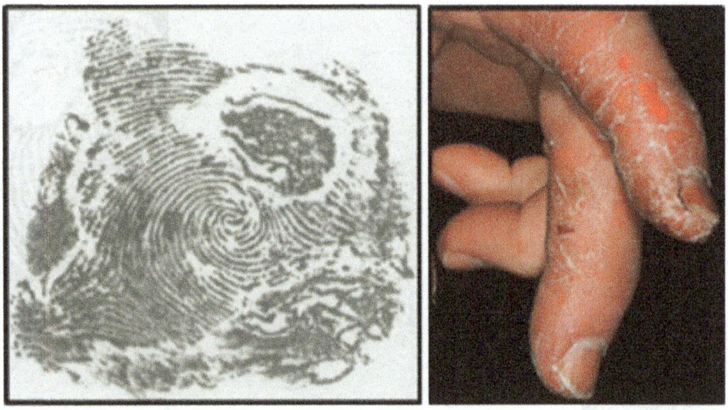

Fig. 13 Psoriasis vulgaris. *Source* Our database and [19]

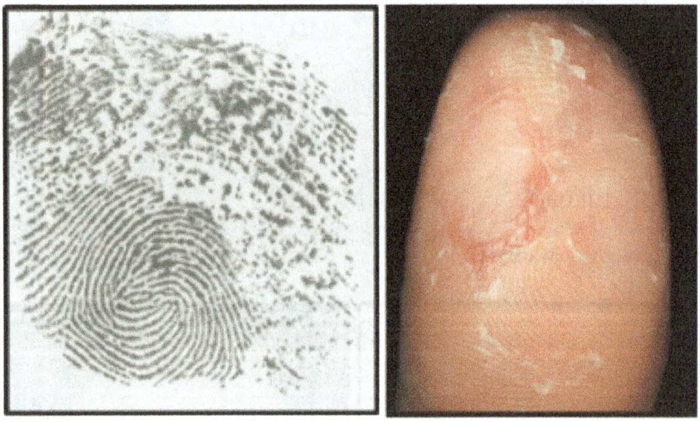

Fig. 14 Hyperkeratotic eczema. *Source* Our database and [19]

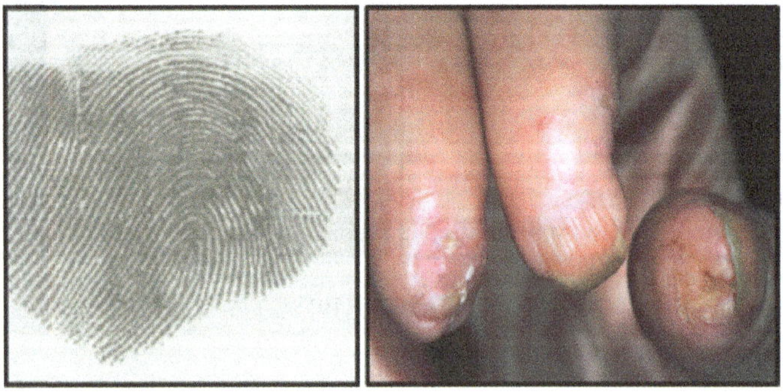

Fig. 15 Systemic scleroderma. *Source* Our database and [19]

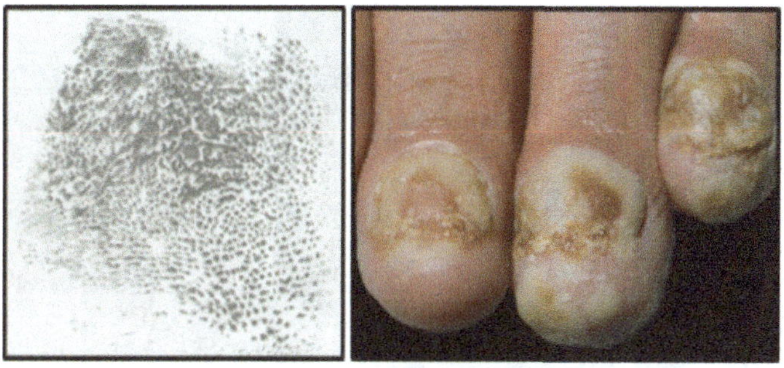

Fig. 16 Acrodermatitis continua. *Source* Our database and [20]

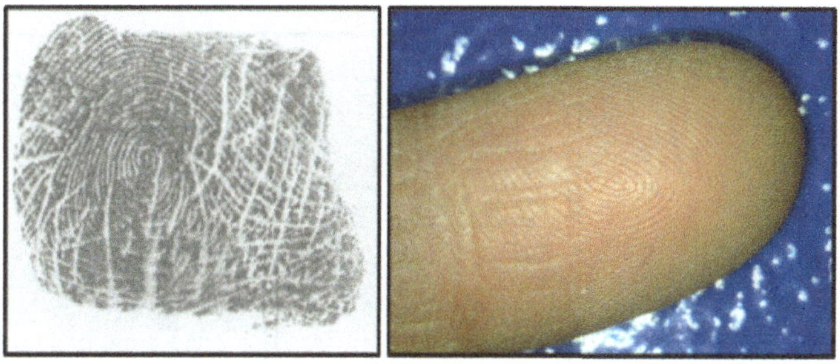

Fig. 17 Colagenosis. *Source* Our database

Almost in all cases, the fingerprint image is completely damaged and cannot be repaired.

Colagenosis (see Fig. 17) is a connective tissue and inflammatory autoimmune disease [19–21]. The only typical feature of fingerprints with this disease is thin lines crossing each other. Under these lines, papillary lines are well visible.

Raynaud's phenomenon (see Fig. 18) is a vascular skin disease that often accompanies an associated disease (most often scleroderma). The fingers have sequential discolorations: they first become pale and cold, then white, blue and finally red. This is caused by constrictions of the small arteries and arterioles in fingers [19–21]. As Raynaud's phenomenon causes discoloration only, fingerprints in the database are always healthy and undamaged.

Although being stated as a disease in the database, *effusion of fingers* is only a syndrome, which manifests itself by a strong swelling. It is one of the symptoms of systemic scleroderma, for instance. Ridges are typically disrupted in many places, and small to medium spots are present. In general, ridges are clearly visible. Sometimes, however, white spots make them unreadable.

A *cut wound* (see Fig. 19) typically cause either a straight line in a fingerprint image or a more blurred white area. The damage is minor and should not be difficult to remove.

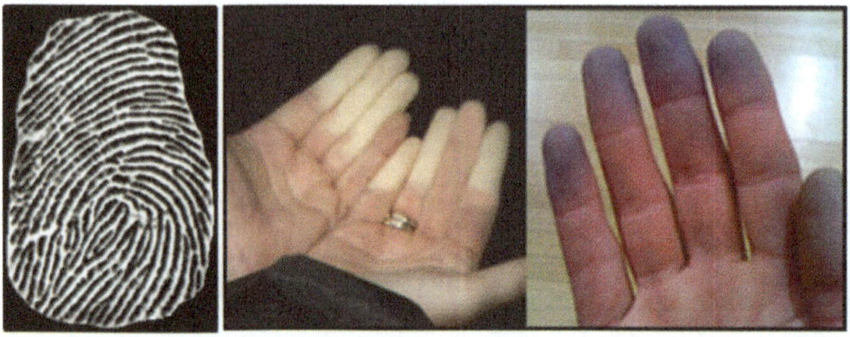

Fig. 18 Raynaud's phenomenon. *Source* Database and [22]

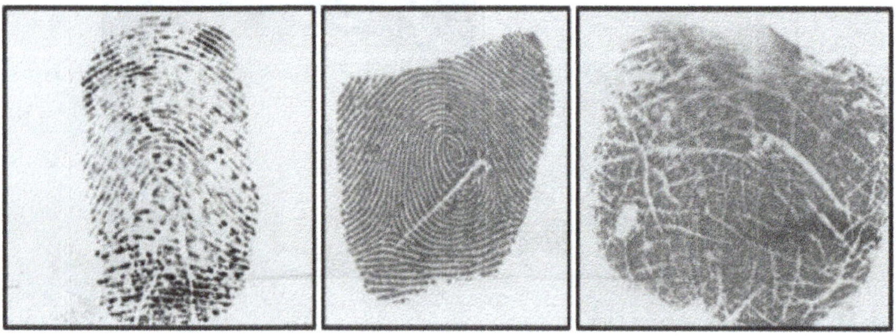

Fig. 19 Effusion of fingers, cut wound and "unknown" disease. *Source* Our database

Fingerprints of the "unknown" or "unnamed" disease (see Fig. 19) are totally covered with lines of different thickness and length and are therefore unreadable. They are very much alike those with fingertip eczema.

4.1 Specific Damages from Diseased Fingerprint Images Database

We analyzed our database with diseased fingerprint in order to find specific properties of each group of diseases. For every particular disease, common signs among all fingerprint images affected by this disease have been found and a general description of each disease and its influences has been defined. Based on these descriptions and sets of common signs and their frequencies, the diseased fingerprint images were classified into five categories. These categories are later used in the actual detection of the damaged areas in a fingerprint image and they help to divide the large detection task into smaller bearable parts [23]. The summarization of fingerprints of each disease is shown in Table 1 [24].

Table 1 Summarization of the content of our database with diseased fingerprints [24]

Disease	No. of fingerprints in the DB	Percentages (%)	No. of patients
Fingertip eczema	1,107	51.132	17
Psoriasis vulgaris	326	15.058	9
Dyshidrotic eczema	247	11.409	4
Hyperkeratotic eczema	18	5.450	2
Verruca vulgaris	96	4.434	4
Scleroderma	0	2.310	1
Acrodermatitis continua	0	1.848	1
Colagenosis	6	1.663	1
Raynaud's phenomenon	9	0.416	1
Effusion of fingers	5	1.617	1
Cut wound	8	0.831	2
"Unknown" disease	3	3.834	1
Total	2,165		44

We analyzed and compared the fingerprint images from the database and found 12 common features, whereas 7 of them are local features [24]: *straight lines* (SL); *grid* (G); *small ridges disruptions* (PRD); *small "cheetah" spots* (CS); *larger round/oblong spots* (ROS); *large irregular spots* (IS) and *dark places* (DP).

The other five global image patterns are [24]: *blurriness of* (parts of) *the image* (B); significantly *high contrast of the image* (HC); the *entire fingerprint area affected* (EA); *total deformation of the fingerprint image* (TD) and *significantly high quality and healthy fingerprint* (HQ).

For every skin disease its image features were summarized (see Tables 2 and 3). Fingerprint images obtained from optical scanners were excluded as their character is significantly dissimilar to the others. The actual number of images taken into account is stated in the column "sum".

4.2 Skin Diseases Detection

In this subsection, the specific algorithms used in the disease detector along with their pros and cons will be discussed, as well as the core methods essential for the program's functionality and data structures used to store and keep important data throughout the process.

For the *detector*, there are three major algorithms that are used for the detection part: *block orientation field*, *histogram analysis* and *flood fill*. Their combination provides valuable information about the fingerprint quality and character of the possible disease.

Table 2 Local features of damaged fingerprint images

Percentages of particular features (%)								
Disease	SL	G	PRD	CS	ROS	IS	DP	Sum
Fingertip eczema	72.03	24.65	15.91	12.24	32.34	16.61	15.73	572
Psoriasis vulgaris	40.37	6.42	2.75	12.84	48.17	32.57	62.84	218
Dyshidrotic eczema	63.11	7.38	14.75	18.03	78.69	29.51	32.79	122
Hyperkeratotic eczema	3.92	0.00	66.67	15.69	74.51	3.92	5.88	51
Verruca vulgaris	3.17	0.00	14.29	12.70	74.60	0.00	25.40	53
Scleroderma	0.00	0.00	0.00	0.00	0.00	0.00	30.43	23
Acrodermatitis continua	14.29	0.00	0.00	85.71	60.00	14.29	65.71	35
Colagenosis	100.00	78.13	0.00	0.00	15.63	0.00	25.00	32
Raynaud's phenomenon	0.00	0.00	100.00	0.00	0.00	0.00	0.00	8
Effusion of fingers	10.00	0.00	73.33	43.33	63.33	6.67	13.33	30
Cut wound	93.75	0.00	0.00	0.00	18.75	0.00	12.50	16
"Unknown" disease	100.00	86.67	0.00	0.00	76.67	30.00	73.33	30

The computation of *block orientation field* is commonly used in the fingerprint recognition process for the purposes of estimating the ridges direction and classifying the fingerprint image into one of the several fingerprint classes [6, 23]. Because a typical fingerprint pattern consists of alternating dark and white lines, this information can be easily processed by a gradient operator that estimates the image gradient for each pixel. This low-level information is gathered and averaged for each $w \times w$ block in the image [24]. The transformation can result in a relatively smooth and continual image of the ridges direction estimates—for a healthy fingerprint of course. If we try to compute the block orientation field for a damaged or a partially damaged fingerprint however, we can easily recognize with the naked eye which areas contain possible damage, because the orientation field in these areas will be discontinuous. These discontinuities can be detected by scanning the field for differences in direction angles. In the program's pipeline, a gradient-based method of block orientation field computation is used [24]. The resulting block orientation field is afterwards analyzed for any discontinuities that may occur. The analysis is done using a row-wise and column-wise scanning approach that reveals areas of possible damage in the fingerprint. Sometimes, the method detects single discontinuities that may be erroneous, and on the other hand, under different circumstances, one unmarked block may appear in the midst of discontinuous blocks. In order to make the algorithm as accurate as possible, although mistakes never disappear completely, these cases are

Table 3 Global features of damaged fingerprint images

Percentages of particular features (%)						
Disease	B	HC	EA	TD	HQ	Sum
Fingertip eczema	18.01	21.50	40.38	36.36	29.02	572
Psoriasis vulgaris	34.86	27.06	61.93	58.72	18.35	218
Dyshidrotic eczema	30.33	30.33	31.97	29.51	9.84	122
Hyperkeratotic eczema	31.37	29.41	9.80	0.00	37.25	51
Verruca vulgaris	19.05	80.95	7.94	7.94	76.19	63
Scleroderma	0.00	0.00	0.00	0.00	100.00	23
Acrodermatitis continua	48.57	25.71	100.00	100.00	0.00	35
Colagenosis	9.38	40.63	0.00	0.00	25.00	32
Raynaud's phenomenon	0.00	0.00	0.00	0.00	100.00	8
Effusion of fingers	23.33	16.67	40.00	16.67	3.33	30
Cut wound	37.50	68.75	0.00	0.00	50.00	16
"Unknown" disease	30.00	20.00	90.00	83.33	0.00	30

Fig. 20 Damaged area detected using the block orientation field

taken into account. The algorithm handles them by copying the properties of their neighboring blocks (marking the single ones either as alright or as a discontinuity, depending on the neighborhood). Example of the detection is shown in Fig. 20.

The method of *histogram analysis* is based on the presumption that a quality fingerprint image consists of equally distributed ridges and valleys. If we assume that

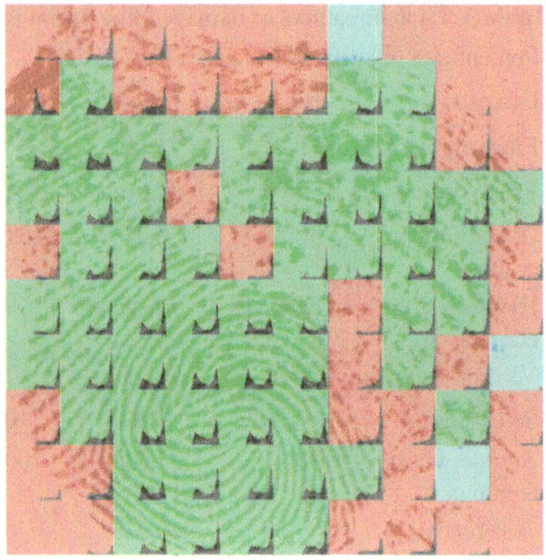

Fig. 21 Histogram analysis result with particular histograms

ridges are roughly the same dark color while valleys are light-colored, a histogram computed from each subfield of the fingerprint's area should ideally consist of two peaks of approximately the same height and one valley between them. Examples of good (bimodal) and bad histograms are presented on Fig. 8. On the other hand, the intensity distribution in a fingerprint image part that belongs to a damaged area is not always as equal as in the quality one. Thus, if a histogram is computed for this subfield, it is very likely that it will not have the ideal bimodal appearance as described above. Experiments showed that the majority of damaged areas break the rules of the bimodal histogram. The lower the quality, the less the histogram resembles the ideal one. A non-bimodal histogram always implies a damaged or low-quality area. Figure 21 shows an example output of this method, along with the particular histograms that were being analyzed. Red background implies an invalid histogram, green means valid and blue stands for background.

Flood fill is a very often used algorithm used for graphical purposes [9] and is especially handy for detecting and filling connected single-colored areas of an image. This characteristics was used in the application in order to find local features of damaged fingerprints, such as straight lines or spots. The *flood fill* algorithm has three parameters: a target color, a replacement color and a start pixel. It is based on examining the color of all pixels in the 4- or 8-neighborhood of the start pixel and changing the color of those pixels that have the target color to the replacement color. Using of either recursion, or stack/queue, the colored pixels become the next start pixels and the process is repeated. In the end, the entire single-colored area is filled. The result from this algorithm is shown in Fig. 22.

All three of the above-described methods detect a different kind of damage in the image and only flood fill provides logically structured results that can be used in classification.

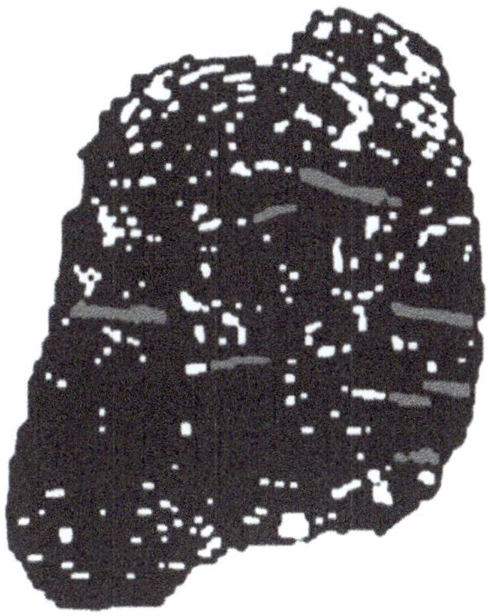

Fig. 22 Extraction of straight white lines

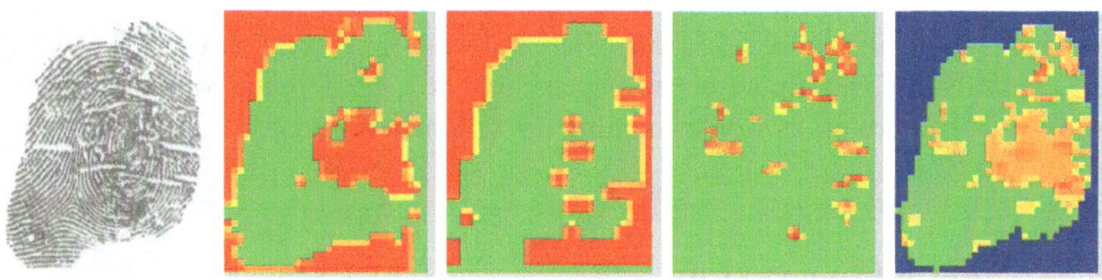

Fig. 23 Example of the final distribution of damage in the image with atopic eczema

However, connecting the three methods together results in a surprisingly accurate description of the extent of damage in an entire area of a fingerprint image. At the end of each detection process, every image pixel is assigned a value between −1 and 1. Negative values stand for background, 0 means a healthy area and positive values indicate damage. The higher the value, the more damaged the area to which the pixel belongs, as explained in the previous parts.

Each of the three detection methods separately provides interesting outputs, but it is their connection that makes the resulting application so notable. Thanks to the connection, very satisfactory results have been achieved for locating the damaged areas—see the results on Fig. 23 (the color representation is as follows—green color marks the healthy areas, blue color highlights the background and for the damaged areas a scale from yellow to red is used; yellow stands for minor damage, whereas red implies extremely damaged places).

The *Classifier* decides based on features extracted by the flood fill algorithm and classifies the fingerprint image, according to the features' numbers, sizes and shapes into one of these 6 categories: acrodermatitis, atopic eczema, psoriasis, verruca vulgaris, unknown disease or healthy (Tables 4 and 5).

Table 4 Rejected and accepted samples

	True positives	False negatives	False positives	True negatives
Acrodermatitis	12	18	103	478
Atopic eczema	134	289	25	163
Verruca vulgaris	23	17	314	257
Total	611	611	611	611

Table 5 Classifier accuracy measures

	FAR	FRR	F1 score [25]	Accuracy
Acrodermatitis	0.1394	0.6667	0.1655	0.8347
Atopic eczema	0.1968	0.7021	0.4300	0.4533
Verruca vulgaris	0.3408	0.7373	0.1956	0.5827
Total	0.2329	0.5000	0.2073	0.7496

The classification accuracy reached high values for acrodermatitis (83.5%) and verruca vulgaris recognition (75.0%), whereas it was lower for atopic eczema (45.3%) and psoriasis (58.3%). Better performance could be gained by improving the classification decision rules, as well as coming up with new types of features detection.

5 Synthetic Fingerprints

When dealing with challenges associated with specifically damaged fingerprints the biggest problem is to get to database of them. Even small databases (thousands of fingerprints) of specific fingerprints is usually not available. On the other hand to get to larger databases of regular fingerprints (we speak of datasets with thousands, or better, tens of thousands of fingerprints) are very hard to get as well. That is because acquisition of databases is very time consuming and expensive. It demands a very trusted organization to attempt to collect a database like that because people tend not to give their fingerprint to everyone. Acquisition of diseased fingerprint also have to be with cooperation of a hospital and doctors. Collecting such a database is also very tiresome, both for the technician, doctors and the users. In this monotonous environment it is easy to make a mistake. Created database is then bound by privacy legislations that protect these kinds of data. Sharing database of diseased fingerprint is basically giving personal and medical information to someone. Obviously from legislative point of view that is not an easy task. When these databases aren't available, algorithms are tested on smaller databases and it is very easy to make them data dependent. So they are very accurate when it comes to a common fingerprint (e.g. loop class) but with an extraordinary fingerprint (like twin loop class) their accuracy collapses [26].

In these cases it would be great to have some generator (application) which would create a large synthetic fingerprint database. If a synthetic database consists of images very similar to human fingerprints, then it can be used instead of a large database of real fingerprints. It also opens space for the testing of just one kind of fingerprint, or else generating fingerprints in very bad conditions to adapt algorithms to their expected workplace. In the scope of diseased fingerprint it opens up the possibility to create damaged fingerprint not only for algorithms but also for doctors and dactyloscopy experts [26].

5.1 Simulation of Fingerprint Generation by Petri Net

Petri nets [27] are specific modelling techniques. They can be defined either by graphs or by a purely mathematical notation. The graphic notation is usually easily understandable, while the mathematical one can be used for various analyses and proofs. In this section the graphical representation will be used for description of synthetic fingerprint generation process. Petri net is a common name for group of similar techniques in scope of these section the P/T Petri net will be used [26].

The definition of Petri net is a sextuplet $N = (P, T, F, W, K, M_0)$ where: P, T, F denote places, transitions and arcs, respectively, $P \cap T = \emptyset$, $F \subseteq (P \times T) \cup (T \times P)$ is a binary relation, P, T, F are finite, $W : F \rightarrow N \setminus \{0\}$ is the weight of each arc, $K : P \rightarrow N \cup \{\omega\}$ is the capacity of each place, $M_0 : P \rightarrow N \cup \{\omega\}$ is the initial marking so that $\forall p \in P : M_0(p) \leq K(p)$. Note that ω is the supremum of the set N with these properties: $\forall n \in N : n < \omega$ and $\forall m \in N \cup \{\omega\} : m + \omega = \omega + m = \omega - m = \omega$. The graphic representation of Petri net is in Fig. 24. Circles denote places, full rectangles denote transitions and arrows denote arcs in their direction. $P0$, $P1$ and $T0$ are labels of each place or transition, numbers above arcs represent weights and the small dot inside place $P0$ denotes a token. The equivalent notation of the Petri net in Fig. 24 is $N = (\{P0, P1\}, \{T0\}, \{(P0, T0), (T0, P1)\}, \{W(P0, T0) = 1, W(T0, P1) = 1\}, \{K(P0) = \omega, K(P1) = \omega\}, \{M_0(P0) = 1, M_0(P1) = 0\})$ [26–30].

·Pro x•cess x• of simulation itself is driven by transition firing. Definition of it can be better described using a preset and a postset. The preset $\cdot x$ is defined as $\forall x \in (P \cup T) : \cdot x = \{y \mid yFx\}$ and the postset $x\cdot$ is defined as $\forall x \in (P \cup T) : x\cdot = \{y \mid xFy\}$. Similarly: for sets $\cdot X$ is the union of $\cdot x$ for each x in X and $X\cdot$ is the union

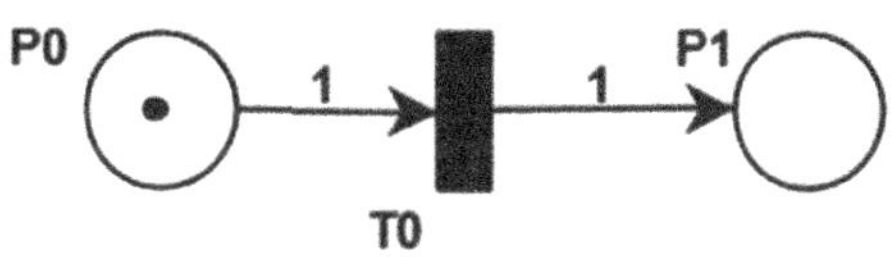

Fig. 24 An example of the graphical representation of Petri net

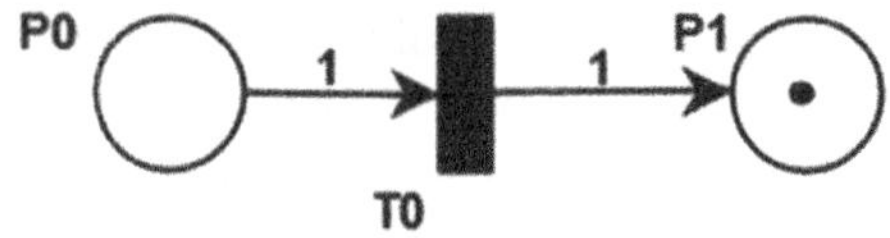

Fig. 25 An example of the graphical representation of the firing of the transition T0 from Fig. 24

of $x\cdot$ for each x in X. Note that the marking M is defined by the same definition as the initial marking M_0. The transition $t \in T$ is enabled from the marking of M when these conditions are met: $\forall p \in \cdot t : M(p) \geq W(p,t)$ and $\forall p \in t^{\cdot} : M(p) \leq K(p) - W(t,p)$. When this transition is fired, one gets the marking of M' which is defined in this way [26–30].

$$\forall p \in P : M'(p) = \begin{cases} M(p) - W(p,t) & if\ p \in {}^{\cdot}t \setminus t^{\cdot} \\ M(p) + W(p,t) & if\ p \in t^{\cdot} \setminus {}^{\cdot}t \\ M(p) - W(p,t) + W(t,p) & if\ p \in {}^{\cdot}t \cap t^{\cdot} \\ M(p) & otherwise \end{cases}$$

The result of the transition $T0$ firing from the net shown in Fig. 24 is shown in Fig. 25.

The goal of this subchapter is to create a Petri net which will simulate the generation of a synthetic fingerprint. It is advantageous to use Petri net as a clear way of showing all possibilities in synthetic fingerprint damage simulation. To make the whole Petri net clearer, the process will be divided into 4 distinct parts. The first part is the master fingerprint generation, the second part simulates the state of the environment, the third part simulates the user and finger condition following damage to the fingerprint, and the last part simulates the sensor condition which affects the fingerprint [26].

The process of master fingerprint generation could be done by various methods. These methods will be described later (in Sect. 5.2). In this example SFinGe generation method is simulated. Master fingerprint is term for synthetic fingerprint in its best quality without any damage. The process starts with the place *Start* where the initial token is placed as can be seen in Fig. 26. After firing the *Input definition* transition, 4 other transitions are enabled. *Class input* is defined (by firing the equivalent transition) by *Class* and then by definition of *Cores and deltas*. After that, the first token is in the *All input data* place. To create the synthetic fingerprint the *Gabor filter* transition needs to be used. But that is possible only when 4 tokens are in the *All input data* place. Another 3 tokens get here by defining *Shape*, *Density* and *Initial seeds*. Then the *Gabor filter* is fired and the token in *Synthetic fingerprint* shows that the generation of a master fingerprint is done [26].

The second part is focused on the influence of the environment and is shown in Fig. 27. The initial marking is set to each of the environment phenomena. After the firing, we should have a value representing each phenomenon. After that, all these values should be saved in the structure representing all phenomena at once, in the place named *All Environmental Effects* [26].

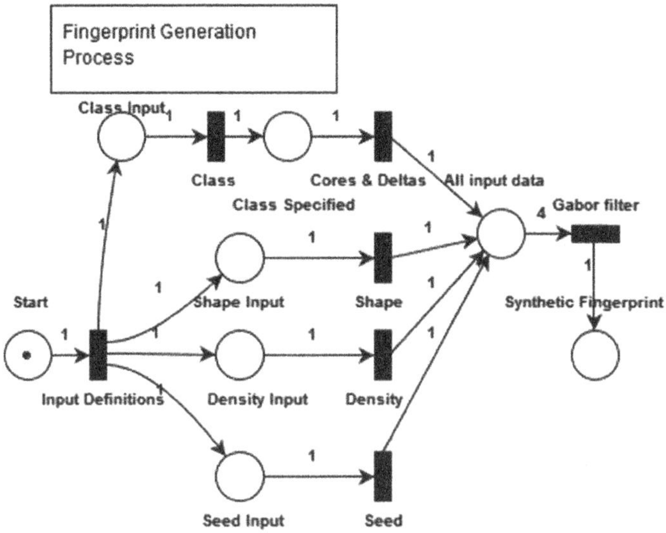

Fig. 26 The first part of the proposed Petri net—the fingerprint generation process

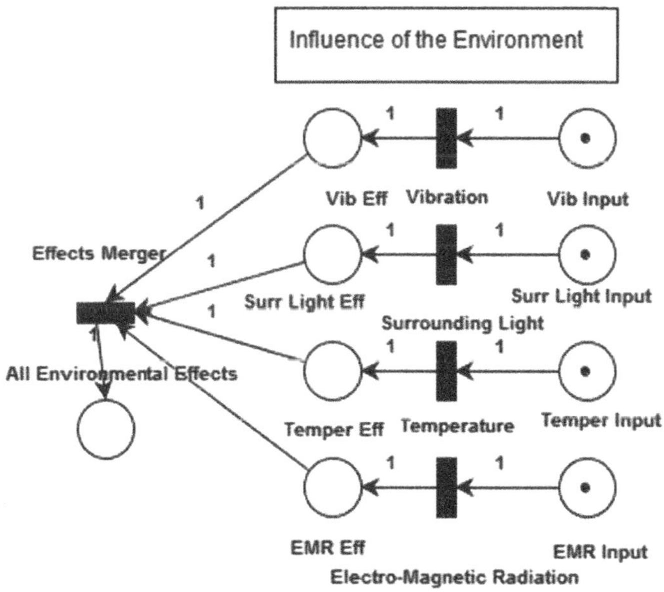

Fig. 27 The second part of the proposed Petri net—the influence of the environment

The third part of the net simulates the condition of user and finger and it is represented in Fig. 28. Transitions in the middle are individual conditions which are connected to the user or the finger. The places near the transitions have two purposes.

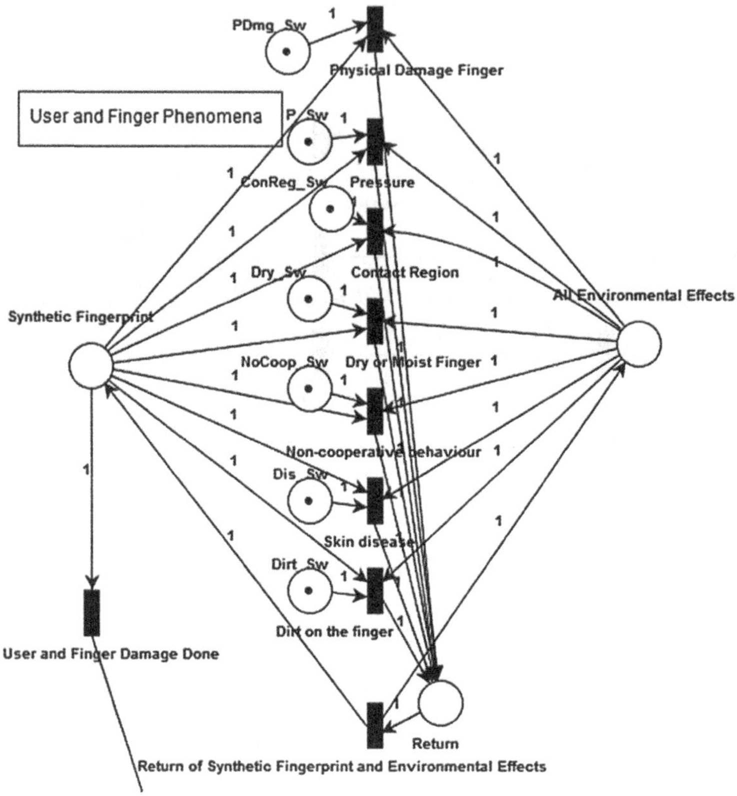

Fig. 28 The third part of the proposed Petri net—the user and finger phenomena

First, they enable or disable this type of damage. Secondly, they ensure that each phenomenon will be used only once. Also notice that the places *Synthetic Fingerprint* and *All Environmental Effects* are the last places of the previous parts. When these previous parts are done, the process continues, after 0 or *n* damage applications to the synthetic fingerprint, by firing the transition *User and finger damage done* and the token is then in the place *Synthetic fingerprint with user and finger phenomena and effects of the environment*, which can be seen in Fig. 29. Note that the place *Return* with the transition *Return of synthetic fingerprint and environmental effects* are returning the damaged synthetic fingerprint and the information about the environment so that 0 or *n* damages can be done. Note that *Skin disease* is one of these conditions [26].

The last part starts where the previous ended and it is focused on the sensor effects. As you can see in Fig. 29, there is a similar structure as in the previous part. Using the damaged fingerprint from the previous steps and the same structure of the environment values, it further damages the fingerprint image. Individual effects of the sensor can be switched on and off and used only once by the switch places near them. From the *Return2* place the token goes by transition back to *All Environmental*

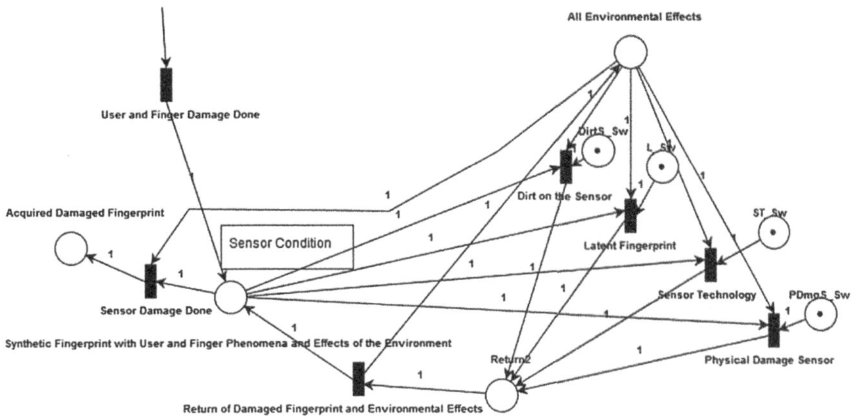

Fig. 29 The fourth part of the proposed Petri net—the sensor condition

Effects and another token representing the generated fingerprint also goes back to its place. Again, the *Sensor Damage Done* transition is fired after 0 or *n* damages and the synthetic fingerprint generation is done. One part of this final step is taking the environment into account. In the rare case when the user, finger and sensor are all in perfect condition then this step is the only way for the environment to influence the final synthetic fingerprint. Generally speaking, the environment influences other damages more often than it has an effect on its own. Figure 30 shows the whole proposed Petri net it is created by connection of all four parts together [26].

5.2 Methods for Generating Synthetic Fingerprints

This subchapter is focused on various algorithms that can be used for synthetic fingerprint generation. The generation is an inverse biometrics problem. According to input variables, it is basically fingerprint recognition process (Sect. 2.2) from the end to the start. Several methods of how to generate a synthetic fingerprint are to be found [31–33]. When we thoroughly study them, we find that they are all based on the same principle. The method used by the SFinGe seems to be the oldest one and also the most commonly known, so it will be described as a template for others. For example, very similar methods are used by Anguli which is an Indian Institute of Science fingerprint generator [34].

For better understanding, one can look at the upper part in Fig. 31 to see the process of generation. The generating part ends with the so called master fingerprint (a perfect fingerprint, equivalent to the phase extracted lines from Fig. 6). Firstly, the fingerprint's shape is determined. The basic shape is oval and each elliptical segment can be changed to create the required shape. The second step is the directional field model. In this step, the fingerprint class is chosen together with the position of cores

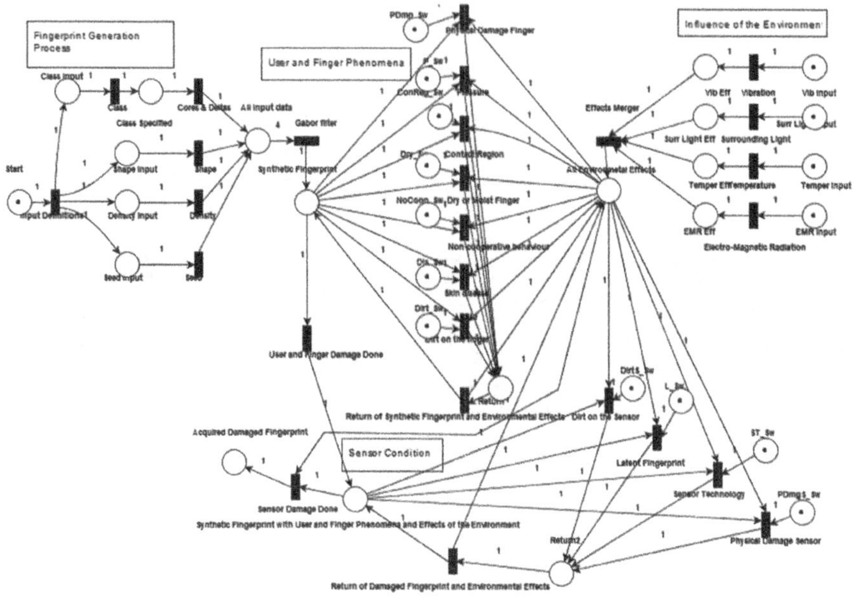

Fig. 30 The whole proposed fingerprint generation Petri net

and deltas. This step uses the Sherlock and Monroe ridge [6] flow model to generate a consistent direction field. The third step creates a density map. Fingerprints do not have equal spread of ridges. After examining several real fingerprints some heuristic criteria about the density could be made. These criteria are based on the position of singularities (cores and deltas) and according to them the density map is generated. The last step is ridge pattern generation. Initial seeds are added to the results from all previous steps. Iteratively, the image with initial seeds is refined with the Gabor filter. The filter orientation and frequency are adjusted according to the directional field and density map. Minutiae are automatically generated at random places with random types (dactyloscopic ones, not only ridge ending and bifurcation). After that phase, the master fingerprint is finished [31–33].

As can be seen, the SFinGe generating process is not precise copy of an inverted recognition process. When inverted process is followed strictly then so called fingerprint reconstruction is performed. These are methods that focus on the creation of a whole fingerprint from only minutiae saved as a template in fingerprint recognition. Another method lies between these two. It says that fingerprint features are dependent on each other [32]. It follows the same scheme but with dependencies on other steps. The orientation field is influenced by singular points. The minutiae density is higher around singularities, and also their appearance isn't random, but rather is statistically driven. The minutiae direction is also dependent on their types and on the orientation of ridges around. This method firstly determines singular points, after that the orientation field, and lastly the minutiae. Each step is dependent on the

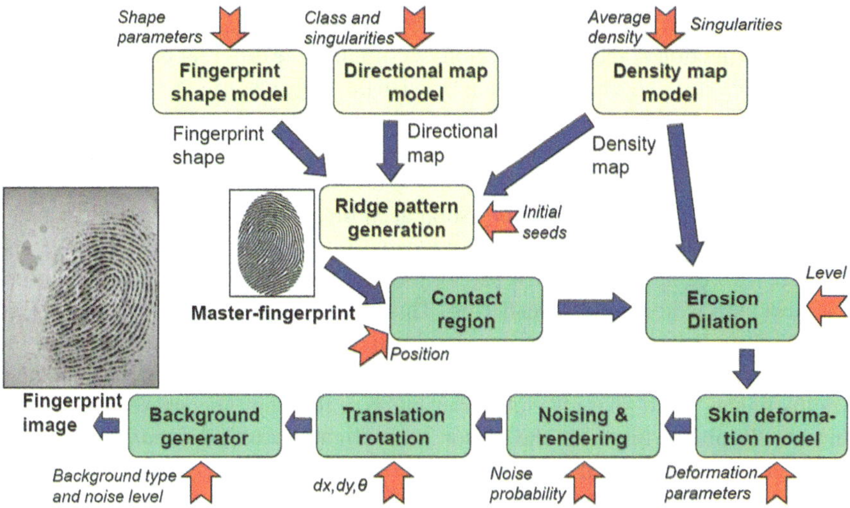

Fig. 31 SFinGe process of fingerprint generation (taken from [33])

previous one. After all the steps are completed, the master fingerprint is made with the use of the AM-FM method [32].

The last method uses user-defined minutiae as an input. The creation of a whole fingerprint is based only on these minutiae. The biggest difference is that the orientation field is generated from minutiae and not only from classes or singular points as it was in the previous methods. It is generated from the minutiae direction and each minutia has a weight based on the distance of it from the point where we are determining the orientation field. The disadvantage of this method is that the final fingerprint could have a class that does not exist in the real world. In this method, the density map can be manually changed. The default state is the uniform density map. After that, using a similar method of Gabor filter as in SFinGe, a master fingerprint is generated. Note that instead of initial seeds this method uses minutiae as these seeds and the generation starts with them, so precisely defined minutiae don't change in the process of generation [31].

5.3 *Impression with Skin Diseases*

The biggest advantage of synthetic fingerprints is that from one master fingerprint it could be made several impressions with various irregularities and damages. As an example of that SFinGe process is used. The process of fingerprint generation is shown in Fig. 31. The upper part, i.e. the part that ends with the generated master fingerprint, is described in Sect. 5.2.

Damages and irregularities are described on the lower part of Fig. 31. The first step is the selection of the contact region. To simulate the different placements of the finger on the sensor area a random translation of the ridge pattern is made. This is done without modifying the global fingerprint's shape and position. The next step is the variation in ridge thickness. The ridge thickness is modified to simulate various skin dampness and finger pressure (by erosion and dilation). The next phase is fingerprint distortion. In this phase, skin deformation according to different finger placements over the sensor is simulated. The skin plasticity (compression or stretching) and a different force applied on each part of the finger creates a non-linear distortion. The next step is noising and rendering. This step is not aimed at specific damage but rather more factors at once are simulated. Fingerprint image is damaged the most by this part of the algorithm. Unfortunately, algorithm was made to fit regular healthy fingers. Next phase is global translation or rotation. This phase simulates an imperfectly placed finger on the sensor. So it slightly translates and/or rotates the whole image. The last step is the generation of a realistic background. At the end of that step, the fingerprint impression is made. For the generation of databases, several impressions are made from one master fingerprint [6, 32, 33].

To create an impression with skin disease it is necessary to implement algorithm that is design to damage the master fingerprint to look like a fingerprint from diseased finger. First method is to base the algorithms on findings from Sect. 4.1. That means simulate the 7 local and 5 global marking which can be found on the diseased fingerprints. After that damage the fingerprint with these markings based on probabilistic distribution of marking in the specific diseased. This way is very dependant on analysis of the available databases but it can create every disease which is in the databases. Second method for creating algorithms that will damage the fingerprints is based on study of the diseases one by one. After that create algorithm for one specific disease (or even a few algorithms based on severity of the disease). Using second method will create more precise results for the damage but only for a few of diseases. Sometimes effects of diseases are difficult to generalize. In that case it might be enough to adapt damage from existing fingerprint image to synthetic. This chapter will focus on algorithms based on this second approach.

5.3.1 Verruca Vulgaris (Warts)

Specific information about the disease itself are in Sect. 4. The most important for generation of disease are images in Fig. 11. As can be seen usually warts creates white oval with irregular border and black dots inside them. Inside of this object there are no traces of former ridges in that place. However outside of warts the ridge flow continues normally. Figure 32 shows a few different effects that can warts have on fingerprint image. Clean ridge structure except for the part where the wart is located. Wart is represented by a white circle-shaped object with several black dots irregularly spread over its surface. Black dots represent the hard and scaly skin of the wart. The ridge structure around the wart is mildly deformed and the ridges are compressed. However, except for the close surroundings of the wart, the ridge

Fig. 32 Different fingerprints affected by warts acquired by Sagem MSO 300

structure of the fingerprint is unaffected. Right image shows at least three large white oval-like objects with irregular borders near each other. These are secondary warts that has spread from the larger one. When the wart is relatively small, usually it contains little or no black dots at all. It has been found that the size of warts on fingers varies from very small ones to ones as large as half of the hypothetical radius of the fingerprint. The location of warts on fingerprint is completely random and often one wart produces other so-called satellite warts in its close surroundings [35].

In order to generate warts into the image with synthetic fingerprint, a centre of the new wart is localised. The point coordinates are randomly generated and they have to be inside of the fingerprint and not too close to the fingerprint border. Also in this step, the size of the generated wart is randomly determined within boundaries set. Each new wart is first drawn separately. A number of small circles of varying radius is generated with their centres being distributed with exponentially large distance from the wart centre point. The next step is to draw the dots with randomly generated coordinates onto the warts surface. The dots are drawn with the same colour as the border of the small circles. Before adding the warts to the image the final colour of each pixel of the wart is determined. Depending on if the pixel is drawn by colour for border or the colour of the inside of the wart, the colour of the neighbouring pixels in the original fingerprint image is acquired (in this case dark pixels for border and light ones for the inside of the wart). The final pixel colour is then determined by one of the two following methods. First method picks random neighbouring pixel and copies its colour. The second method computes the mean colour of all the neighbouring pixels and then the mean colour is computed and applied to the pixel. Afterwards the whole wart image is blurred slightly and blend into the original fingerprint image. Eventually, secondary warts are drawn into the fingerprint if required following the same steps of the algorithm as for the main wart. Example of wart generation can be seen in Fig. 33 and final image with wart in Fig. 34 [35].

5.3.2 Fingertip Eczema (Atopic Dermatitis)

Specific information about the disease itself are in Sect. 4. The most important for generation of disease are images in Fig. 11. From these images the two types of

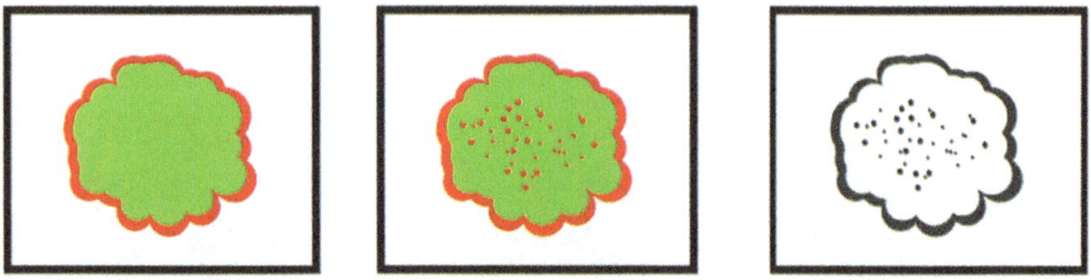

Fig. 33 Wart in distinctive colors (left), with dots added (middle), in final colors (right)

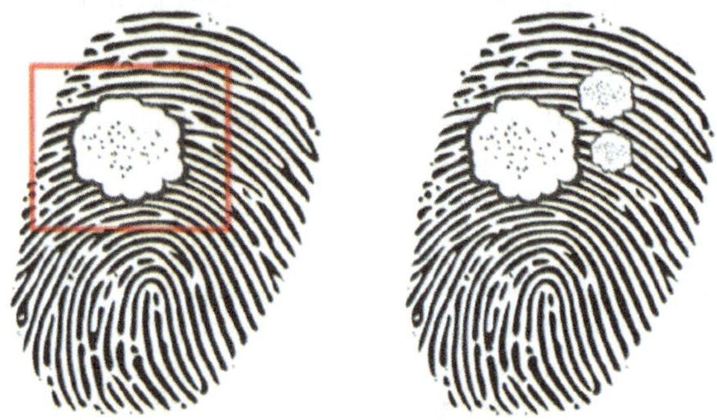

Fig. 34 Wart blended into synthetic image (left), same synthetic image with secondary warts (right)

damage by atopic dermatitis are abnormal white lines and light and dark patches. The lines are mostly horizontally oriented. The finger is dry and ridge structure is in some areas of the fingerprint image less visible than in an image of a healthy finger. On the other hand, other parts of the fingerprint show unusually dark areas with a damaged ridge structure. According to Lee et al. the patches represent dystrophy of the skin and the median percentage of the surface area of dystrophy in their study was 22.80% [36]. The abnormal white lines usually run in horizontal or vertical direction and their length ranges from very short up to lines running throughout the whole fingerprint. According to the study of Lee et al. the median number of white lines per fingerprint was 12 and short horizontal lines prevailed (with occurrence in 73.0%), followed by short vertical lines (56.5%), long horizontal lines (52.5%), and long vertical lines (18.0%) [36]. The only difference is Fig. 35c which contains many large white-only patches with no ridge structure whatsoever. Overall the fingerprint image is damaged to the point where its usage is hardly imaginable [35].

Next the generation of the fingerprint eczema is described. First the light and dark patches are drawn into separate "diseased image". The number of patches is generated randomly within set boundary values. Afterwards the type (light, dark), size, and a center point for each patch is determined. Color of dark or light patches is set the same way as the color of the warts. After this, the patches are blurred. Drawing of the patches is shown in Fig. 36. The second significant part of the algorithm is drawing of the white lines. The length of the line is determined within set boundary values and the line direction (either vertical or horizontal) is set. Line points are generated from the starting point. Other leading points are generated based on the

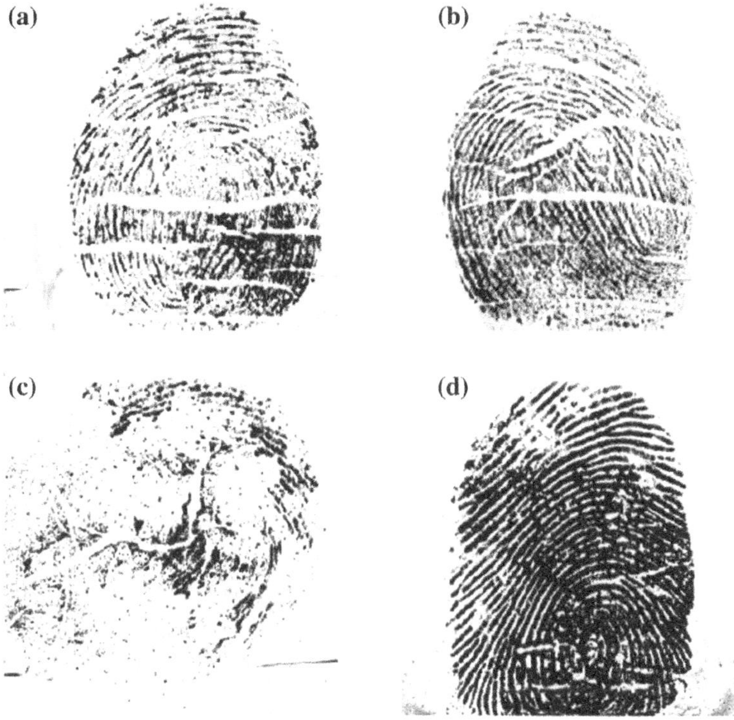

Fig. 35 A set of fingerprints affected by atopic eczema acquired by Sagem MSO 300

length of the line, the direction of the line, and a random generated angle within a pre-defined range. To make the lines look more realistic, the line leading points count is doubled and spline interpolation of the first order is applied. The thickness of the line is set and the line is drawn in several steps starting with the whole length drawn in the smallest thickness. Then the first and last leading points are removed and the line is drawn over with a higher thickness. This process repeats until the final set thickness is reached. This ensures that the lines width decreases towards line ends. Lines generation is shown in the Fig. 37. When all the lines are prepared the "diseased image" is blurred and after that blended to the original fingerprint image [35].

5.3.3 Psoriasis Vulgaris

Disease itself is described in Sect. 4 and can be seen in Fig. 13. This is the type of disease when effects can be hardly generalized. But we still can extract specific objects from diseased image and introduce them into the synthetic images. Extraction can be done manually or semi-automatically by color differentiation, canny edge detector with help of other morphological operators, etc. Example of object of interest

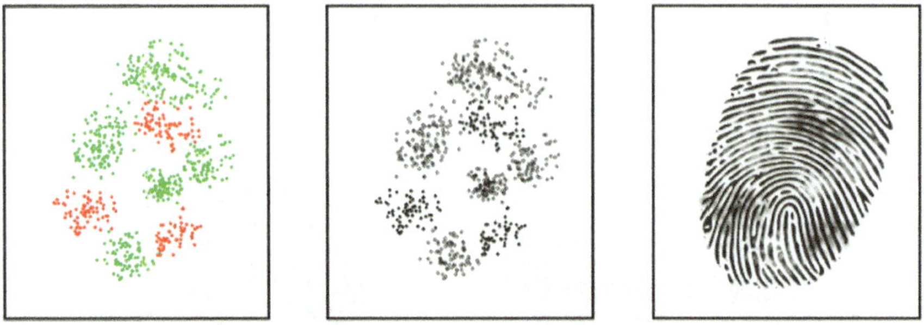

Fig. 36 Eczema patches generation—in distinctive colors (left), in final color (middle), blended to synthetic fingerprint image (right)

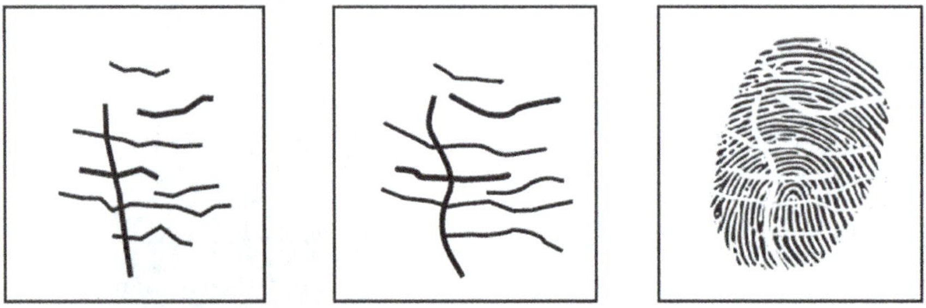

Fig. 37 Eczema lines generation—line leading points (left), interpolated line leading points (middle), lines blended to synthetic fingerprint image (right)

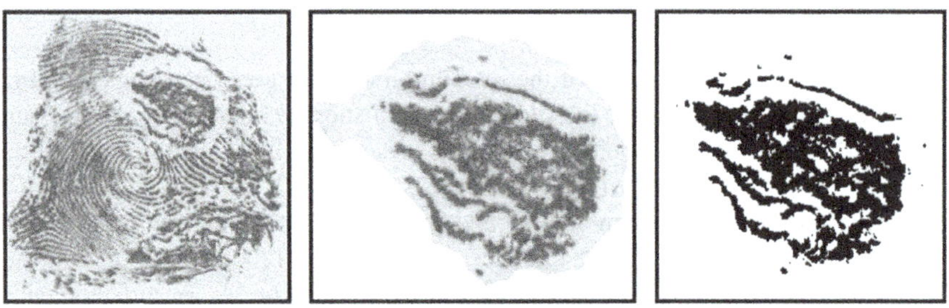

Fig. 38 Object extraction and preparation—original image (left), object extracted (middle), object ready for blending into synthetic fingerprint image (right)

from diseased image is shown in Fig. 38. Similarly to previous methods colors of extracted image has to be altered to fit new image. Using a database of these objects can ensure creation of realistically looking synthetic fingerprint image. On the other hand this method is rather semi-automatic and getting large database of different impressions is time consuming.

6 Conclusion

The area of fingerprints is one of the most explored and experienced biometrics. Liveness detection technology have to be integrated in the fingerprint sensor, of course. Most manufacturers are already doing this, and liveness detection really becomes an obvious part of fingerprint reader hardware.

This chapter includes a description of fingerprints which are common in a population, however cause big troubles in automatic processing and recognition. Before the methods for detection and extraction of diseases were designed, the faculty's database of fingerprint images affected by skin disease was thoroughly analyzed and, based on this foundation, possible algorithms for damaged area detection were tested. The best results were achieved in *Detector* by connecting the methods (block orientation field, histogram analysis and flood fill) together.

The *Classifier* makes decisions based on statistics that resulted from testing the algorithms on the whole fingerprint database. Using the methods described in this work, the program reached an accuracy of 83.5% for acrodermatitis, 45.3% for atopic eczema, 58.3% for psoriasis and 75.0% for verruca vulgaris.

There is a great potential for improvements and enhancements, and it is assumed that the research will continue. At the moment we work on algorithms for detection of concrete troubles in fingerprints and other skin diseases. Furthermore we work on development of algorithms for simulation of diseases and damages into synthetic fingerprints. In this chapter basis of the synthetic fingerprint generation and creation of diseased impression was described. Methods for simulation of verruca vulgaris and fingerprint eczema were shown. Other methods based on general damages done by diseases or semiautomatic damage extraction were proposed.

Acknowledgements This work was supported by The Ministry of Education, Youth and Sports of the Czech Republic from the National Programme of Sustainability (NPU II); project IT4Innovations excellence in science—LQ1602.

References

1. W.J. Herschel, *The origin of finger-printing* (Oxford University Press, H. Milford, 1916)
2. F. Galton, *Fingerprints* (Macmillan and Co., London, 1892)
3. F. Galton, Hereditary talent and character. Macmillan's Mag. **12**(157–166), 318–327 (1865)
4. Web page. https://www.fbi.gov/services/cjis/fingerprints-and-other-biometrics/ngi. Accessed on 5 Jan 2018
5. A. Jozefek, *Principy některých daktyloskopických klasifikačních systémů* (*Principles of some dactyloscopic classification systems*), Ústav kriminalistiky PF UK, 1972
6. D. Maltoni et al., *Handbook of Fingerprint Recognition* (Springer Science & Business Media, 2009). ISBN 978-1-84882-254-2
7. E. Auksorius, A.C. Boccara, Fingerprint imaging from the inside of a finger with full-field optical coherence tomography. Biomed. Opt. Express **6**, 4465–4471 (2015). https://doi.org/10.1364/boe.6.004465
8. D.A. Godse, *Computer Graphics* (Technical Publications, 2008). ISBN 9788189411589

9. J.C. Russ, F.B. Neal, *The Image Processing Handbook*, 7th edn. (CRC Press, 2016), p. 1035, ISBN 978-1-4987-4028-9
10. L. Hong, Y. Wan, A. Jain, Fingerprint image enhancement: algorithm and performance evaluation. IEEE Trans. Pattern Anal. Mach. Intell. **20**(8), 777–789 (1998)
11. R. Cappelli, M. Ferrara, D. Maltoni, Minutia cylinder-code: a new representation and matching technique for fingerprint recognition. IEEE Trans. Pattern Anal. Mach. Intell. **32**(12), 2128 (2010). https://doi.org/10.1109/tpami.2010.52
12. A.M. Bazen, G.T.B. Verwaaijen, S.H. Gerez, L.P.J. Veelenturf, B.J. Zwaag van der, A correlation-based fingerprint verification system, in ProRISC 2000 Workshop on Circuits, Systems and Signal Processing, Veldhoven (NL) (2000)
13. I. Koichi, N. Hiroshi, K. Koji, A. Takafumi, H. Tatsuo, A fingerprint matching algorithm using phase-only correlation. IEICE Trans. Fundamentals **E-87A**(3), 682 (2004)
14. V. Krivec, J. Birchbauer, W. Marius, H. Bischof, *A Hybrid Fingerprint Matcher on Card*, in Proceedings of the 1st Conference on Biometrics and Electronic Signatures of the GI Working Group BIOSIG, 2003
15. F. Liu, D. Zhang, L. Shen, Study on Novel Curvature Features for 3D Fingerprint Recognition. Neurocomputing **168**, 599–608 (2015). https://doi.org/10.1016/j.neucom.2015.05.065
16. C. Hong, T. Jie, Y. Xin, *Fingerprint matching with registration pattern inspection*, in Proceedings of the 4th International Conference on Audio- and Video-Based Biometric Person Authentication, Guildford (UK), 2003
17. N. Ratha, R. Bolle, *Automatic Fingerprint Recognition Systems* (Springer, New York, 2004)
18. M. Drahanský, O. Kanich, E. Březinová, *Is Fingerprint Recognition Really So Reliable and Secure? Challenges for fingerprint recognition—Spoofing, Skin Diseases and Environmental Effects*, Handbook of Biometrics for Forensic Science. Springer, p. 21, ISBN 978-3-319-50671-5
19. T.P Habif, *Clinical Dermatology*, 4th edn. (Mosby, China, 2004), p. 1004, ISBN 978-0-323-01319-2
20. K. Wolff, R.A. Johnson, D. Suurmond, *Fitzpatrick's Color Atlas and Synopsis of Clinical Dermatology*, 5th edn. (McGraw-Hill, USA, 2005), p. 1085, ISBN 0-07-144019-4
21. W.D. James, T.G. Berger, D.M. Elston, *Andrews' Diseases of the Skin Clinical* Dermatology, 10th edn. (Elsevier Inc., Canada, 2006). ISBN 0-7216-2921-0
22. Jr., W.C.S.: *Raynaud's Phenomenon*. 2016. Retrieved from on 2017-11-01: http://www.medicinenet.com/raynauds_phenomenon/page4.htm
23. M. Drahanský, R. Pernický, O. Kanich, Š. Barotová, *Verarbeitung von beschädigten Fingerabdrücken in der polizeilichen Praxis*. DuD - Datenschutz und Datensicherheit. **41**(7), 407–414 (2017). ISSN 1614-0702
24. Š. Barotová, *Detector of Skin Diseases by Fingerprint Technology*. Bachelor's thesis. FIT BUT, 2017
25. D. Powers, *Evaluation: From Precision, Recall and F-measure to ROC*, Informedness and Correlation. J. Mach. Learn. Technol 2, 37–63(2011). ISSN 2229-3981. Retrieved from on 05 Sept 2017. http://www.bioinfo.in/contents.php?id=51
26. O. Kanich, M. Drahanský, *Simulation of Synthetic Fingerprint Generation Using Petri Nets*, IET Biometrics, 13 (2017). ISSN: 2047-4938
27. W. Reisig, *Understanding Petri Nets* (Springer, New York, 2013), p. 230. ISBN 978-3-642-3327-7
28. E. Badouel, L. Bernardinello, P. Dorondeau, *Petri Net Synthesis* (Springer, New York, 2015), p. 339. ISBN 978-3-662-47966-7
29. L. Popova-Zeugmann, *Time and Petri Nets* (Springer, New York, 2013), p. 209. ISBN 978-3-642-41114-4
30. M. Češka, *Petriho sítě*. Akademické nakladatelství CERM Brno, 1994. ISBN 8-085-86735-4
31. R. Chaloupka, *Generátor otisků prstů*. Master's thesis. FIT BUT, 2007
32. Q. Zhao, A.K. Jain, N.G. Paulter, M. Taylor, *Fingerprint Image Synthesis Based on Statistical Feature Models*, in 2012 IEEE Fifth International Conference on Biometrics: Theory, Applications and Systems (BTAS), pp. 23–30. ISBN 978-14-673-1384-1. http://ieeexplore.ieee.org/stamp/stamp.jsp?tp=&arnumber=6374554&isnumber=6374538

33. R. Cappelli, *SFinGe: an Approach to Synthetic Fingerprint Generation*, in BT 2004—International Workshop on Biometric Technologies. Calgary, Canada: 2004, pp. 147–154
34. A.H. Ansari, *Generation and Storage of Large Synthetic Fingerprint Database*, Master's thesis, Indian Institute of Science, 2011
35. M. Bárta, *Generation of Skin Disease Into the Synthetic Fingerprints*, Master's thesis. FIT BUT, 2016
36. C. Lee, C. Chang, A. Johar, O. Puwira, B. Roshidah, *Fingerprint Changes and Verification Failure Among Patients with Hand Dermatitis*. JAMA Dermatol. **149**(3), 294–299 (2013). ISSN 2168-6068

Iris Recognition in Cases of Eye Pathology

Mateusz Trokielewicz, Adam Czajka and Piotr Maciejewicz

Abstract The purpose of this chapter is to provide an insight on iris recognition system performances for users suffering from some ocular pathologies. Such situations are analyzed in terms of influences and possible means are discussed to take this case into account during the phase of iris samples matching.

1 Introduction

1.1 Iris Recognition

Automatic iris recognition has emerged as an important biometric identity recognition method more than two decades ago, although the concept of recognizing people by their irises is known to have a history of 80 years. Frank Burch was the first to suggest that iris texture, and not the eye color, can be an important human identifier [1]. This idea was later reproduced in ophthalmology textbooks to finally fall upon a breeding ground in 1993, when the first, automatic iris recognition method based

M. Trokielewicz (✉)
Biometrics Laboratory, Research and Academic Computer Network (NASK),
Kolska 12, 01-045 Warsaw, Poland
e-mail: mateusz.trokielewicz@nask.pl

M. Trokielewicz
Institute of Control and Computation Engineering, Warsaw University of Technology,
Nowowiejska 15/19, 00-665 Warsaw, Poland

A. Czajka
University of Notre Dame,
384, Fitzpatrick Hall of Engineering, Notre Dame, IN 46556, USA
e-mail: aczajka@nd.edu

P. Maciejewicz
Department of Ophthalmology, Medical University of Warsaw,
Lindleya 4, 02-005 Warsaw, Poland
e-mail: piotr.maciejewicz@wum.edu.pl

A. Nait-Ali (ed.), *Biometrics under Biomedical Considerations*,
Series in BioEngineering, https://doi.org/10.1007/978-981-13-1144-4_2

on 2D Gabor wavelets was proposed by Daugman [2]. Numerous algorithms were proposed to date, most of them inspired by the original Daugman's invention. A common pipeline of these systems begins with image acquisition in near infrared light, which is used to make this process comfortable for users, as near infrared light is almost invisible to humans. Since this kind of illumination is hardly absorbed by melanin pigment present in the iris, as opposed to visible light, the iris texture information can be efficiently extracted even for highly pigmented, dark irises. Occlusions such as eyelids, eyelashes and specular reflections are automatically detected. The remaining, non-occluded image of iris texture is processed with various filters (e.g., 2D Gabor) and filtering results are quantized to two values depending only on their signs, ending up with a binary code as a feature vector. Since iris images are normalized prior to filtering, the iris codes have identical structure for all irises. This makes it possible to use the exclusive disjunction (XOR) to calculate the number of disagreeing bits, and hence the dissimilarity score at a very high speed. While false rejections in iris recognition are mostly due to variations in the acquisition process, impostor score distributions are not influenced by such factors making the probability of false acceptance to be predictable and relatively low. Good surveys of various iris recognition methods and important research problems related to iris recognition may be found in Bowyer et al. [3–6].

The iris, placed anteriorly in the human eye and protected by the cornea, consists of a stroma, which holds a fibrovascular mesh, and of a layer of interworking muscles, whose role is to control the amount of light getting into the eyeball. The iris is relatively easy to be observed and measured. Only the structural layout of the iris' trabecular meshwork is analyzed for the purpose of extracting individual features. Neither the color, nor other iris global features (such as tissue density) are used in biometric recognition. It is believed that high degree of structural richness and uniqueness observed in iris patterns are caused by limited dependence on human genotype (a.k.a. low 'genetic penetrance'). Consequently, iris recognition is often envisioned as almost 'ideal biometrics', characterized by low error rates, robustness against variations in time, immunity to diseases, resistance to forgeries, and neutrality in terms of social, religious and ethical aspects. Numerous large-scale installments, such as NEXUS program [7], which offers dedicated processing lanes to pre-screened travelers when entering the United States and Canada, or AADHAAR [8], which applies biometrics for de-duplication of unique personal IDs in India, and evaluation programs such as IREX [9], focused on interoperability of iris recognition, present its high usability in operational scenarios.

More than two decades of operational practice and experience allowed to identify new research challenges in iris recognition. Due to successful spoofing attacks targeted at commercial iris sensors, numerous liveness detection methods were proposed at the sensor and software levels [10]. Temporal stability of iris patterns seems to be lower than initially believed [11–15] and hard to be estimated as we may observe even contradictory conclusions drawn from the same datasets [16, 17]. New research in post-mortem iris recognition [18, 19] reveals that it is possible to use this biometric characteristic in forensic applications. This chapter is dedicated to yet another important aspect of iris recognition reliability and presents how its reliability is influenced by different eye diseases.

1.2 Possible Influence of Ocular Disorders

In the ISO/IEC 29794-6 standard that provides requirements for iris image quality, two possible scenarios of eye disease impact on iris recognition are distinguished [20]:

Scenario 1 in which injury or illness occurs when the affected person is already using a biometric system as a registered user (e.g., a medical procedure affecting the eye is performed or an illness influencing the iris takes place); in this case, we may observe a degradation in recognition system's performance when pre-disease images are compared to those obtained afterwards;

Scenario 2 in which injury or illness is present before the enrollment and therefore an overall performance of a biometric system may be worse than when presented with healthy eyes; in severe cases the eye may not be suitable for iris recognition at all (e.g., person suffers from a congenital disease called aniridia, in which only the iris tissue is absent, often leaving an irregularly shaped pupil [21]).

For both scenarios the ISO/IEC standard specifies medical conditions that may apply. The first case includes excessive dilation or constriction of the pupil (associated with disease, trauma or abuse of drugs and alcohol), illnesses that affect the iris itself and the cornea, behind which the iris is located (e.g., iritis, micro- and megalocornea, keratitis, leukoma), the aforementioned congenital diseases, such as aniridia or iris hypoplasia (underdevelopment), surgical procedures (cataract surgery, laser therapy in glaucoma treatment, iridectomy) along many other pathologies—disease, age or injury-related.

It is also possible to imagine a third scenario, that is not explicitly defined by the ISO/IEC standard:

Scenario 3 in which patient registered in a biometric system while already suffering from certain eye disease undergoes a treatment (e.g., a lens replacement surgery in cataract conditions); this may lead to worse recognition when compared to condition before the treatment.

Although ISO/IEC standard focuses on evaluating the impact of certain eye disorders on iris recognition, it is extremely difficult to gather data that represents diseases that are isolated from others, as different eye disorders often occur simultaneously. Therefore, a different approach to classifying disease influence may be proposed. This incorporates assembling selected medical conditions into groups that represent certain types of impact on the eye structures, and thus may affect iris recognition performance in different ways. These include eyes with no visible change, but with disease present, eyes with geometrically distorted pupils and irises, eyes with irises altered or damaged or, finally, eyes with irises and pupils obstructed by pathological objects located in front of them. Such classification approach can stimulate more practical conclusions, such as putting forward changes in the eye that lead to worse recognition, which would be easy to identify during visual inspection by an expert.

1.3 Terms and Definitions

FNMR **False Non-Match Rate**—an estimator of the false non-match probability, defined as considering a sample as not matching a given template when only samples matching this template are presented (i.e. *genuine* attempts—presenting same-eye iris images); FNMR is a function of an acceptance threshold τ:

$$FNMR(\tau) = \frac{\#\, of\ attempts\ not\ matching\ the\ template\ (\tau)}{\#\, of\ all\ genuine\ attempts}$$

FMR **False Match Rate**—an estimator of the false match probability, defined as considering a sample as matching a given template while only samples not matching this certain template are presented (i.e. *impostor* attempts—presenting iris images of different eyes); FMR is a function of an acceptance threshold τ:

$$FMR(\tau) = \frac{\#\, of\ attempts\ matching\ the\ template}{\#\, of\ all\ impostor\ attempts}$$

EER **Equal Error Rate**—an acceptance threshold τ value, at which both False Match Rate and False Non-Match Rate are equal:

$$EER = FMR(\tau) = FNMR(\tau)$$

FTE **Failure to Enroll**—a proportion of samples that could not be enrolled to the overall number of samples, from which there was an enrollment attempt:

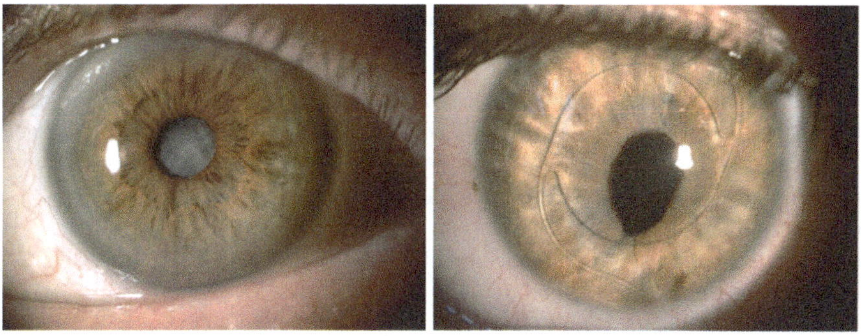

Fig. 1 **Left** cataract-induced lens clouding. **Right** artificial lens placed in the anterior chamber instead of behind the iris

$$FTE(\tau) = \frac{\#\ of\ failed\ attempts\ to\ create\ a\ template}{\#\ of\ all\ attempts\ to\ create\ a\ template}$$

2 Description of Medical Conditions

2.1 Cataract

Disease characteristics. Cataract is one of the most common eye pathologies, being a complete or partial lens opacification leading to loss of transparency and ability to properly focus light onto the retina, Fig. 1. This results in a blurred and dimmed vision. Cataract accounts for as much as 30% of blindness and visual impairment worldwide, mostly in developing countries.

Cataracts may be classified twofold: congenital cataracts, that occur even before the birth or developing in the first few years of life, caused by congenital metabolic disorders, ionizing radiation, certain drugs (sulfonamides, corticosteroids), infections in the uterus. The second type are the more common, usually age-related cataracts or secondary cataracts, induced by systemic diseases, accompanying other eye pathologies, or as a result of trauma, ionizing radiation, or toxins [22]

Treatment incorporating lens replacement. Cataract treatment focuses solely on a surgical procedure of lens extraction (called *phacoemulsification*) using aspiration, during which a small probe is inserted through an incision in the side of the cornea and then through a circular hole in the lens capsule to reach the lens. The probe then emits ultrasound waves to break the opacified lens, which is later removed using suction. The resulting condition, *aphakia*, can be compensated using either prescription glasses, contact lenses, or an intraocular implant (artificial lens) with individually calculated focusing power [23].

Influence on the remaining eye structures. Cataract itself should not have any significant impact on other parts of the eyeball, including the iris. However, there are many other pathologies that accompany cataract and may cause damage or alteration to the iris [23], for instance:

- acute glaucoma (possible flattening of the iris and pupil distortion),
- anterior and posterior synechiae (iris adhered to either the lens or to the cornea, respectively, Fig. 4),
- rubeosis iridis (pathological vascularization in the iris),
- iris atrophy,
- pseudoexfoliation syndrome (accumulation of protein fibers inside the eye),
- conditions after iris laserotherapy,
- post-traumatic cataract.

In addition to this, in some cases the surgical cataract extraction incorporates placing the lens implant not behind, but in front of the iris. This happens when there is not enough capsular support remaining after the lens extraction and in such cases the implant has to be attached to the circumferential part of the iris or supported on the iridocorneal angle using claw-shaped hooks, Fig. 1. This may affect the look of the iris and even distort the circular shape of the pupil [23].

2.2 *Glaucoma*

Disease characteristics. Glaucoma is a group of ocular disorders, typically described as damage to the optical nerve followed by visual impairment as a result of increased intraocular pressure, however, in certain cases these may happen with low or normal pressure levels. Therefore, glaucoma should be given a description of a multi-factorial optical nerve neuropathy, with increased intraocular pressure being a risk factor [22]. Glaucoma is a second-leading cause of visual impairment and blindness worldwide, secondary to cataracts only [24].

Glaucoma and its impact on iris. This disease can be subdivided into two main categories, namely the open-angle and the closed-angle glaucoma. The former type is chronic and often does not cause any pain as it progresses slower than the latter. The second type of glaucoma, known as the acute glaucoma, usually happens suddenly when the angle between the iris and the cornea closes completely, preventing the aqueous humor from flowing through the trabecular meshwork towards the inside of the eyeball, causing a sudden increase of the intraocular pressure and severe pain to the patient. Aqueous humor pushing against the iris and the lens may cause certain flattening of the iris and distort the shape of the pupil, as shown in Fig. 2. While the first symptom is often not visible, the other one is significant and can contribute to a decrease in iris recognition performance as it affects the segmentation procedures that often approximate iris and pupil boundaries with circles. Also, an increase in the intraocular pressure can cause a corneal edema reducing the cornea's transparency, and thus making it difficult to get a clear image of the iris behind it.

Treatment. Glaucoma treatment, usually associated with reducing the pressure inside the eyeball, incorporates either a surgically performed, triangle-shaped inci-

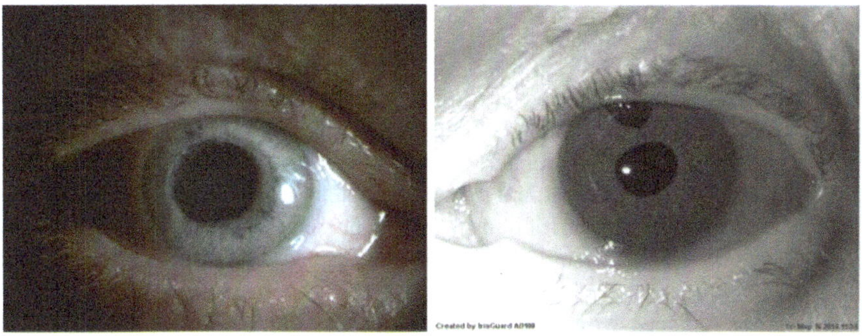

Fig. 2 **Left** a distortion in pupil shape resulting from acute (closed-angle) glaucoma. **Right** iris tissue damage due to the iridectomy procedure

sion in the upper part of the iris (iridotomy) or making a small, circular puncture also in the upper part of the iris (iridotomy). Obviously, both these procedures affect the look of the iris. However, the incision created during iridotomy is usually very small and hidden under the upper eyelid, and thus rarely visible. The triangle-shaped cutout, however, is rather large and may significantly affect the look of the iris, Fig. 2.

2.3 Posterior and Anterior Synechiae

Synechiae occur when the iris becomes attached either to the lens (posterior synechiae) or to the cornea (anterior synechiae). They may be caused by various conditions, in most cases by ocular trauma, iritis (inflammation of the iris) or other forms of uveitis (inflammation of the uvea). **Posterior synechiae** become clearly visible when pharmacological pupil dilation is performed and can alter the shape of the pupil significantly, as well as bring opacification to the surface of the lens, making it brighter than normally, Fig. 3. When synechiae accompany cataract, after the lens extraction there may be deformation left, even though the iris is no longer attached to the lens, Fig. 4. In some cases, this condition may lead to glaucoma, as the aqueous humor flow is partially blocked. **Anterior synechiae**, when the iris adheres to the cornea, often lead to glaucoma due to closing the angle between the iris and the cornea and blocking the flow of the aqueous humor through the trabecular meshwork. This condition, however, does not directly affect the look of the iris or the pupil.

2.4 Retinal Detachment

Subdividing into many conditions, **retinal detachment** can be roughly summarized as a disorder, in which the retina—a light-sensitive part of the internal membrane of the eyeball—detaches from the underlying layers, creating a severe visual impairment leading to complete blindness if not treated immediately [22]. The illness itself does not have a direct impact on the iris or on the lens. The treatment, however, incorporates filling the eyeball with silicon oil to push the detached retina towards the back of the eyeball and then fusing it in place using laser photocoagulation. The oil is then removed and replaced with artificial fluid that substitutes the vitreous humor. In certain cases, the oil may flow back to the anterior chamber of the eye, Fig. 5, left. This creates an obstruction that alters the look of the iris. We also came across one case in which the retinal detachment was accompanied with wide, oval, non-reacting pupil with partial iris atrophy, Fig. 5, right. These are obvious candidates to cause potential trouble for iris recognition.

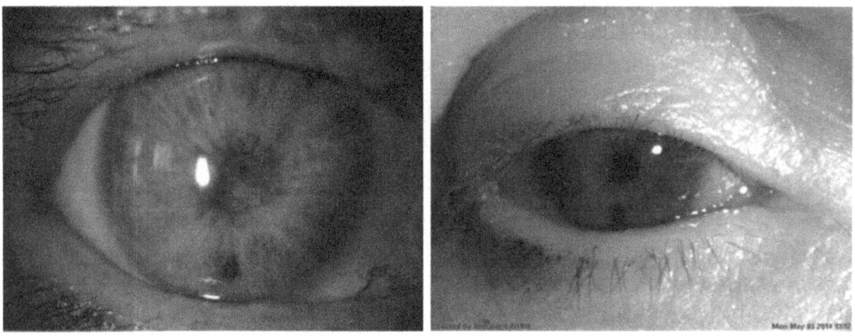

Fig. 3 **Left** posterior synechiae seen in visible light. **Right** same, but in NIR illumination. The pupil appears dark in the NIR image, in contrast to the visible light image

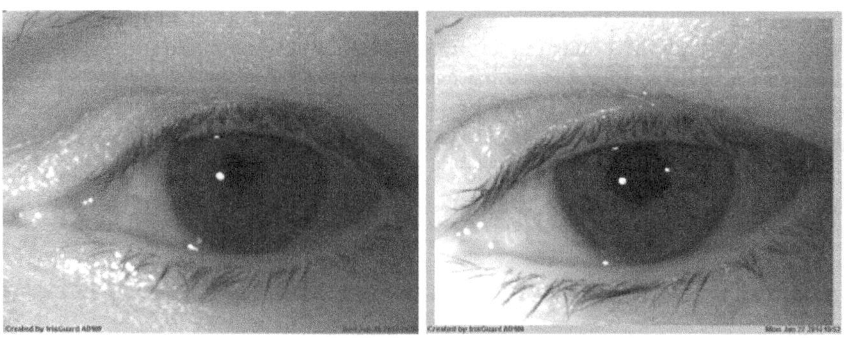

Fig. 4 Cataract eye with posterior synechiae before (**left**) and after the procedure of lens removal (**right**). Distortion of the inner iris boundary remains even after the lens has been extracted and the synechiae have been removed

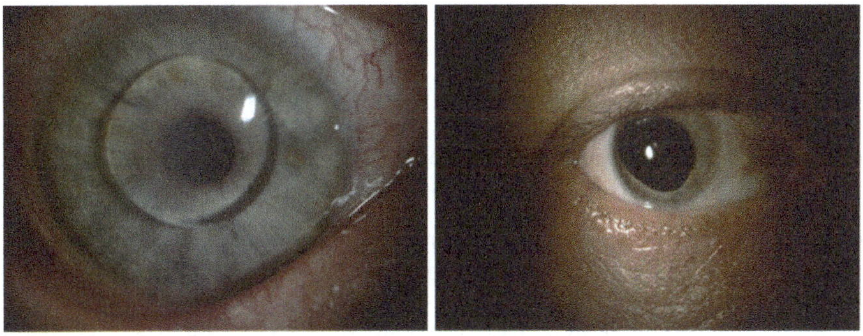

Fig. 5 Conditions that may accompany retinal detachment or occur during treatment: silicon oil in the anterior chamber of the eye (**left**), distortion of the pupil (**right**)

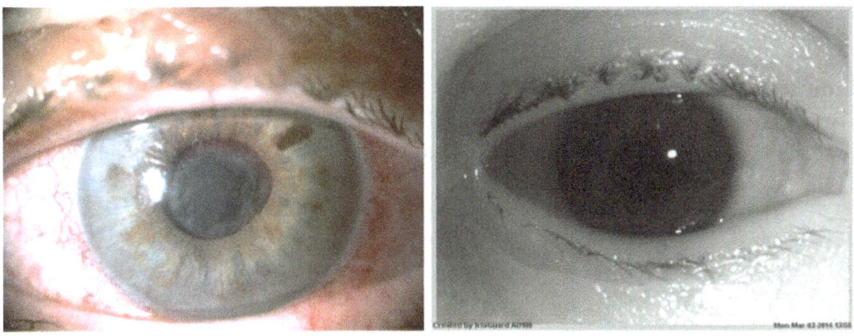

Fig. 6 Pathological angiogenesis in the iris tissue—rubeosis iridis. Visible light (**left**) and NIR (**right**) illumination images show differences in the appearance of iris tissue populated with blood veins. The NIR image is much less influenced by this type of pathology

2.5 Rubeosis Iridis

Rubeosis is a pathological vascularization process afflicting the surface of the iris, often as a result of disease present in the retina, Fig. 6. Ischemic retina can release vessel growth factors to meet its oxygen needs, however, such vascularization is not a normal condition and may clog the iridocorneal angle, causing an increase in intraocular pressure [25]. This alters the look of the iris surface with the potential to affect iris recognition. However, it is worth noticing that NIR light transmission through blood is different than that of visible light, therefore the NIR-illuminated images may show this pathology as far less obtrusive than the visible light photographs. See Fig. 6 for comparison of images taken with those two types of illumination.

2.6 Other Eye Pathologies

During the process of data collection, we came across several less common eye pathologies, illnesses or conditions, often accompanying other disorders, but sometimes occurring independently. Those are divided in respect to the part of the eye they influence most. Disorders affecting the cornea were:

- pathological **vascularization (angiogenesis)** in the corneal tissue, causing occlusions that partially prevent the light from entering the eye, and also obstructing the view of iris pattern, Fig. 7,
- **corneal haze, ulcers or opacities** of different origin, with consequences similar to the ones associated with pathological angiogenesis (obstruction of the iris pattern),
- **corneal grafting** with grafts sutured and therefore obstructing the view of the iris, Fig. 8.

Disorders affecting the iris were:

- **sutures** in the iris,
- **iris dialysis** with a piece of iris tissue detached from the base of the iris,
- **damage or atrophy to the iris tissue** resulting in missing fragments of its structure, Fig. 9.

3 Related Work

3.1 Cataract and Phacoemulsification Procedure Influence

A study by Roizenblatt et al. [26] involved 55 patients suffering from cataract. Enrollment for each eye was performed using an iris biometric system (LG IrisAccess 2000)

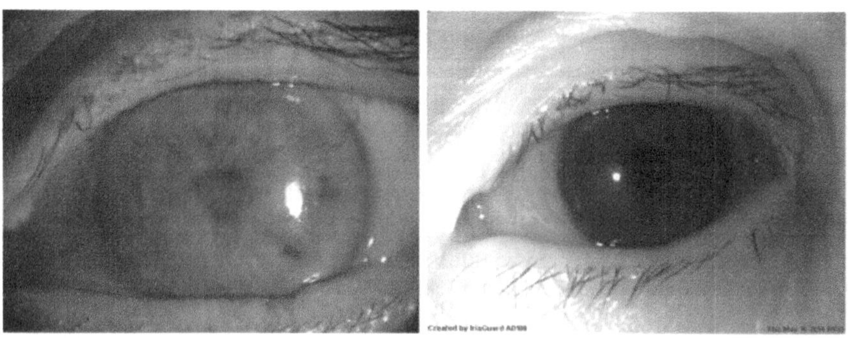

Fig. 7 Pathological angiogenesis in the cornea creates a haze obstructing the view of the iris below it. Visible (**left**) and NIR (**right**) illumination comparison shows much less impact of this type of pathology with the latter

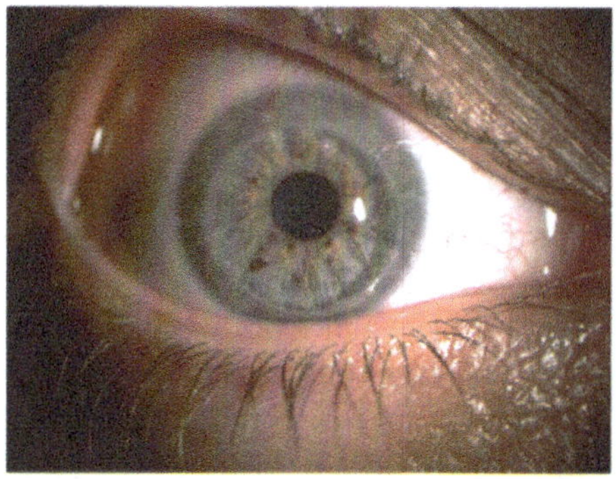

Fig. 8 Grafted cornea with visible sutures obstructing the view of the iris. Also, the newly implanted cornea is far more transparent than the remainings of the original tissue

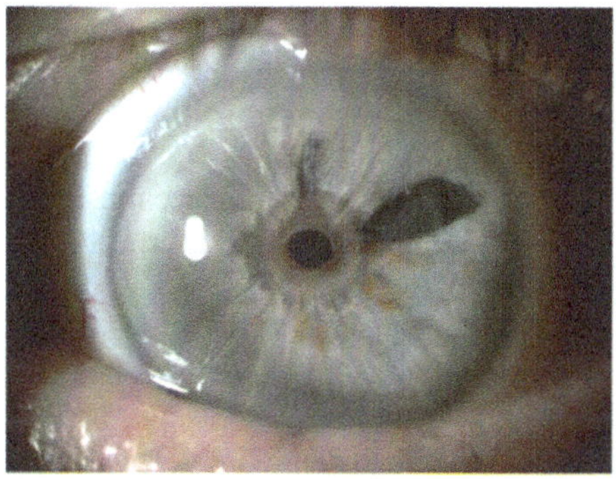

Fig. 9 Rupture-like damage to the iris tissue revealing the lens underneath it and altering the iris pattern by removing significant portions of the iris tissue

before cataract extraction. Then, verification was performed, three times before the surgery, and three times after the cataract extraction (the latter three verification trials were performed 30 days after the procedure and 7 days after stopping the administration of pupil-dilating drugs, i.e., a time period, after which healing should be finished). After such time period, pupils are expected to revert to their normal size and reaction to light, which is later confirmed by authors' observations, revealing differences in the size no larger than 1.5 mm when compared with images collected before the treatment. The verification experiments showed an increase in Hamming distance when average HD score obtained in post-surgery trials (HD = 0.2094) is compared with average HD score obtained in pre-surgery trials (HD = 0.098). The biometric system failed to recognize 6 out of 55 eyes (FNMR of approx. 11% is reported). It is worth noticing that authors used a very liberal acceptance threshold of 0.4. To come up with possible explanation of worse performance, the authors assigned a visual score between 0 and 4, given by an ophthalmology surgeon, to each of the eyes that underwent the cataract extraction procedure. One point was assigned for each of the following ocular pathologies: depigmentation, pupil ovalization, focal atrophy with and without transillumination. Statistical analysis revealed a correlation between iris pattern deviation intensity and HD shift towards worse (i.e., higher) scores. Authors suggest that probe manipulation and energy released in the eyeball during the procedure may cause atrophic changes to the iris tissue. They also make a claim to be able to predict trouble with iris recognition based on a visual inspection of the eye that underwent the procedure. In such cases, re-enrollment is suggested.

Phacoemulsification refers to the extraction of the lens through aspiration. The procedure involves the insertion of a small probe through an incision in the side of the cornea. The probe emits ultrasound waves that break the opaque lens which is later removed using aspiration [23]. Seyeddain et al. [27] conducted a study on the effects of phacoemulsification and pharmacologically induced mydriasis on iris recognition. The experiment aimed to determine whether irises, following phacoemulsification or drug induced mydriasis (preventing the dilated pupil from reacting to light stimulation) perform worse when compared to the same irises before the procedure or before the drug-induced pupil dilation. They revealed that 5.2% of the eyes subject to cataract surgery could no longer be recognized after the procedure. In the pupil dilation group, this portion reached as high as 11.9%. In both cases the authors suggest re-enrollment for patients whose eyes were not successfully identified after the surgery or instillation of mydriatics. No false acceptances were observed in either case.

In our previous work [23] we presented an experimental study revealing weaker performance of the automatic iris recognition methods for cataract-affected eyes when compared to healthy eyes. A database of 1288 eye images coming from 37 ophthalmology patients has been gathered. To assess the extent of recognition accuracy deterioration, three commercial and academic iris recognition methods were employed to calculate genuine match scores for healthy eyes and those with cataracts. A significant degradation in recognition accuracy was shown for all three matchers used in this study (12% of genuine score increase for an academic matcher, up to 175% of genuine score increase obtained for an example commercial matcher). False-non match rates were affected by cataract-induced changes in two out of three iris matchers.

In a study by Dhir et al. [28], the effects of the use of mydriatics accompanying cataract surgery, as well as the effects of the procedure itself, are examined. A group of 15 patients had their eyes enrolled prior to cataract extraction surgery. Four verification attempts were then performed: 5, 10, and 15 min after the application of pupil-dilating drugs (but still before the surgery), and finally—14 days after the surgery. Pupil dilation due to mydriatics use caused Hamming distances to gradually increase as time after drug instillation elapsed. This led to FNMR of 13.3% (6 out of 45 verification attempts failed). Surprisingly, none of the eyes were falsely rejected in the verification attempt conducted 14 days after the cataract removal procedure. However, the authors excluded from the dataset eyes with pre-existent corneal and iris pathologies, or those with iris tissue damaged during the surgery. This brings a certain bias to the analysis, thus the results may not reflect a real-world application. The study suggests that decreased recognition accuracy after the cataract surgery originates from a slight shift of the iris towards the center of the eyeball as a result of implanting an artificial lens that is thinner than the natural one. Specular reflections from the implant may also contribute when certain localization methods are employed. Authors also warn that excessive pupil dilation can be exploited by criminals in order to enroll under multiple identities to deceive law enforcement.

Recently, Ramachandra et al. [29] conducted experiments regarding iris biometrics in the context of cataract surgery, when verification is performed using pre-surgery gallery samples and post-surgery probe samples, coming from the same individual. Studies were carried out using a database of iris images acquired from 24 h pre-surgery and 36–42 h post-surgery from 84 subjects. Recognition accuracy is reported to drop significantly, reaching genuine match rate of 85.19% @ FMR = 0.1 and EER = 7%, when compared to performance achieved using pre-surgery images only.

3.2 *Refractive Surgeries*

Laser-assisted refractive correction surgeries and their possible impact on the accuracy of iris biometrics is studied by Yuan et al. [30]. These procedures involve making an incision in the cornea to create a flap with a hinge left on one side. The flap is then lifted to expose the middle part of the cornea, which is then ablated using short pulses of a 193 nm excimer laser to achieve a finely-tuned shape that depends on the treated condition. The corneal flap is then folded back and eye is left to heal itself. These methods are commonly known as *Laser-Assisted in Situ Keratomileusis* method, abbreviated *LASIK*. *LASIK* procedures are widely used across the globe to improve life quality for patients suffering from myopia, hypermetropia, or astigmatism. An experiment is thus carried out to find out whether such manipulation may result in false non-matches when iris biometrics is employed. Using Masek's method, 13 eyes out of 14 were correctly recognized after the procedure. However, the one eye that was falsely rejected had a significant deviation in circularity of the pupil and increased pupil diameter. The authors argue that refractive correction procedures have little effect on iris recognition. Nonetheless, more experiments involving larger datasets are called for.

3.3 *Other Ocular Pathologies*

Aslam et al. [31] conducted one of the most extensive studies in terms of disease representation in the database. 54 patients suffering from various diseases had their eyes imaged with an IrisGuard H100 sensor and then enrolled in a biometric system based on Daugman's idea, before they underwent any treatment. Then, verification attempts were carried out after the treatment and Hamming distances (HD) between iris codes obtained before and after the treatment were calculated to tell whether medical procedures applied had any impact on recognition accuracy. Iris recognition method employed by the authors turned out to be resilient for most illnesses, i.e., glaucoma treated using laser iridotomy, infective and non-infective corneal pathologies, episcleritis, scleritis and conjunctivitis. However, 5 out of 24 irises affected by anterior uveitis, a condition in which the middle layer of the eye, the uvea, which includes the iris and the ciliary body, becomes inflamed [32], were falsely rejected after the treatment, hence FNMR = 21% in this particular subset. Rejected eyes had earlier been administered with mydriatics and thus had pupils significantly dilated. In addition, two eyes suffered from high corneal and anterior chamber activity, while the remaining three had posterior synechiae that caused deviation from the pupil circularity. Hypothesis stating that the mean HD in the anterior uveitis subset is equal to mean HD in the control group consisting of healthy eyes has been rejected with $p < 10^{-4}$, while there were no statistically significant differences between scores obtained from other disease subsets when compared to the control group. For pathologies related to corneal opacities, Aslam tries to explain the lack of recognition performance deterioration by the fact that NIR illumination used in iris biometrics is more easily transmitted through such objects and therefore allows correct imaging of underlying iris details. Laser iridotomy also showed little influence, as the puncture in the iris tissue made by laser appears to be too small to alter the iris pattern significantly. However, certain combinations of synechiae and pupil dilation can affect the look of the iris texture vastly enough to produce recognition errors. A deviation in pupil's circularity caused by synechiae may also contribute to segmentation errors.

Borgen et al. [33] conducted a study focusing on iris and retinal biometrics in which they use 17 images chosen from the UBIRIS database and then digitally modified to resemble changes to the eye structures caused by various ocular illnesses: keratitis and corneal infiltrates, blurring and dulling of the cornea, corneal scarring and surgery, angiogenesis, tumors and melanoma. High FNMR values (32.8–86.8%) are reported for all modifications, except for the pathological vascularization (6.6%), changes in iris color (0.5%) and iridectomy, for which FNMR = 0%. Faulty segmentation is supposed to be the main cause, especially in cases with present corneal clouding. Authors, however, do not acknowledge the fact that near-infrared illumination enables correct imaging even in eyes with corneal pathologies such as clouding or other selected illness-related occlusions.

McConnon et al. [34] studied disorders causing pupil/iris deformation, pupil/iris occlusion and those with no iris or with a very small iris, to estimate the impact they may have on the reliability of iris image segmentation. Encoding and matching are not executed in their study. However, due to lack of publicly available datasets, the authors are forced to use images coming from the Atlas of Ophthalmology. These are photographs obtained in visible light, and thus not always suited for iris recognition, which typically uses NIR-illuminated, 8-bit grayscale images. The dataset was downsampled to 320×240 resolution and manually segmented to obtain the ground truth iris localization. The authors then performed the automatic segmentation using Masek's algorithm to find out if the results would vary from those obtained when segmenting the images manually. The results suggest that segmentation stage can be influenced by the presence of the foretold pathologies, as automatic segmentation deviated, when compared to manual segmentation, by two or more pixels in 46 and 55% of images for the limbic and pupillary boundaries, respectively.

Our most recent work devoted to this subject expands earlier experiments regarding cataract-related effects and brings the novelty of assessing which types of eye-afflicted damage caused by diseases have the greatest impact on the accuracy of biometric systems employing iris recognition. Changes to the iris tissue and geometrical distortions in the pupillary area are shown to have the highest chance of degrading genuine comparison scores for three different iris recognition algorithms employed in that research [35]. This study was later extended with more data collected from ophthalmology patients over a longer period of time, as well as with experiments that revealed an increased chance of failure-to-enroll errors when iris biometric systems are presented with images obtained from patients with diseased eyes [36]. The samples that were used for these experiments were selected so that they comply with the ISO/IEC 29794-6 standard for iris image quality. Segmentation errors were suggested as the most probable source of deteriorated matching accuracy. The datasets of eye images affected by ocular disorders, used in the preliminary and the expanded research, are publicly available to all interested researchers [37, 38]. Further extension of this research, including experiments conducted with the use of an additional iris recognition method can be found in [39].

3.4 Post-mortem Iris Recognition

Post-mortem iris recognition in humans can certainly be classified as the most extreme case of 'ocular pathology'. To establish whether iris can still serve as a biometric identifier after death is crucial from the forensics point of view: '*Can it provide a tool for forensics examiners in cases when other biometric methods may be unavailable or too burdensome?*', as well as because of security issues: '*Can the iris be stolen after death and used in presentation attacks?*'.

The authors' studies regarding this important, yet maybe unpleasant topic, revealed that images obtained 27 h post-mortem can still be successfully recognized when matched against samples obtained shortly after demise (with accuracy reaching 70%), despite common statements about iris recognition being impossible to employ after death [19]. This study was later expanded to examine whether this can still be considered true when images obtained after even longer periods of time are considered. Paper [18] extends the earlier work with images obtained up to 17 days post-mortem, revealing that iris recognition can still occasionally work even as long as 407 h after a person's death. Moreover, the dataset collected for the purpose of these studies is made publicly available to all interested researchers to facilitate research in this field [40]. Other reports on post-mortem iris recognition can be found in Saripalle et al. [41] and Sansola [42].

4 Database of Iris Images Collected from Patients with Ocular Disorders

We had a rare opportunity of close cooperation with an ophthalmologist's office, which provided us with the dataset that represents various medical conditions affecting the iris and its surrounding structures. This database is detailed in the following section.

4.1 Data Collection Process

The data collection process was carried out during approximately 16 months of typical patients' visits to the ophthalmologist's office. During its course, all patients attending visits had their eyes photographed with a professional iris recognition camera operating in near-infrared spectrum (IrisGuard AD100) and, in selected cases, also with two cameras operating in visible spectrum (Canon EOS 1000D with EF-S 18–55 mm f/3.5–5.6 lens equipped with a Raynox DCR-250 macro converter and a ring flashlight for macro-photography, and an ophthalmology slit-lamp camera Topcon DC3). The two latter cameras are employed in particularly interesting cases to perform visual inspection of the illnesses' impact in samples showing significant changes to the eye structures.

During the first visit, a typical ophthalmology examination was performed by an ophthalmologist, and the patient was enrolled anonymously into the system. Then, at least six photographs of each eye have been captured. Images within a single acquisition session were acquired in separate presentations, as recommended by ISO/IEC 19795-2, i.e., the patient was asked to lift his/her head from the chin and forehead rests after each capture. This is to purposefully introduce some noise in the intra-session sample sets. This procedure has been then repeated with future visits, each of which had a separate metadata, e.g., to distinguish between pre- and post-surgery examinations when such treatment is applicable and has been performed.

4.2 Database Description and Statistics

The entire dataset comprises 2996 images collected from 230 distinct irises. Each class, corresponding to each iris, is represented by NIR images collected with the IrisGuard camera, while some classes are also represented by color images. Regarding session count, for 184 image classes there are samples collected in one acquisition session. 38 classes contain images collected in two sessions, 6 classes contain images collected in three sessions, and finally for 2 classes there are four different image acquisition sessions. The second and subsequent sessions are usually conducted after some kind of medical procedures or treatment (a cataract extraction surgery, for instance).

4.3 Access to the Database

Papers [35, 36] present the described dataset as a package publicly available to interested researchers in version 1 and an expanded version 2, respectively. The released database also contains an extensive medical description for each eye represented in the data, including all diseases present in the eye, medical procedures, examinations, and other remarks.

5 Experimental Study and Results

5.1 Data Subsets Creation Methodology

Criteria A: medical conditions. In the first part of this study we attempted to divide the data with respect to the type of illness or other medical condition present in the eye. Only one particular disorder is represented by enough data for a thorough analysis, thus in this section we discuss only the cataract eyes (before any surgical treatment)—referred to as Cataract subset later on.

Criteria B: influence on different eye structures. Having a database that gathers various eye pathologies does not necessarily mean that each eye is affected by one illness only. In most cases, there are two or even more conditions present in one eye. Some of them do not affect the iris at all, some impact pupillary regions, other target the cornea or the iris tissue itself. This abundance of various and often unrelated medical conditions makes precise analysis of illness impact on iris recognition very difficult. Therefore, we come up with a solution of subdividing the dataset into several groups, connected not by the illnesses present in the eye themselves, but by the type of impact they have on iris or other structures of the eyeball, regardless of their medical origin. Such approach provides a possibility to analyze the impact of certain types of disorders on the recognition performance irrespective of an actual medical condition. It also makes it easier to propose countermeasures based on visual inspection of samples.

Five major subsets can be distinguished on that basis: healthy eyes (referred to as *Healthy* partition, serving as a control group), eyes with no visible changes (referred to as *Clear* partition), eyes with changes in pupillary regions, such as deviation from pupil's circular shape (referred to as *Geometry*), eyes with visible alterations to the iris tissue (referred to as *Tissue*) and eyes with iris covered by obstructions located in front of it (referred to as *Obstructions*), Fig. 10. Table 1 provides detailed description and number of samples in each group.

5.2 Iris Recognition Tools

This subsection provides a brief overview of iris recognition methods employed for the purpose of the experiments. We took effort to proceed with this study using four well-known, both commercial and academic iris recognition solutions.

OSIRIS (*Open Source for IRIS*) [43] is an open-source method developed in the framework of the BioSecure project, which follows Daugman's original concept of

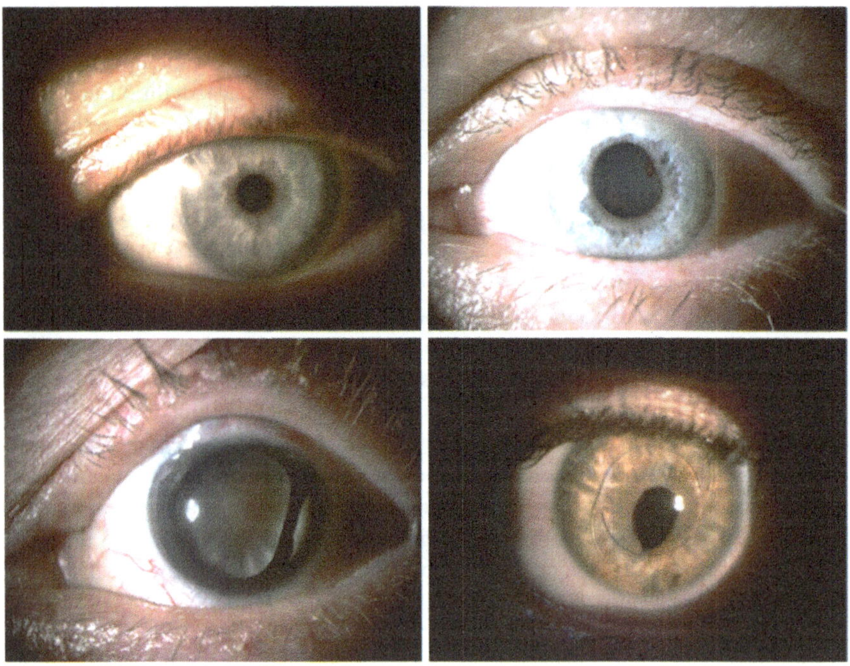

Fig. 10 Images obtained using Topcon DC3 slit-lamp microscope camera. **Top left** a sample from the Clear subset without any visible changes that may affect iris recognition. **Top right** a sample from the Geometry subset with posterior synechiae disturbing the shape of the pupil and the iris. **Bottom left** a sample from the tissue subset with severe changes to the iris tissue. **Bottom right** a sample from the obstructions subset lens implant placed in the anterior chamber of the eye, in front of the iris

Table 1 Database partitions created in respect to the type of illness impact on certain eye structures

Influence type *data partition*	Description	Number of eyes	Number of NIR samples
Healthy eyes (*Healthy*)	Eyes with no illness or pathology present	35	216
No visible changes (*Clear*)	Eyes with present medical condition, but not affected visually	87	568
Changes in pupil/iris geometry (*Geometry*)	Distortions in pupil and iris circularity due to various medical conditions	53	312
Iris tissue alterations (*Tissue*)	Damage or atrophy of the iris tissue itself	8	50
Obstructed iris (*Obstructions*)	Objects or opacities in the eye structures located in front of the iris (i.e., cornea and anterior chamber) that prevent proper imaging of the iris	36	207

Only NIR sample count is shown as these are used in experiments with automated iris recognition methods.

iris recognition based on the quantization of Gabor filtering outcomes to create a binary iris code. Iris codes are then compared against each other using the exclusive or (XOR) operation to yield a dissimilarity metric in a form of fractional Hamming distance normalized to a range of [0; 1], where values close to 0 are expected when samples coming from the same eye are compared, and values close to 0.5 would typically be expected for different-eye comparisons, as in the outcome of a series of independent coin tosses. However, as a result of the iris code shift to find the best match, intended as a countermeasure against eyeball rotation, impostor comparison scores distribution is usually centered around 0.4–0.45 values.

VeriEye is a commercially available product, offered by Neurotechnology [44]. Its encoding methodology is not disclosed by the manufacturer, apart from the claim that the method employs non-circular approximations of the iris boundaries using active shape modeling. This matcher gives comparison scores in the form of a similarity metric, i.e., the greater the score, the better the match. Scores near zero (typically lower than 40) are considered different-eye comparisons, while scores above 40 are by default classified as same-eye comparisons. The highest recorded in our experiments score (when two identical images are compared) is 1557.

IriCore is another commercially available matcher, offered by IriTech Inc. [45]. Similarly to the VeriEye method, its manufacturer does not divulge the implementation details. Comparison scores when same-eye images are matched against each other are expected to fall in the range of 0–1.1, while different-eye comparisons should yield scores between 1.1 and 2.0.

MIRLIN is a method also offered on the market in the form of an SDK [46] by Fotonation Ltd. (formerly Smart Sensors Ltd.). The underlying methodology of this product is said to incorporate discrete cosine transform (DCT) applied to overlapping iris image patches in order to calculate binary code out of iris features [47]. Binary templates are later on compared with each other to return fractional Hamming distance as a dissimilarity metric.

5.3 Examination of Cataract Influence

For both the *Cataract* and *Healthy* subsets, all possible genuine and impostor comparisons were performed to come up with comparison score distributions in the form of cumulative distribution functions (CDFs), which are to reveal whether cataract-afflicted eyes are expected to perform worse than their healthy counterparts, Figs. 11 and 12. Notably, for each of the four iris matchers we can observe a visible shift of the genuine score distribution obtained from the *Cataract* subset towards worse scores (i.e., lower for VeriEye, and higher for the remaining three methods). Regarding impostor scores, the differences in score distributions are smaller and uneven across matchers (highest for IriCore, negligible for MIRLIN and OSIRIS).

To examine whether the observed differences can be considered statistically significant, a two-sample Kolmogorov-Smirnov test is applied with the significance level $\alpha = 0.05$ (further referred to as K-S test). One-sided variant of the test is used when comparing genuine score distributions, and a two-sided variant for the impostor score distributions. The K-S test makes no assumptions on the distributions (apart from their continuity) and the test statistics simply quantifies the distance between two empirical cumulative distribution functions F(x1) and F(x2) of the random variables x1, x2 being compared. To alleviate the issue of statistical dependencies between comparison scores that are introduced when performing all possible comparisons

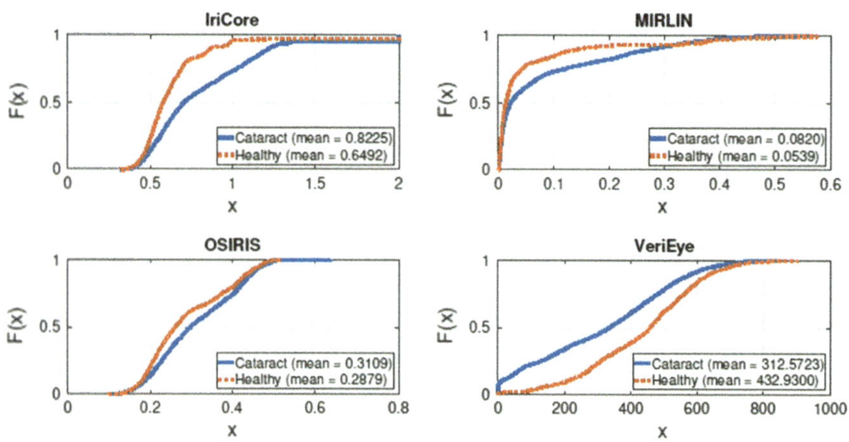

Fig. 11 Cumulative distribution functions for **genuine comparisons** obtained for all four iris recognition methods denoting the performance of these systems when cataract eyes are enrolled compared to a control group of healthy eyes. Mean values are provided in brackets

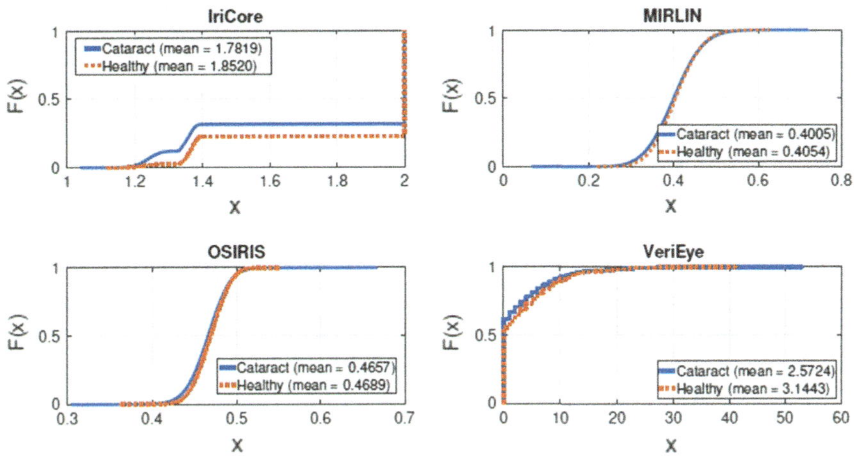

Fig. 12 Same as in Fig. 11, except that graphs for **impostor comparisons** are presented, together with respective mean values in brackets

Table 2 Kolmogorov-Smirnov statistical testing results for the cataract experiment

	Genuine comparisons *Cataract* (g_c) versus *Healthy* (g_h) H_1: $F(g_c)<F(g_h)$	Impostor comparisons *Cataract* (i_c) versus *Healthy* (i_h) H_1: $F(i_c) \nsim F(i_h)$
OSIRIS	<0.0001	<0.0001
MIRLIN	<0.0001	<0.0001
IriCore	<0.0001	<0.0001
	Cataract (g_c) versus *Healthy* (g_h) $H1$: $F(g_c)>F(g_h)$	*Cataract* (i_c) versus *Healthy* (i_h) $H1$: $F(i_c) \nsim F(i_h)$
VeriEye	<0.0001	<0.0001

The null hypothesis H_0 in all tests state that the scores originating from two compared partitions are drawn from the same distribution. Alternative hypothesis H_1 for genuine scores state that scores obtained from *Cataract* set are **worse** than those obtained from *Healthy* partition, while for impostor scores H_1 state that scores obtained from *Cataract* set are **different** than those obtained from *Healthy* partition One-sided test is used for genuine comparisons and two-sided test for impostor comparisons.

between samples, we resample with replacement each set of comparison scores 1,000 times for genuine scores and 10,000 times for impostor scores, providing sets of statistically independent scores. These resampled sets of scores are later used for performing the K-S tests, whose results are presented in Table 2.

In addition to cumulative distribution function graphs, we also present Receiver Operating Characteristic (ROC) curves which demonstrate the accuracy of these iris recognition systems when they are presented with eyes affected by cataract, compared to a scenario, in which healthy eyes are used. The ROC curve presents a relation of true positive ratio to false positive ratio obtained by a given decision system and is therefore helpful for assessing its expected behavior. ROCs for respective iris

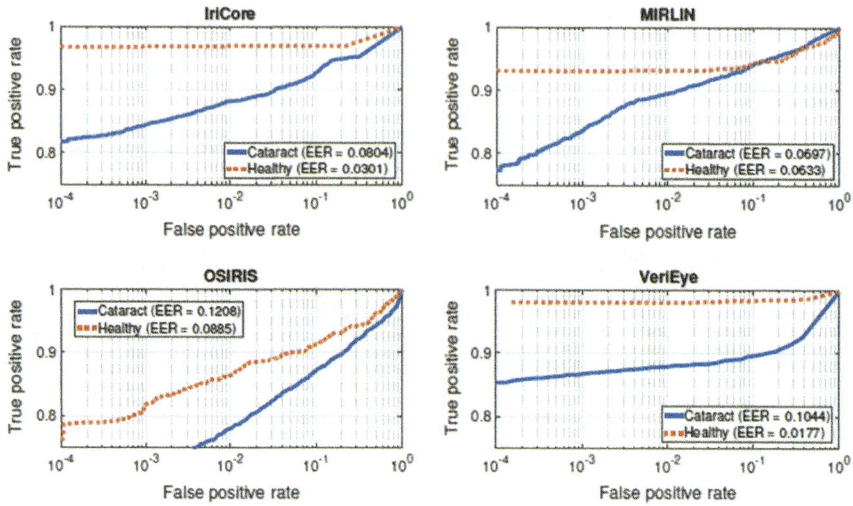

Fig. 13 Receiver operating characteristic (ROC) curves obtained for all four iris recognition methods denoting the performance of these systems when cataract eyes are enrolled compared to a control group of healthy eyes. Equal error rate values are provided in brackets

matchers are shown in Fig. 13. Equal Error Rate values are also provided as another metric for quantifying a system's performance. For all four employed iris recognition methods, the *Cataract* subset gives worse ROC-wise performance that the *Healthy* subset except for the MIRLIN matcher, which gives similar EER values for both partitions.

5.4 Examination of Disease Influence by Types of Eye Damage

Accordingly with the testing procedure described for the *Cataract* subset experiments, in this section we present CDF graphs complemented by the results of the K-S statistical testing, together with ROC curves for all four iris recognition methods employed. The CDFs for genuine comparison scores are presented in Fig. 14. Similarly across all methods, the *Geometry* and *Obstructions* subsets present the worst scores, with the CDF shifted to the right for VeriEye, and to the left for the remaining matchers, when compared to the *Healthy* subset serving as a control group. Uneven behavior can be observed for the *Clear* subset, which gives worse scores for most matchers, except for the OSIRIS matcher, in which its CDF intertwines with the CDF corresponding to the *Healthy* subset. Surprisingly, the *Tissue* subset displays behavior that is either similar to this of the *Healthy* subset, or even slightly better. These fluctuations can be, however, attributed to the small number of samples in the

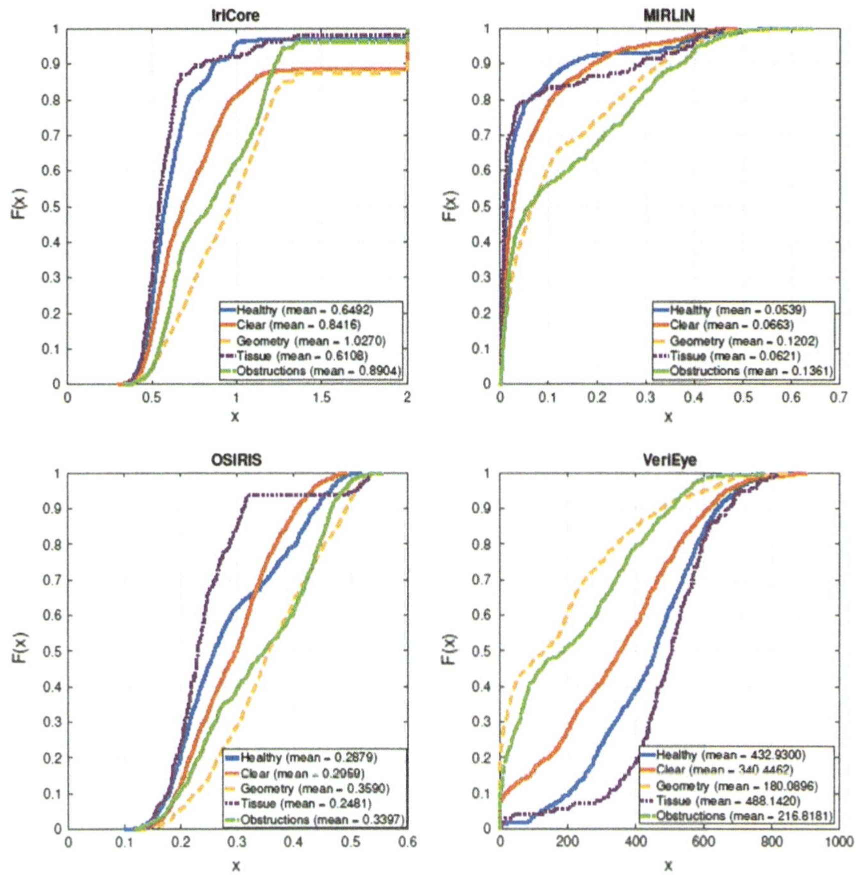

Fig. 14 Cumulative distribution function plots for **genuine comparisons** obtained for all four iris recognition methods denoting the performance of these systems for five data subsets. Mean values are provided in brackets. Note that these graphs may differ from those in [36], as here we use samples that are not required to comply to the ISO/IEC 29794-6 iris image quality standard

Tissue subset, which makes it easier for given samples to influence the performance of the whole subset. Table 3 presents the results of K-S testing procedure for genuine comparisons, which confirms that the differences between the *Clear*, *Geometry*, and *Obstructions* subsets when compared against the *Healthy* subset are statistically significant. As for the impostor-related CDFs, again rather large differences may be observed for the IriCore matcher, and smaller for the remaining three methods. The K-S statistical testing again confirms that there are statistically significant differences between each of the four subsets comprising diseased eyes and the *Healthy* partition, Table 4 (Fig. 15).

Figure 16 present ROC curves plotted collectively for all five data subsets representing different types of damage inflicted to the eye. This is repeated for all four of

Table 3 Kolmogorov-Smirnov statistical testing results for the disease influence type experiment, genuine comparisons

		Clear (g_c) versus *Healthy* (g_h)	*Geometry* (g_g) versus *Healthy* (g_h)	*Tissue* (g_t) versus *Healthy* (g_h)	*Obstructions* (g_o) versus *Healthy* (i_h)
OSIRIS	H_1	$F(g_c)<F(g_h)$	$F(g_g)<F(g_h)$	$F(g_t)<F(g_h)$	$F(g_o)<F(g_h)$
	p-value	<0.0001	<0.0001	0.1071	<0.0001
MIRLIN	H_1	$F(g_c)<F(g_h)$	$F(g_g)<F(g_h)$	$F(g_t)<F(g_h)$	$F(g_o)<F(g_h)$
	p-value	<0.0001	<0.0001	0.0039	<0.0001
IriCore	H_1	$F(g_o)<F(g_h)$	$F(g_o)<F(g_h)$	$F(g_o)<F(g_h)$	$F(g_o)<F(g_h)$
	p-value	<0.0001	<0.0001	0.3325	<0.0001
VeriEye	H_1	$F(g_o)>F(g_h)$	$F(g_o)>F(g_h)$	$F(g_o)>F(g_h)$	$F(g_o)>F(g_h)$
	p-value	<0.0001	<0.0001	0.6942	<0.0001

The null hypothesis H_0 in all tests state that the samples originating from two compared partitions are drawn from the same distribution. Alternative hypotheses are detailed in rows labeled H_1. One-sided test is used. $F(g_k)$ denotes the cumulative distribution function of g_k, where g_k denotes the genuine scores calculated to the k-th partition.

Table 4 Same as in Table 3, except that impostor comparison scores are analyzed and two-sided Kolmogorov-Smirnov test (for the resampled data) was applied

	Clear (i_c) versus *Healthy* (i_h) H_1: $F(i_c) \nsim F(i_h)$	*Cataract* (i_c) versus *Healthy* (i_h) H_1: $F(i_c) \nsim F(i_h)$	*Tissue* (i_t) versus *Healthy* (i_h) H_1: $F(i_c) \nsim F(i_h)$	*Obstructions* (i_o) versus *Healthy* (i_h) H_1: $F(i_c) \nsim F(i_h)$
OSIRIS	<0.0001	<0.0001	<0.0001	<0.0001
MIRLIN	0.0078	<0.0001	<0.0001	<0.0001
IriCore	<0.0001	<0.0001	<0.0001	<0.0001
VeriEye	0.0051	<0.0001	<0.0001	<0.0001

$F(i_k)$ denotes the cumulative distribution function of i_k, where i_k denotes the impostor scores calculated to k-th partition.

the iris matchers involved in this study. Here as well, the *Geometry* and *Obstructions* subsets are giving the worst performance. Surprisingly, for the IriCore and VeriEye matchers, the *Clear* subset also performs poorly, while the *Tissue* subset is behaving similarly or better than the *Healthy* subset.

5.5 Recognition Errors Analysis

As finding the actual reasons behind erroneous performance is crucial for getting a complete picture of the studied phenomenon, we performed a careful visual inspection of the samples that generated exceptionally poor comparison scores. As impostor

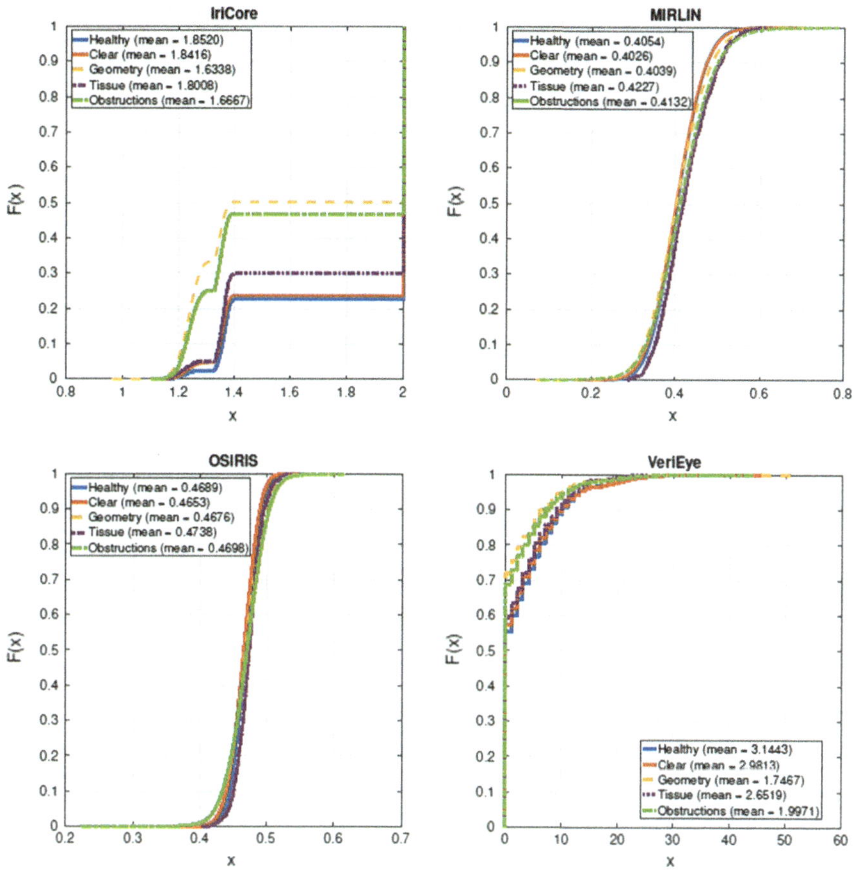

Fig. 15 Same as in Fig. 14, except that CDFs for **impostor comparisons** are presented, together with respective mean values provided in brackets

comparison scores are not impacted in a significant way, this is done only for the genuine scores. Since bad performance in iris recognition typically originates from incorrect execution of the segmentation stage, we employed two of the iris matchers that are capable of showing image segmentation results: MIRLIN and OSIRIS, to generate iris images with denoted iris localization results. This analysis confirmed that failed iris localization is the most prevalent source of bad iris matcher performance. Segmentation errors that we have come across were most likely caused by some artifacts, such as distortions in the pupil boundary, obstructions such as corneal hazes, or damages to the iris tissue being interpreted by image segmentation algorithms as the pupil itself. Thus, the following matching stage, which is executed after the segmentation stage, could not be performed correctly, but instead was performed using the non-iris portions of the image. This is especially true for *Geometry* and *Obstructions* subsets of the data, which is coherent with exceptionally

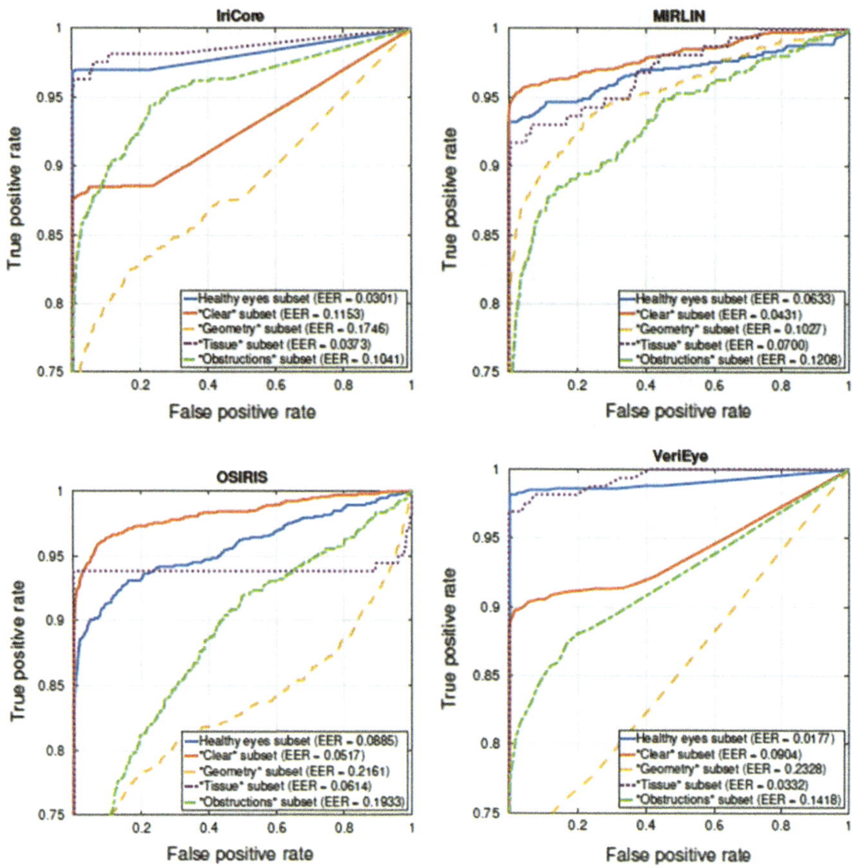

Fig. 16 Receiver operating characteristic (ROC) curves obtained for all four iris recognition methods denoting the performance of these systems for five data subsets. Equal error rate values are provided in brackets

poor ROC-wise performance of the data belonging to these subsets. VeriEye and IriCore algorithms do not provide a way to read the segmentation results, however, an examination of those samples that perform the worst when using these method, easily identifies conditions responsible for errors, namely: significant geometrical distortions, severe corneal hazes, blurred boundary between the iris and the pupil, letting us hazard a guess that segmentation issues are the ones responsible for errors here as well.

6 Conclusion

This book chapter summarizes the Authors' knowledge regarding iris recognition behavior under conditions involving ophthalmic disorders, including both mild illnesses and severe eye pathologies. Together with an extensive literature review concerning this subject, two extensive experiments are described with important results delivered.

The first experiment related to probably the most proliferated eye illness worldwide, the cataract, proves that despite usually not affecting the eye and the iris in a significant way, this pathology is capable of causing serious negative impact on the performance of state-of-the-art iris recognition technologies used today. With Equal Error Rates for the cataract-affected eyes being a few percent higher than those obtained with data corresponding to healthy eyes, one may expect recognition accuracy to be noticeably lower for people suffering from this illness. Combining this with high number of cataract occurrences, especially in third-world countries leads to a conclusion that this issue should be seriously taken into consideration when building future, large-scale biometric applications employing iris recognition.

The latter of the two experiments deals with a more broad and universal approach to the problem and tries to predict the recognition accuracy deterioration with respect to the type of damage inflicted by pathological processes in the eye, regardless of the actual medical origin and disease taxonomy. Ophthalmological disorders that are expected to cause the highest performance drops are those causing the pupil to be irregularly shaped and those introducing obstructions that make correct imaging of the iris texture difficult or impossible. An attempt to explain the underlying reasons of such poor performance is carried out, pointing to image segmentation errors as the predominant source of performance deterioration. However, knowing which types of eye damage are the ones most likely to cause recognition errors, one can employ visual inspection of the eyes of a person under enrollment to assess whether iris recognition can be reliably used to manage this person's identity.

With such knowledge in place, more research in this field with more data is necessary to better quantify the effect ophthalmic disorders have on iris recognition systems on larger scales and to be able to propose appropriate countermeasures against the reported drops in recognition accuracy to make this biometric technology an even better and more reliable solution to use globally without excluding subjects suffering from eye illnesses.

Acknowledgements All patients who participated in this study have been provided with detailed information regarding the research and an informed, written consent has been obtained from each volunteer. Ethical principles of the Helsinki Declaration were carefully followed by the authors.

References

1. J. Daugman, Iris recognition. Am. Sci. **89**, 326–333 (2001)
2. J.G. Daugman, High confidence visual recognition of persons by a test of statistical independence. IEEE Trans. Pattern Anal. Mach. Intell. **15**(11), 1148–1161 (1993)
3. K.W. Bowyer, M. Burge, *Handbook of Iris Recognition*, 1st edn. (Springer, London, 2013)
4. K.W. Bowyer, M. Burge, *Handbook of Iris Recognition*, 2nd edn. (Springer, London, 2016)
5. K.W. Bowyer, K. Hollingsworth, P.J. Flynn, Image understanding for iris biometrics: a survey. Comput. Vis. Image Underst. **110**(2), 281–307 (2008)
6. K.W. Bowyer, K.P. Hollingsworth, P.J. Flynn, A survey of iris biometrics research: 2008–2010, in *Handbook of Iris Recognition* (Springer, London, 2013), pp. 15–54
7. Canadian Border Services Agency, *CANPASS, Air*. http://www.cbsa-asfc.gc.ca/prog/canpass/canpassair-eng.html. Accessed 11 Aug 2015
8. Unique Identification Authority of India. *AADHAAR*. http://uidai.gov.in. Accessed 11 Aug 2015
9. National Institute of Standards and Technology (NIST), *IREX*. http://www.nist.gov/itl/iad/ig/irex.cfm. Accessed 20 May 2016
10. A. Czajka, Pupil dynamics for iris liveness detection. IEEE Trans. Inf. Forensics Secur. **10**(4), 726–735 (2015)
11. A. Czajka, Template ageing in iris recognition, in *6th International Conference on Bio-inspired Systems and Signal Processing, February 11–14, 2013, Barcelona, Spain* (2013)
12. S.P. Fenker, K.W. Bowyer, Experimental evidence of a template aging effect in iris biometrics, in *IEEE Computer Society Workshop on Applications of Computer Vision*, pp. 232–239 (2011)
13. S.P. Fenker, K.W. Bowyer, Analysis of template aging in iris biometrics, in *Proceedings of CVPR Workshops 2012*, pp. 45–51 (2012)
14. S.P. Fenker, E. Ortiz, K.W. Bowyer, Template aging phenomenon in iris recognition. IEEE Access **1**, 266–274 (2013)
15. M. Trokielewicz, Linear regression analysis of template aging in iris recognition, in *IEEE 3rd International Workshop on Biometrics and Forensics (IWBF 2015), 3–4 March 2015, Gjovik, Norway*. https://doi.org/10.1109/iwbf.2015.7110233 (2015)
16. K.W. Bowyer, E. Ortiz, Critical examination of the IREX VI results. IET Biometrics **4**, 192–199(7) (2015)
17. P. Grother, J.R. Matey, G.W. Quinn, Irex vi: mixed-effects longitudinal models for iris ageing: response to bowyer and ortiz. IET Biometrics **4**, 200–205(5) (2015)
18. M. Trokielewicz, A. Czajka, P. Maciejewicz, Human iris recognition in post-mortem subjects: study and database, in *8th IEEE International Conference on Biometrics: Theory, Applications and Systems (BTAS 2016), September 6–9, 2016, Buffalo, NY, USA* (2016)
19. M. Trokielewicz, A. Czajka, P. Maciejewicz. Post-mortem human iris recognition, in *9th IAPR International Conference on Biometrics (ICB 2016), June 13–16, 2016, Halmstad, Sweden* (2016)
20. ISO/IEC 29794-6, *Information Technology—Biometric Sample Quality—Part 6: Iris Image Data (FDIS)* (August 2014)
21. American Association for Pediatric Ophthalmology and Strabismus, *Aniridia*. http://www.aapos.org/terms/conditions/26. Accessed 10 May 2016
22. M.H. Nizankowska, *Podstawy okulistyki*, 2nd edn. (Volumed, Wroclaw, Poland, 2000)
23. M. Trokielewicz, A. Czajka, P. Maciejewicz, Cataract influence on iris recognition performance, in *Proceedings of SPIE 9290, Photonics Applications in Astronomy, Communications, Industry, and High-Energy Physics Experiments 2014, 9290* (2014)
24. O. Seyeddain, H. Kraker, A. Redlberger, A.K. Dexl, G. Grabner, M. Emesz, Reliability of automatic biometric iris recognition after phacoemulsification or drug-induced pupil dilation. Eur. J. Ophthalmol. **24**(1), 58–62 (2014)
25. M. Dhooge, J.J. de Laey, The ocular ischemic syndrome. Bull. Soc. Belge Ophtalmol. **231**, 1–13 (1989)

26. R. Roizenblatt, P. Schor, F. Dante, J. Roizenblatt, R. Belfort Jr, Iris recognition as a biometric method after cataract surgery. BioMed. Eng. Online **3**(2) (2004)
27. O. Seyeddain, H. Kraker, A. Redlberger, A.K. Dexl, G. Grabner, M. Emesz, Reliability of automatic biometric iris recognition after phacoemulsification or drug-induced pupil dilation. Eur. J. Ophthalmol. **24**(1), 58–62 (2014)
28. L. Dhir, N.E. Habib, D.M. Monro, S. Rakshit, Effect of cataract surgery and pupil dilation on iris pattern recognition for personal authentication. Eye **24**, 1006–1010 (2010)
29. R. Ramachandra, K. Raja, V. Vemuri, S. Kumari, P. Gacon, E. Krichen, C. Busch. Influence of cataract surgery on iris recognition: a preliminary study, in *9th IAPR International Conference on Biometrics (ICB 2016), June 13–16, 2016, Halmstad, Sweden* (2016)
30. X. Yuan, H. Zhou, P. Shi, Iris recognition: a biometric method after refractive surgery. J. Zhejiand Univ. Sci. A **8**(8), 1227–1231 (2007)
31. T.M. Aslam, S.Z. Tan, B. Dhillon, Iris recognition in the presence of ocular disease. J. R. Soc. Interface **2009**, 6 (2009)
32. American Optometric Association, *Anterior Uveitis*. http://www.aoa.org/patients-and-public/eye-and-vision-problems/glossary-of-eye-and-vision-conditions/anterior-uveitis?sso=y. Accessed 10 May 2016
33. H. Borgen, P. Bours, S.D. Wolthusen, Simulating the influences of aging and ocular disease on biometric recognition performance, in *International Conference on Biometrics 2009, LNCS 5558*, vol. 8, no. 8, pp. 857–867 (2009)
34. G. McConnon, F. Deravi, S. Hoque, K. Sirlantzis, G. Howells, Impact of common ophthalmic disorders on iris recognition, in *2012 5th IAPR International Conference on Biometrics Compendium, IEEE*, pp. 277–282 (2012)
35. M. Trokielewicz, A. Czajka, P. Maciejewicz, Database if iris images acquired in the presence of ocular pathologies and assessment of iris recognition reliability for disease-affected eyes, in *2nd IEEE International Conference on Cybernetics (CYBCONF 2015), Special Session on Reliable Biometrics (BIORELIABILITY 2015), June 24–26, 2015, Gdynia, Poland* (2015)
36. M. Trokielewicz, A. Czajka, P. Maciejewicz, Assessment of iris recognition reliability for eyes affected by ocular pathologies, in *7th IEEE International Conference on Biometrics: Theory, Applications and Systems (BTAS 2015), September 8–11, 2015, Arlington, USA* (2015)
37. Warsaw University of Technology, *Warsaw-BioBase-Disease-Iris-v1.0*. http://zbum.ia.pw.edu.pl/en/node/46 (2015)
38. Warsaw University of Technology, *Warsaw-BioBase-Disease-Iris-v2.1*. http://zbum.ia.pw.edu.pl/en/node/46 (2015)
39. M. Trokielewicz, A. Czajka, P. Maciejewicz, Implications of ocular pathologies for iris recognition reliability. Image Vision Comput. (2016)
40. Warsaw University of Technology, *Warsaw-BioBase-PostMortem-Iris-v1.0*. http://zbum.ia.pw.edu.pl/en/node/46 (2016)
41. S.K. Saripalle, A. McLaughlin, R. Krishna, A. Ross, R. Derakhshani, Post-mortem, iris biometric analysis in *Sus scrofa domesticus*, in *IEEE 7th International Conference on Biometrics Theory, Applications and Systems (BTAS)* (2015)
42. A. Sansola, Postmortem iris recognition and its application in human identification. Master's Thesis, Boston University (2015)
43. G. Sutra, B. Dorizzi, S. Garcia-Salitcetti, N. Othman. *A Biometric Reference System for Iris. OSIRIS version 4.1*. http://svnext.it-sudparis.eu/svnview2-eph/refsyst/irisosirisv4.1/. Accessed 1 Oct 2014
44. Neurotechnology. *VeriEye SDK, version 4.3*. www.neurotechnology.com/verieye.html. Accessed 11 Aug 2015
45. IriTech Inc. *IriCore Software Developers Manual, version 3.6* (2013)
46. Smart Sensors Ltd. *MIRLIN SDK, version 2.23* (2013)
47. D.M. Monro, S. Rakshit, D. Zhang, DCT-based iris recognition. IEEE Trans. Pattern Anal. Mach. Intell. (Special Issue on Biometrics: Progress and Directions) **29**(4), 586–595 (2007)

Facial Rejuvenation Modeling

Elham Farazdaghi, F. Majidzadeh and Amine Nait-Ali

Abstract Face modeling is discussed in this chapter entitled "Facial Rejuvenation Modeling by using biometric features" describes the process of normal facial growth trajectory within the categories of different age groups. The de-aging study is based on a 2D generic face model highlighted by providing numerous illustrations showing digital face rejuvenation from adulthood to childhood appearance. At the end of this chapter, an extension of a 3D generic model is proposed.

1 Introduction

As we saw in the previous chapter, digital facial ageing modeling has been an active research topic. This may be explained by the fact that numerous potential applications have recently been developed in the field of forensics, biometrics, security applications, etc. However, one can notice that digital face image rejuvenation, called in this chapter, backward prediction or de-ageing, has not been deeply considered. So what is the reason? Probably, this is due to a lack of some concrete applications.

Actually, based on some published works from the literature, digital backward facial ageing models have been considered, in particular, to rejuvenate old adult face images into young adult face images [1, 2]. This is regarded as soft rejuvenation. Therefore, when considering rejuvenation process into very young ages (e.g. 3–4 years old), changes in terms of face shapes and textures are obviously more contrasted and more important compared to the forward process changes. Basically, the natural ageing affect texture changes due to wrinkles appearance and slight face shape feature changes, due to elasticity loss, weight gain or weight loss.

Within the context of this book, we distinguish potentials biometric security and healthcare biometrics applications, as follows:

A. **Biometric Security and Forensics**

E. Farazdaghi · F. Majidzadeh · A. Nait-Ali (✉)
Université Paris-Est, LiSSi, UPEC, 94400 Vitry-sur-Seine, France
e-mail: naitali@u-pec.fr

A. Nait-Ali (ed.), *Biometrics under Biomedical Considerations*,
Series in BioEngineering, https://doi.org/10.1007/978-981-13-1144-4_3

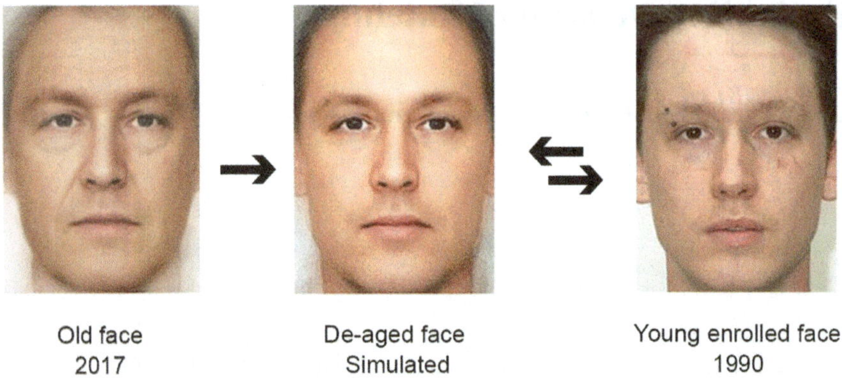

Fig. 1 An example of a missing child in 2 years old on the left and the age-progressed simulation on the right (www.fbi.gov)

When considering face recognition systems, one of the issues related to this modality is the fact that, among other factors, face traits are subject to changes due to ageing phenomena. Therefore, when unrolling a person at a given moment, recognition process, may totally fail. Consequently, when dealing with verification, updating regularly the face template, is commonly recommended to decrease the False Rejection Rate. However, in some face identification applications, recognition systems become less reliable because, matching faces is not straightforward when it comes to objectively compare two different faces at different ages (see Fig. 1). Certainly, the performances depend on the recognition algorithm used, but it is obvious that, if the algorithm is sensitive to some specific texture or shape features, matching faces fails for the same person (considered as different persons).

As illustrated through Fig. 1, the advantage of using age rejuvenation methods is to objectively compare the aged face image (shifted to a younger age), to the reference face image, corresponding to the same person. This comparison scenario can be interesting if the rejuvenation model is accurate and reflect the real ageing process.

As another application, closed to Biometrics, once can consider the process of facial rejuvenation to automatically compare some face images to reference face images, corresponding to reported missing persons. Commonly, face images of reported missing persons are processed using some ageing algorithms to shift the initial reference face image, to an older simulated face image (see Fig. 2). Using, the rejuvenation approach, the process is inverted. This means that from a given real face image (input image), one can simulate the face appearance at a younger age (output) to be compared to the real young face image. In this scenario, the process is considered as a verification. The advantage of inverting this process is the possibility to capture the input image, either in 2D or 3D. In case of 3D capture, one can adjust the pose after applying the rejuvenation algorithm. Therefore, resulting image can be compared to the reference by considering the same conditions.

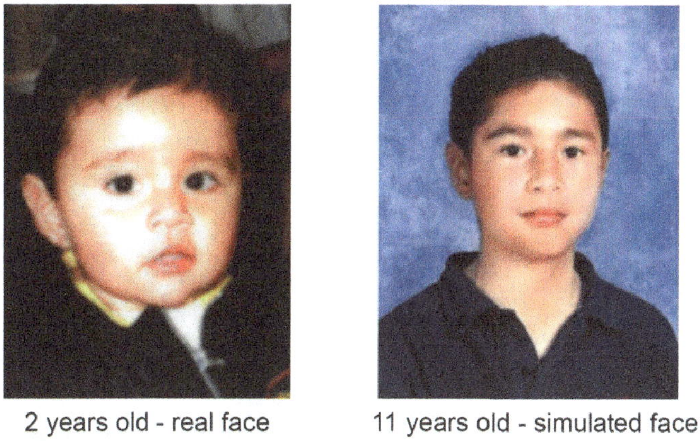

Fig. 2 An example of a missing child in 2 years old on the left and the age-progressed simulation on the right (www.fbi.gov)

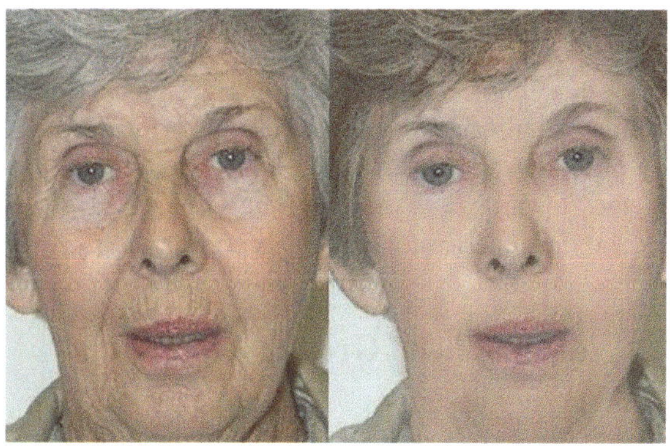

Fig. 3 Face lift surgery simulation. Original image is from PAL database [3]

B. Biomedical (Healthcare Biometrics)

In healthcare biometrics, digital rejuvenation can be used as an automatique advanced simulation tool. For example, in plastic surgery, it can be used to de-age digitally a face image from adult to adult in order to simulate different type of face lift surgery (see Fig. 3). We understand by this application that there is no need to shift the rejuvenation up to childhood age. However, in other potential applications, adult-childhood de-aging can be considered as a useful tool to analyse, to simulate and to model abnormal facial growth caused by some pathologies in comparison to normal facial growth. For example, one can mention some genetic disorders such as Progeria, angelman, Treacher collins and acromegaly which are genetic disorders which produce abnormal facial deformation due to (see Fig. 4). Creating typical face models require a normalisation towards a specific age in order to allow objective analysis.

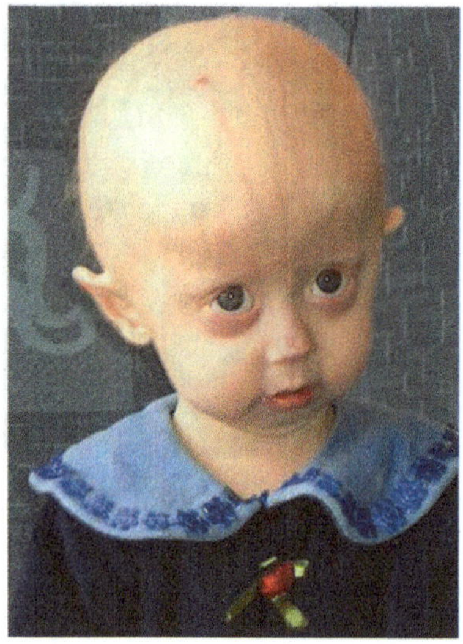

Fig. 4 HGPS is a childhood disorder caused by mutations in one of the major architectural proteins of the cell nucleus

In this chapter, we will describe the process of normal facial growth trajectory within different age group categories. Afterwards, 2D generic model is presented by providing numerous illustrations showing digital rejuvenation from adult face images towards 3–4 years old face images [4, 5]. At the end of this chapter, we extend the generic model to 3D representation.

2 Normal Facial Growth Trajectory

Child face is not a miniature model of the adult face. The process of its evolution is truly complex. To study the craniofacial growth process, variety aspects of the face should be considered. In this procedure, changes in the hard tissue structure and soft tissue take place at different times, in different directions and with different rates [6]. To model the face growth process or, as in this chapter, to rejuvenate the face image down to childhood, it is important to peruse craniofacial morphology and be knowledgeable of the several aspects of the growth.

First, let us define "facial growth". Growth is the increase in size and change in facial proportions over the time as Wilton M. Krogman defined. Therefore, it is the quantitative and the measurable sight of the biologic life. Growth might be the change in proportions, size, texture, and complexity and in general, change in quantity. Development, however, includes all the changes from a single cell till death. It can be said, growth is an "anatomic phenomenon" whereas development is a "physiological and behavioural phenomenon". According to this, the focus in this study is nearly facial growth [7, pp. 46–47].

Facial geometry growth is affiliated with adaptability of soft tissue with hard tissue growth. Moreover, when dealing with facial growth, there are usually two

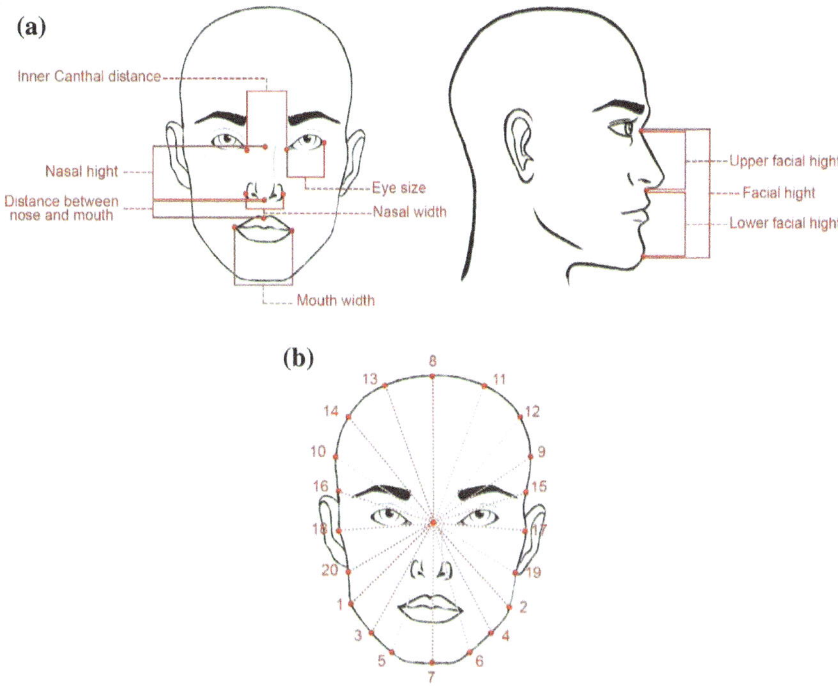

Fig. 5 **a** Illustration of the vertical and horizontal facial measures' definition. **b** Illustration of the 20 radiuses with their given numbers, which define the face shape and scale

important matters to consider; intensity and direction of the growth. In the following, both direction and intensity of growth-related changes, in hard and soft tissues, for different facial landmarks are briefly discussed.

Eyes: The eye is a complex organ, although it is small. A newborn's eye has approximately 70% of the size of an adult's eye [8, p. 24]. That is why eyes seem larger in proportion to the head. Over the first three years of life, the eyes grow rapidly and continue to grow through the childhood until they reach almost their adulthood length by age of 13; however, the changes are dedicated and insensible. The eye socket grows as the eyeball grows and gives the appropriate space to the eye to grow in different directions [9, Chap. 1]. The visible part of the eyeball has only 1/6 of the eye's total surface area, and the rest are hidden behind the eyelids. In this study, eye size is the visible part of the eye that is the palpebral fissure length, which is the distance from the endocanthion to the exocanthion landmarks (see Fig. 5a).

Nasal: A young child has a small rounded nose which is vertically short and shallow, with a small protrusion. In childhood, the nasal bridge is quite low and the nasal profile is concave. The human nose continues to grow in downward and forward directions until early adulthood. The length of the nose, in average, increases about 1–1.5 mm per year. The vertical dimension of the nose has more share of growth than the anteroposterior projection. Increase in width of the nose is about 0.5 mm per year. The inclination of the nose remains constant. The nose profile changes is due to increments in nasal length [6, 10].

Lips: Lips growth follows the general body growth procedure with soft tissue and muscles growth. The lips grow at the gradual rate till 15 years old. It was observed that the maxillary (upper) and mandibular (lower) lips increase in both dimensions under the influence of growth. The upper lip grows away from the palate and the lower one grows away from the chin. The upper and lower lips have their most share of the increment from age 6–12. In general, the lower lip grows more than the upper lip and upper lip grows slightly with age. From childhood to adolescence, increase in the thickness of the upper and lower lisps is obvious [6, 7, pp. 153–154].

Face: Face continues its growth downward and forward until the adult face shape and size is reached. Skull growth is complex and inhomogeneous. Different parts of the skull are differently matured in terms of size, direction, and rate.

In childhood the head is big, so it seems that it does not have a considerable growth until adulthood. The fact is that, during the growth process, the size of the body will increase three times but this ratio is two for the head. Bellow, different parts of cranium affecting frontal face shape and size will be discussed.

Forehead: The brainpan has much greater extent than facial complex in childhood. It grows earlier and faster so that cranial cavity reaches 90% of its size by 5 years of age [6]. The forehead in young children is bulged and seems very large and high as the frontal lobe of the brain is huge in size, and also because the face beneath it is still comparably small.

The large head-small face is noticeable up to ages 7 and 8. In the process of growth, the face enlargement is much more than forehead so that the proportionate size of the forehead becomes reduced [11, 12].

Anterior facial height: Anterior facial height is the sum of upper and lower facial height which from now on in this study is called facial height (see Fig. 5a). Vertical growth of the child face occurs due to the both respiratory needs and tooth eruption [6]. These growth spurts take place during the third and fourth year, from the seventh to eleventh year, and again between the sixteenth and the nineteenth year [13].

Face width: Child face is wide and less in height. Facial width growth is associated with mandibular, maxillary and zygomatic growth. Facial growth arises mostly in the vertical direction compared to the horizontal one. However, Studies show that the ratio facial height/facial width increased from childhood to adulthood [14, 15, Chap. 1].

3 Age Groups and Sub-groups Definition

Globally, one can consider four main age groups in the periods of human life: infancy (1 month to 2 years old), childhood, adulthood and old ages. The general features of a person's face in the infancy is quite different from the same individual's grown face and is governed almost purely by genetic factors. Moreover, the general structure of the face has not yet formed and there are only minimal variations in facial structure of their faces. Relatively large-appearing eyes, a tiny nose, a small mouth, puffy cheeks, a high forehead, a low nasal bridge and overall wide and short proportions. As a result, the face in less than 3-year-olds cannot be very useful for identification and subsequently for many applications that can be imagined for this work. Therefore, childhood (3–16), adulthood (16–30) and old ages (after 30: With the appearance of wrinkles) are arbitrarily considered as the main age groups.

We also consider four age subgroups: 3–4, 7–8, 12–13 and 16 years old for adulthood, since these are the formative years in which most of the changes in the process of growth occur [16]. Considering the difficulty of predicting very close age, sub-age groups are defined in a way that their age differences are more distinguishable. For example, distinction between a 4-year-old and a 5-year-old child is very difficult but differences between a 4-year-old child and a 7-year-old one are entirely visible and noticeable. In fact, sub-groups have been set up in a way that transformations are clearly visible especially in the face's geometry. Therefore, age clusters for digital face rejuvenation are; Cluster 3–4, Cluster 7–8, Cluster 12–13 and Cluster Adult for 16 years old and later (until when that person can be considered as young adult).

4 2D Facial Rejuvenation Modeling

4.1 Facial Features Assortment

Human face is a complex combination of many information which represents identity, gender, emotional state, and age. Many kinds of useful features can be extracted from a face image. By categorising these features, a better perception of the face is achieved and faces can be assorted based on age, gender, ethnic, and expressions. Furthermore, a model can be obtained according to the different features corresponding to the different assortments.

When dealing with face changes over the time, numerous aspects should be taken into account. For instance, in some age ranges, one can notice important face variations in both its shape and its skin texture. Therefore, facial changes can be divided as geometrical (anthropometric) and textural. In the growth process (childhood), most of the changes are geometrical, however, texture undergoes subtle changes. In the ageing trajectory (adulthood) is inverse. Based on these, facial features can be divided into two different categories:

A. **Geometrical features** which are concerned with changes in craniofacial morphology and anthropometry of an individual. These changes can occur in the bony portions of the head and soft tissues of the face [15]. It can be said that the geometrical features have two subdivisions:

    1. The information which is connected to the facial components' (eyes, nose and lips) size and their distribution.

2. The information about the face shape (the contour), consisting of the upper face, its mandible size and its maturation process.

B. **Textural features** that are concerned with the skin changes. These changes consist of the thickness and the colour of the skin, elasticity of the facial muscles and the appearance of the wrinkles.

4.2 Geometrical Measurements

Changing the facial age geometrically requires the transformation of the facial components' size, distances between them, face shape, and its size. Choosing the practical scales and distances has commenced with the study of the craniofacial morphology and information collection dealing with craniofacial growth. One part of this information was about development and understanding how organs change during the ageing process. The result was just useful to have a better perception of the facial growth so the focus was shifted to the changes in the magnitude of facial landmarks. After collecting all the craniofacial measurements, they were filtered: First, to those which were useful for 2D facial modeling. Therefore, information such as the head length and width are eliminated. Second, to those which can present a component or its location in the face. For example from inner Canthal distance, outer Canthal distance and interpupillary distance of the two eyes, only the first one was retained as a measure for eyes distribution in the face since the other two are redundant for this study. Some of the measurements such as skull height or forehead height are included in the face shape and size.

To apply the geometrical face transformation, craniofacial measurements in different ages are needed. First, the required components' size and distances should be defined. In the following, you can find their definition by using the 79 previously extracted landmark points related to the face components and the contour.

Eye size: Distance between the inner and outer canthus of one eye. The head should be held erect with the eyes open and facing forward.

Inner canthal distance: The distance between the innermost corners of one eye to the innermost corner of the other eye, in a straight line.

Nasal height: The distance from the deepest depression at the root of the nose to the deepest concavity at the base of the nose, in a vertical axis.

Nasal width: Distance between the most lateral aspects of the alae nasi.

Distance between nose and lips: Distance between the base of the nose and the border of the upper lip, in the midline.

Mouth width: Measure from one corner of the mouth to the other. The head should be held erect with the eyes facing forward and the mouth held close and in a neutral position.

Face shape: 20 facial feature points are the face contour indicator as cited in the previous chapter and nasal bridge is considered as the origin of the face. Accordingly, the face shape and its scale are defined using these points (Fig. 5b).

Fig. 6 **a** Facial components' size and distances—both genders, four age clusters—extracted by Judith G. Hall [15]. **b** Facial components' size and distances—both genders, four age clusters—extracted from calculated average size and distance of all face images in each age cluster

(a)

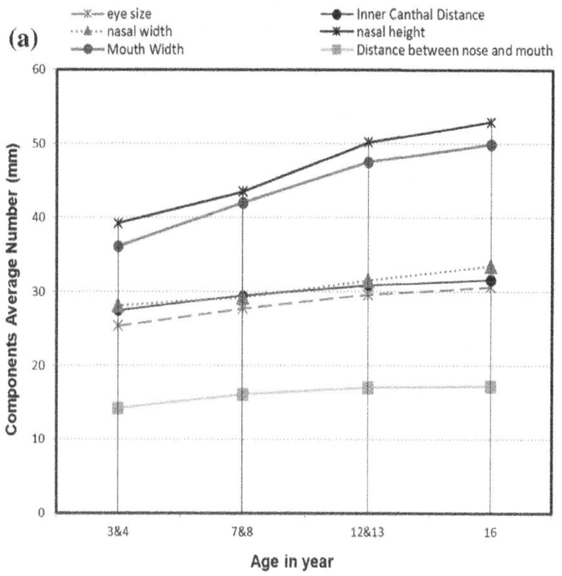

(b)

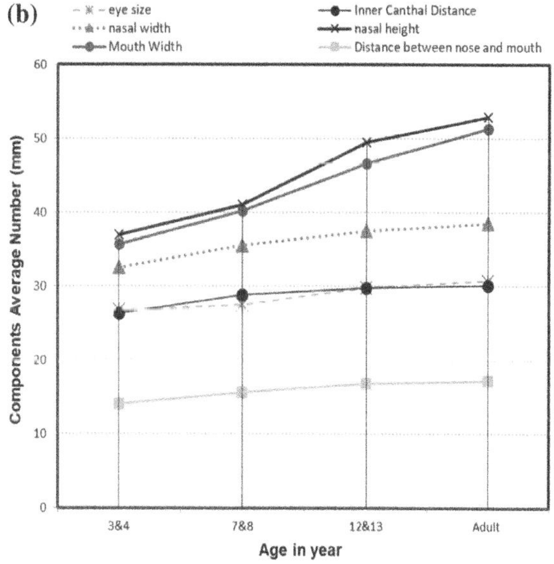

This measurements have already been done by Judith G. Hall from the birth to 16 years old [16]. Among these numbers, those corresponding to formative years (3–4, 7–8, 12–13 and 16) are considered in this study (Fig. 6a).

In addition to that, having the calculated average size and distance of all face images in each age cluster in our database, the same chart has been extracted (Fig. 6b).

As it can be seen, these two charts are very close, so, using these results, the percentage of the scale and the distance changes has been extracted. Final percentages and orders are obtained by comparing these two charts. Figure 7a shows the percentage number of all components' size in different age clusters compared to the adult components' size. Reference measures are the measures of the early years of adulthood.

Note that an adult, in terms of geometry, means a person with more than 16 years of age because the geometrical growth of the face winds up at this age.

4.3 Rejuvenation Model Development

Information from research resources [16, 17] show that there are two periods in facial ageing:

1. From birth to adulthood, that can be named formative years, in which most of the changes related to the geometry appear. There are minor changes in the skin colour, facial hair colour, and the density. It can be said that the textural variations tend to be subtle during this period.
2. Adulthood which is the end of growth until old ages. The majority of changes in this period are related to the skin texture. The skin becomes thinner and darker with less elasticity; this is the time for wrinkles to appear. Some small craniofacial changes such as the shape change of the face and eyes also occur.

B-FAM is dealing with the childhood face alterations, so the geometrical variations have more important roles.

In this work, the procedure is separated into two parts: geometrical and textural. At the end, the final result is visualised through merging these two sections.

4.3.1 Geometrical Model

Since the geometrical features have been divided into two parts, the facial transformation will be applied in two steps:
Components' scale and their distribution, which are defined by some modifications such as eye size, inner Canthal distance, nasal height, nasal width, the distance between nose and lips, and mouth width, all previously measured. Each face components is considered as a discrete object.

For mathematical description of the component's shape, a finite number of points along the boundary are located in the feature extraction phase and concatenated to a shape vector. Face components are each modified according to a special scale factor that is the percent of the component's size in an age cluster. Besides, their distances are modified, if required, according to a translation factor.

$$S_i = \alpha S_j + d(i < j, 0 < \alpha < 1) \tag{1}$$

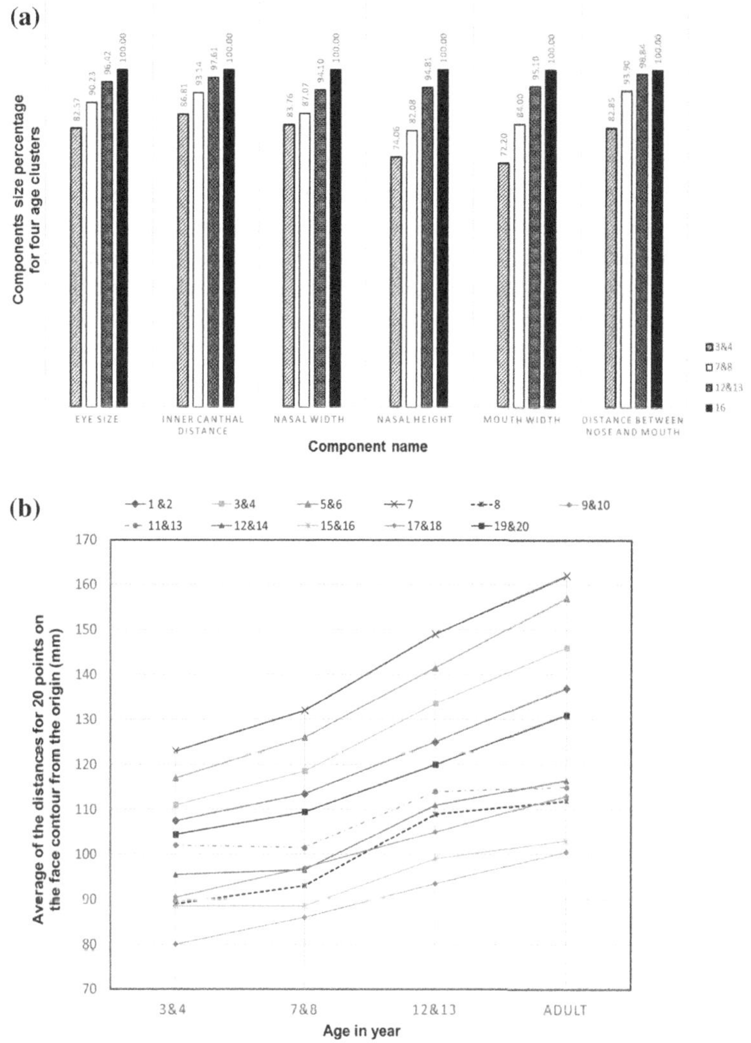

Fig. 7 **a** Percentage of the components' size in different age clusters compared to the adult components' size—both sexes, four age clusters. **b** Average distance of each point on the face contour from the origin—both genders, four age clusters—extracted from calculated average reduces of all face images in each age cluster

where S_i is the target size, α is the scale factor and S_j is the current size and d is the translation vector. j is the input age and i is the target age and for backward modeling, $i < j$. For adult to child rejuvenation, α is always between zero and one since all the face components' sizes in childhood are smaller than their sizes in the adulthood. Note that, the order of the changes is important. It means that at first the size changes, and then translation is applied to modify the distances and decomposition of the facial component distribution.

Face contour: the nasal bridge is considered as the origin of the face and 20 points for the contour get extracted in the feature extraction phase. The face shape can be defined by these points and its scale is defined by 20 radiuses obtained from those and the origin point. Studies show that in the process of facial growth, these radiuses do not change similarly; the upper face and the mandible mature unevenly. For each radius, one different coefficient is obtained. For example, the forehead has larger share of the face in childhood compared with adulthood. Yet as the face is symmetric, some of them can share the same coefficient.

Since holding the same angles in the radius proportion is important, in order to apply the changes the following method is used. Where $O(O_x, O_y)$ is the origin of the face, $C_j(C_{xj}(n), C_{yj}(n))$ is the facial contour point in the input age j, and $D_j(n)$ is the distance between them for n equals 1–20,

$$D_j(n) = \left(C_{yj}(n) - O_y\right)^2 + \left(C_{xj}(n) - O_x\right)^2 \quad (2)$$

For calculating the angle $\theta(n)$:

$$\tan\,\theta(n) = C_{yj}(n) - O_y C_{xj}(n) - O_x \quad (3)$$

$$D_i(n) = D_j(n)\beta \quad \beta > 0 \quad (4)$$

$C_i\left(C_{xi}(n), C_{yi}(n)\right)$ is the point of interest which is the point corresponding to the target age face contour, $D_i(n)$ is the new radius. β is the coefficient which differs from one age cluster to another and there are 11 different β to change a face in age j to the face in age i (see Fig. 8a).

$$C_{xi}(n) = C_{xj}(n) + D_i(n) * \cos\theta(n) \quad (5)$$

and

$$C_{yi}(n) = C_{yj}(n) + D_i(n) * \sin\,\theta(n) \quad (6)$$

The final result of geometrical changes is shown in Fig. 8b.

As the mathematical aspect is done, the result should be visualised. To visualise the geometrical correction, image warping should be applied. As completely explained in Chap. 2, image warping changes the input into an output image according to a mapping between the input and output space. For this reason, the points in the image, yet not deformed, should be mapped to the deformed image. For example, in Fig. 8b the facial points for the adult face image (green circles) should be mapped to the points for the 3-4-year-old facial points (red crosses).

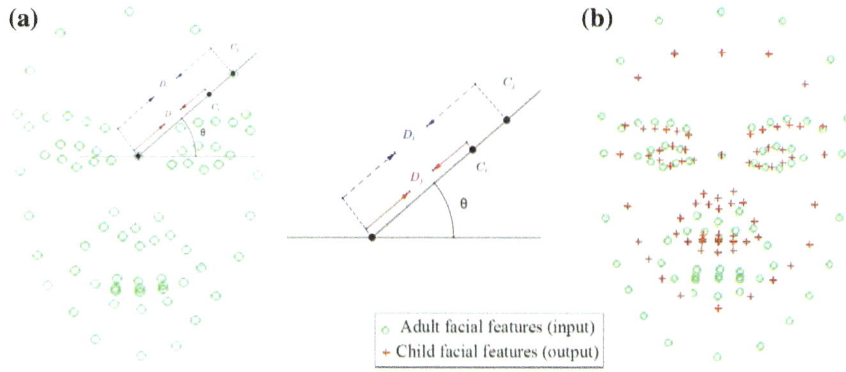

Fig. 8 **a** How adult facial contour radiuses are changed—C_j is changed to the C_i by having the β coefficient and keeping the angle θ. **b** Result of geometrical changes for both components and facial contour from adult (green circles) to 3–4 years old child (red crosses)

4.3.2 Reconstructive Face Templates

Since growth process depends on the genetic and the complex environmental interactions, growth related effects on the face are different from one person to another. However, the facial growth pattern similarities among the people in the same age cluster cannot be disregarded.

To study these common characteristics in an age range, an average face is calculated (explained in Chap. 2). In fact, the average of the facial images in the same age cluster can demonstrate; first, the sum of the important and distinctive similar information for that specific age; that is why it has been named *Face Template*. Second, the typical differences between two age clusters. Of course the more these two ages are far from each other the better the differences are distinctive.

Initially, the Face Templates are constructed using 5 points for which all images in an age cluster are selected and the average was calculated based on steps mentioned in Sect. 2.4. The results in the first row in Fig. 9 show the Face Templates resulted from this method, which are not satisfying because of the blurriness in some parts of the face, such as facial components boundaries and especially the face contour. Moreover, extracting the facial features and the information of the texture is not easy.

Therefore, this time Face Templates are constructed using all of the extracted feature points according to the different phases, fully described in Chap. 2, by employing feature extraction, image warping and averaging among all the faces in each age cluster.

It can be said, for each facial image f(k, ρ), *k* describes the value of a pixel position ρ. μ(λ, ρ) is the Face Template for one age cluster, described by the Face Template colour λ, which is the average of the pixel values for the same pixel position ρ from all facial images in an age cluster. In fact, this is a fusion method (simplest method) to fuse multiple images in order to combine their related information.

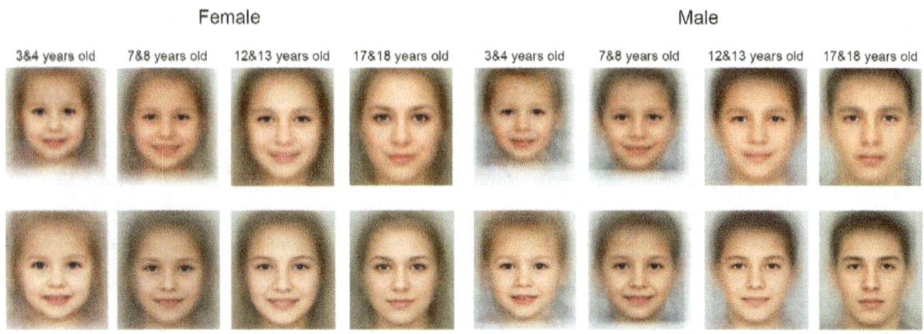

Fig. 9 Illustration of the failures in creating the Reconstructive Face Templates—both sexes—four age clusters. The first row includes the Face Templates constructed by using only 5 points. The second row includes the Face Templates contracted using all the extracted facial feature points that are mostly with a smile which should be eliminated

$$(\lambda, \rho) = \sum_{i=1}^{n} [fi(k, \rho)] \tag{7}$$

n is the number of facial images in the age cluster.

These results are sharper; the face contour and facial components' borders are vivid. The features and the textures are also clearer as it can be seen in the second row of the Fig. 9.

However, this time, the weakness of the calculated Face Templates is that the Face Templates are with smile. The reason is, in the collected database most of the face images are faces with a smile on them. This affects the quality of the results because of the changes made in the size of the mouth and the existence of laugh lines on the two sides of the lips. To have the better Face Templates and solving these issues in the final results, the database is revised and the faces with big smiles are pretermitted and replaced. As it can be seen in Fig. 10, the final *Reconstructive Face Templat*es are in a better condition and the obtained results using them are more satisfying.

4.3.3 Textural Model

Although the geometrical results have geometrical characteristics related to the associated age groups, they are not visually sufficiently realistic and convincing. Therefore, in order to have a better textural aspect, they should be fused with the Reconstructive Face Templates belonging to the same age cluster.

To apply the fusion, the geometrically changed face is fused with the associated Reconstructive Face Template, by engaging the 79 facial feature points and the image warping. In fact, first, the average of the coordinates of these two face images is calculated. Then, the facial feature points in the Face Template and the geometrically changed face image are warped to the calculated average coordinates. At the end,

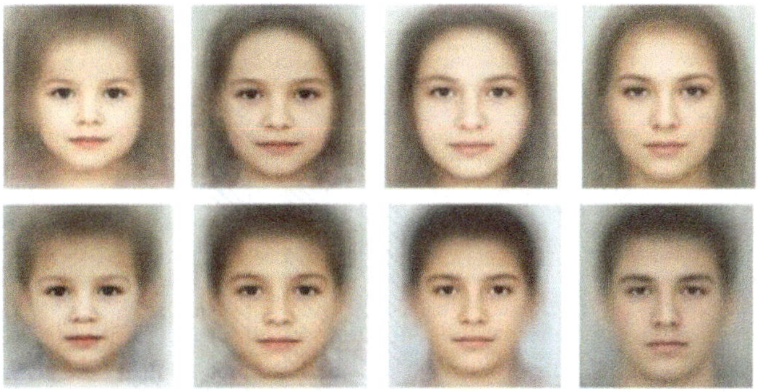

Fig. 10 Illustration of the final reconstructive face templates, without smile—both sexes—four age clusters

these two face images are fused by using the texture averaging, which is a basic fusion method.

In applying the Textural Model, there are some intricacies that significantly change the quality of the results:

Keeping the original eye shape: As mentioned before, to apply the texture variation, both face images (Reconstructive Face Template and the geometrically changed face image) should be warped to the calculated mean coordinates of each feature point. By paying more attention to the primary results and taking into consideration that in many cases, people are recognised by their eyes, the original shape of the eyes is preserved. For this reason, when the averages of coordinates are calculated, the feature points corresponding to the eyes will not be considered. Instead, the eyes in the Face Template image will be warped to the eyes coordinated in the geometrical result. Otherwise, the eyes area in the result will be blurred.

Weighted average calculation: In order for the result to have a texture most proximate to its related age group, the main face image and the Reconstructive Face Template should undergo an average calculation procedure. The given weight to each input is very crucial. It means that it can be decided which face image has more share in the result; the main one (Geometrically changed) or the Face Template. Of course, the more shares the main face image has, more similarity in the result is obtained. Most often, a percentage between 50 and 70 for the weight of the main face image, gives an acceptable result.

These results also contain all the expected geometrical characteristics defined for a face in the target age. In the meantime, their texture is adapted under the influence of the Reconstructive Face Template. Thus, these results are visually realistic and acceptable (Fig. 13—column b).

For weighted average which is in fact the weighted sum:

$$T(\lambda, \rho) = \sum_{i=1}^{n} [fi(k, \rho)] \tag{8}$$

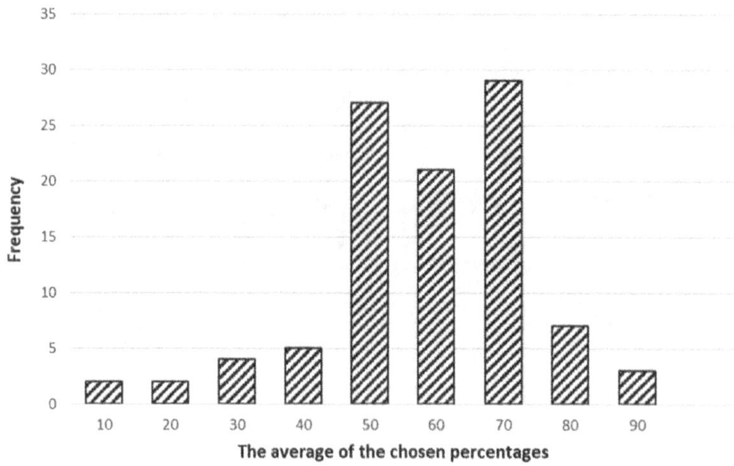

Fig. 11 The histogram of the average of the chosen percentage based on the highest similarity to the childhood reference image by all the respondents

w is the weight given to each face image. As a matter of fact, in this case, we have only two images so n = 2. Hence, it can be said, Face Template construction for instance is a special case of the weighted average where all the face images have equal weights.

To achieve the percentage range between 50 and 70, a survey has been made on the results of 34 adult input images. For each input image, the percentages from 10 to 90 were given to the main face to achieve a range of results. Note that, 0% for the main face means that only the Reconstructive Face Template is used and 100% means that the effect of the Face Template is neglected, so these two conditions are passed up. Then, we asked 100 people to choose the percentage that gives the most likely result to a reference image from the input face's childhood. Figure 11 shows the average of the responses for all 34 sets of results. As it can be seen most of the chosen percentages are between 50 and 70. The input images are selected from Set 1 and childhood face images used as reference are from Set 2 of FaceTiM database.

Mask creation to keep the natural eye colour: With a closer examination of the obtained results, it is undeniable that in order to have better results in the images of people who have coloured eyes, some pre-processing should be done before the average calculation. To this purpose, pupil, iris and sclera should not be considered in calculating the average (see Fig. 12).

To do this, a binary mask is overlaid onto the eyes area in the Face Template, to prevent the masked area from getting included in the calculation. The area is selected by using double thresholding with pixel values and morphology-based opening and closing, which are derived from the operations of erosion and dilation [18, 19, Sect. 2.5].

In fact, we use opening to preserve the region that have the similar shape to the structuring element or can completely contain structuring element while eliminating

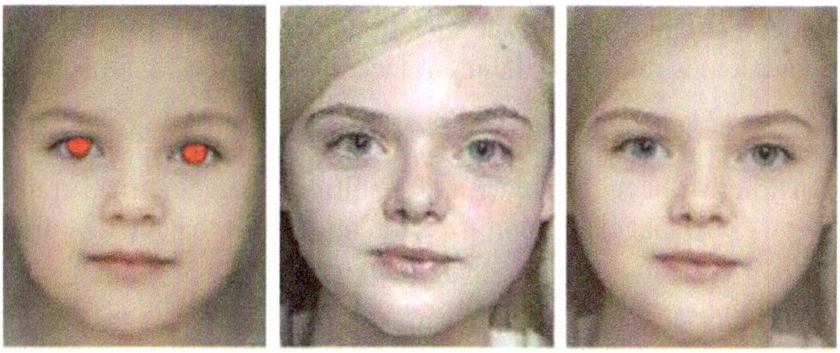

Fig. 12 Mask creation for keeping the original eyes colour. From left to right, created eye mask, geometrical result and rejuvenated face image

all other regions of foreground pixels. If I is the eye region from the facial image and H is the structure element that arbitrary defined, opening is:

$$I \circ H = (I \ominus H) \oplus H \tag{9}$$

which is the morphological operation of "an erosion followed by a dilation". Opening removes the isolated foreground structures that are smaller than the H structure element, and keeps the larger structures. Closing is:

$$I \bullet H = (I \oplus H) \ominus H \tag{10}$$

which is the morphological operation of "a dilation followed by an erosion". Closing removes the holes in the foreground that are smaller than the H structure element [18, p. 209].

If the selection is not perfect then a manual selection can be created.

It is worth mentioning, in calculating the average in this section, eyes are weighted separately.

4.4 Backward Facial Ageing Model (B-FAM)

Based on the steps mentioned above, B-FAM is proposed for digital facial image rejuvenation. In this model each rejuvenated face, in one of the formative years, is the sum of the Reconstructive Face Template correspond to that specific age and its special geometrical characteristics, facial differences and details:

$$F_{(i)} = w_1 T_r(i) + w_2 \varphi(f(j)), \{i < j 0.5 < w_2 < 0.7 \quad w_1 + w_2 = 1\} \tag{11}$$

where $F(i)$ is the face in the age i, $T_r(i)$ is the Reconstructive Face Template in the same age and $\varphi(f(j))$ is the transformation function applied on the face f in age j

to adapt the geometrical aspect of the face to the target age i, by keeping the entire differences and details related to the input face. j is the input age, i is the output age and $i < j$.

This sum is the weighted sum in which w_1 and w_2 are weights given respectively to the Reconstructive Face Template and result of the geometrical transformation function. $50 < w_1 < 70$ that can be confirmation for the fact that most of the alterations during formative years are geometrical and texture undergoes light changes.

This ageing model can even be carried out for different ethics or genders. For example, a Mongolian face in the age *i* can be presented with the sum of a typical Mongolian face in that specific age and its particular details and differences. Of course, the more face images used in the database in creating the Face Template, the more exquisite the results are.

4.5 Some Experimental Results

In order to study the facial changes during the growth ages and construct the model, Backward Facial Ageing Collection (BFAC) album from the Facial Ageing Modeling Database (FAMD), explained in Chap. 2, is used. This album contains almost 700 facial images in 4 different age clusters, Cluster 3–4, Cluster 7–8, Cluster 12–13 years old and the Cluster Adult, separated by gender.

For testing the model, 112 adult face images from Face Time-Machine Database (FaceTiM V1.0), Set 1, are used, which are frontal and almost without any facial expressions. Besides, 112 face images from Set 2, that contains childhood face images belongs to the individuals in Set1, are employed as the reference to compare the simulated results.

According to the presented model, the input is an adult face image and the result is how it might have looked in one of the defined age clusters. The reference face images have been used to compare the results and to evaluate them. Some of the results are shown in Fig. 13.

The model can generate faces in different specified age clusters (Cluster 3–4, Cluster 7–8, Cluster 12–13 and Cluster Adult). However, we have changed the faces only to some age clusters that we actually have references for them, in order to be able to evaluate the results.

Figure 14 illustrates the rejuvenation results on one subject facial image in different age clusters.

5 3D Facial Rejuvenation Modeling

In this section, the purpose is to extend the previous digital rejuvenation work to third dimension. This totally makes sense, since faces are basically tri-dimensional. The depth (third dimension) provides additional information to represent accurately

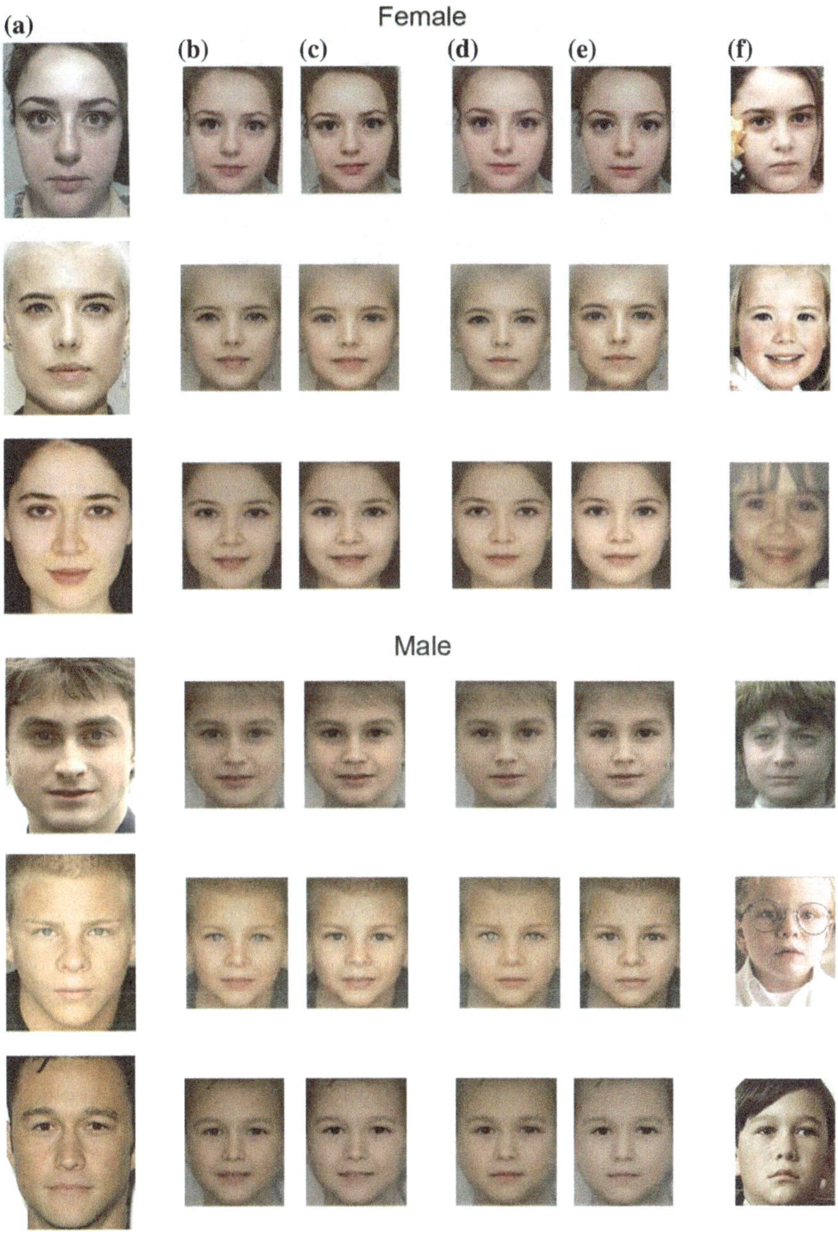

Fig. 13 Results of applying B-FAM. **a** Original image—young adult. **b** Result using Face Templates with smile and natural eye colour. **c** Result using Face Templates with smile and calculated average eyes. **d** Result using reconstructive face templates without smile and with natural eye colour. **e** Result using reconstructive face templates without smile and with calculated average eyes. **f** Reference image in approximately the same age as the result, for further comparison

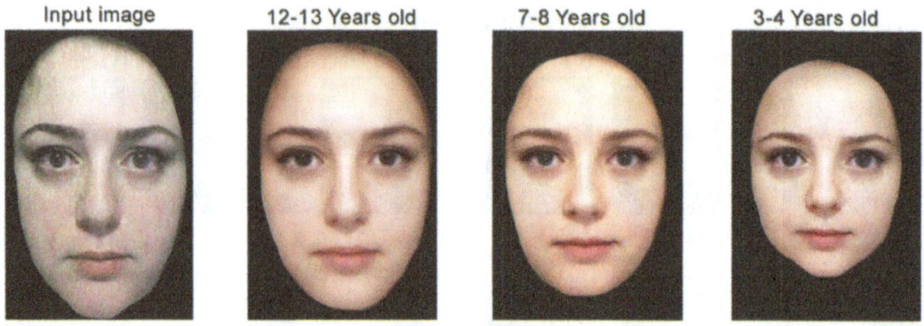

Fig. 14 Illustration of the rejuvenation results in different age clusters

faces. Therefore, when dealing with 3D face mages, the following questions are raised.

How to visualize a facial surface in three-dimension environment without losing track of data cohesion? How to easily identify correlations between large three-dimensional data volumes and how to make changes in them in order to change the facial appearance according to its age?

The aim in this study is to deal with some of these challenges and extend the visualization possibilities of gerontology physical changes (Ageing and Rejuvenation) for analysis 3D facial data.

Visualization plays an important role in analyzing, exploring and presenting data. However, visualizing large amounts of data is often challenging, especially for multivariate data spanning across several dimensions.

Three-dimensional (3D) models represent a physical object using a collection of points in 3D space, connected by various geometric entities such as triangles, lines, curved surfaces, etc.

There are three popular ways to represent a 3D geometrical model:

1. Polygonal modeling,
2. Curve modeling,
3. Digital Sculpting.

In this study, the polygonal modeling is considered so that the characteristics of the 3D facial image data called "Geometry Mesh Topology" is used in order to represent the 3D face image. Generally, we need at least three characteristics for 3D graphics:

1. A number of points in 3D space (called vertices)
2. Information on how these points connect to each other to form triangles (called faces)
3. Information on how to draw the 3D model (that means: colors, images to be used, etc. This is summarized in the term material).

According to the previous chapter regarding 2D face models, we learned that Face perception is powerful channel to human social interaction [20]. Faces visualize many different types of important and vital clues and the processes and mechanisms involved in extracting them are complex [21] and it can become more challenging by going beyond and add one more dimension to represent the face with a massive amount of data points in 3-dimension.

To help readers become more familiar with the keywords in the context of the 3D facial models, the terms "3D Feature Extraction", "Reference Model", "Geometrical Model" and "Textural Model" are thoroughly discussed below.

5.1 3D Facial Features Extraction

A 3D face model represents a 3D object using a massive collection of points in 3D space, connected by various geometric entities of 3D file formats. Talking about face models, it is complicated to extract reliable facial features in 3D environment as they are numerous; however, considering the depth information and pose variation gives 3D face model the advantage. In this research, we extract the geometrical information (i.e., the geometric vertex and its 3D coordinates of 3D face image) and the textural information (i.e., the UV map of its texture material). The polygon mesh data containing the geometric vertices and texture information are extracted and stored, Fig. 15.

5.2 Reference Model

Anthropologist Sir Francis Galton who was also the cousin of Charles Darwin pioneered the method of "composite portraiture" in the 1880s. He superimposed multiple portraits of individuals' faces together to create an average. In the psychological science, it has been assumed that the distinctiveness of a face is related to its distance from the average face in a generic face space [22]. Based on this, the average face of individuals in the same age group is considered which represents the sum of the information of maximal importance that can be distinctive in that specific age. The model proposed by Heravi et al. [23] starts out with the recognition that there are a finite number of typical faces in different age groups. In each age group, typical face categories exist as average female face, average male face. In other words, in each of the age categories, there are certain characteristics which change in almost all faces of a given type, e.g., the types of wrinkles which appear, the loss of facial elasticity, etc. Based on this premise, the "Reference Face" is constructed for each age group representing the average typical face with the common characteristics of each specific age class [24, 25]. Then the features are extracted from all 3D images of each age group for both genders, then the average of each is calculated. Consequently, the geometrical and textual reference models are constructed based on the study on the common traits of each age group by using the average of all geometrical and textural information in every age class. [26].

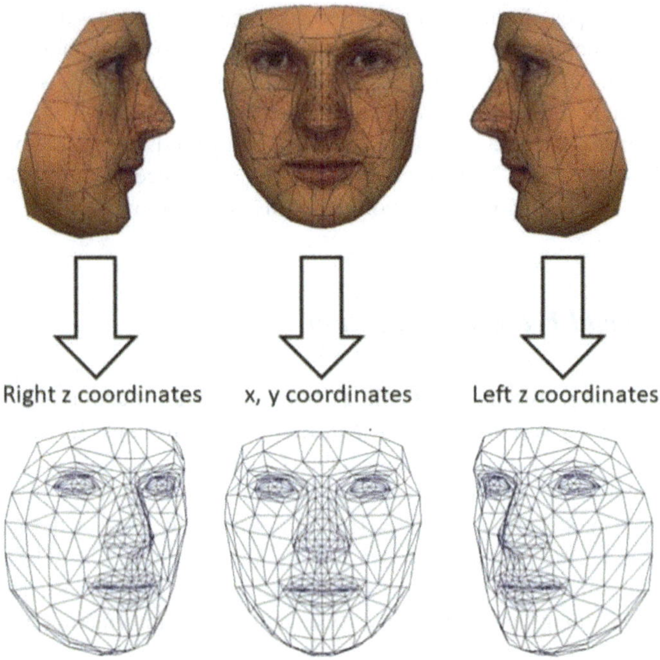

Fig. 15 The 3D face model and geometrical view

Various aspects should be taken into account when dealing with face changes over the time. It is noticeable that face variations take place in both geometrical shape and skin texture. Therefore, two sub-models of geometrical and textural are constructed.

5.3 Geometrical Model

Undoubtedly, altering a face age down to its childhood or up to its adulthood requires a reliable geometrical model as the craniofacial morphology growth pattern is specific for each age. The polygon mesh (shape) of a model is often stored as a set of 3D points (or vertices) (Fig.16a). The surface of the model is then stored as a series of polygons (faces) that are constructed by indexing these vertices (Fig.16c). The number of vertices the face may index can vary, though triangular faces with three vertices are common. The geometry information in ASCII format consists of 5904 points specifying Vertex data (specifies a vertex by its three coordinates), Vertex texture (the UV mapping determines how to map the texture), Vertex normal (specifies a normal vector), and the faces (specifies a polygon made from the vertices listed). For each vertex, there is an associated Vt, says how to map the texture at this point.

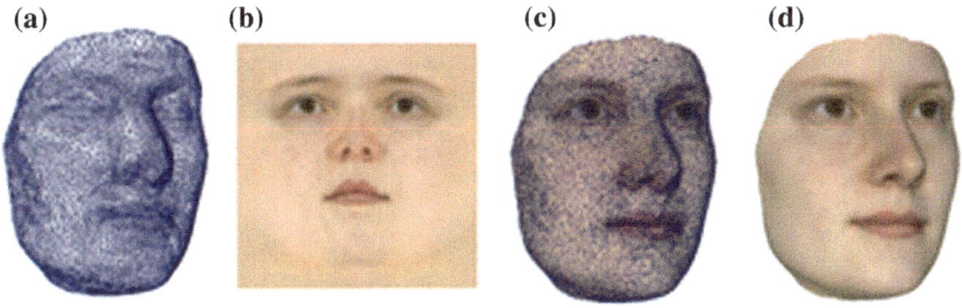

Fig. 16 Illustration of a 3D polygon mesh (**a**), the texture UV map (**b**), the warped texture UV map on 3D mesh (**c**) and the 3D face (**d**)

5.4 *Textural Model*

To make the geometrical model visually realistic, the textural aspect has been considered in the context of textural model for both genders in each age class. There exists many methods to represent the surface properties of a 3D object in computer graphics. In case of the original Morphable Model [27] per vertex, a texture information was used. Texture UV maps are commonly used in computer graphics to wrap a 2D image texture along "U" and "V" axis of a plane on a 3D mesh with "X", "Y" and "Z" coordinates in the 3D space and are an appropriate way to represent the skin surface color of a face. The advantage is that the resolution is independent of the geometry. For this reason, we defined a single 2D texture UV map for the facial regions of the head. Figure 16c. All the UV map regions of importance are normalized to a size of 256 * 256 pixels. The texture average is considered to calculate a single mean texture template for each age as a UV texture map.

5.4.1 3D Rejuvenation Model (RM)

Based on the above mentioned pre-processing steps, the 3D dynamic RM (Rejuvenation Model) is proposed for digital facial image rejuvenation using 3D data:

$$F_{RM}(i) = \left\{T_{RM/input}(j) + wT_R(i); \underline{R} + \sum \alpha V_{RM}\right\} \tag{12}$$

where F_{RM} (i) is the target rejuvenated output of the 3D face at the given age, $\underline{R}$ is the Reference 3D image which consists of R_M and R_T, the reference Mesh and Texture respectively. Coefficient α is specified such that the personal characteristics similarity measurement between current age and target age can be kept as much as possible. In other words, to synthesize the different measures using the geometrical and textural models, it is necessary to calculate varying model coefficients. The approximation of the coefficients is not the same for shape and perception since most of the changes in the polygon mesh of the face occurs until the age of 18 [28], while the changes in

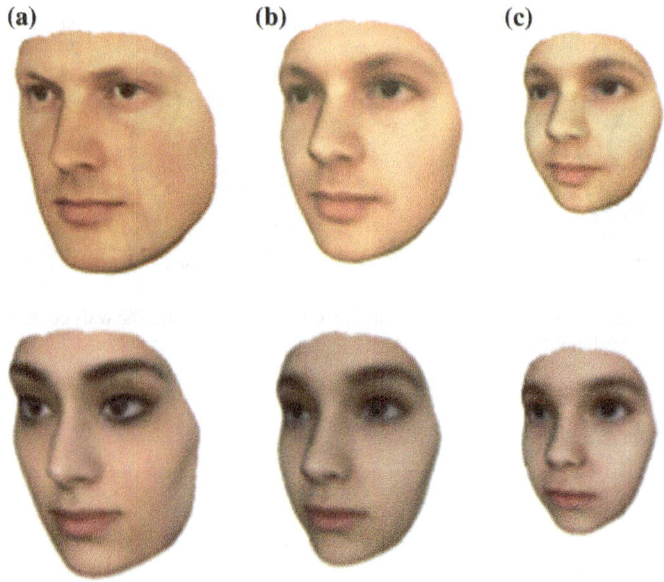

Fig. 17 Simulation of 3D facial rejuvenation trajectory in three different ages, adulthood (**a**), teenager (**b**) and childhood (**c**)

texture take place when facial muscles begin to lose their elasticity and the wrinkles appear.

The Rejuvenation Training Database (RTD) is designated to rejuvenation trajectory containing the 3D faces features (Mesh and Texture) between 3 to 18 years old. M and T are representatives of shape (Mesh) and perception (Texture) of the face. Using PCA, the model is proposed in which w is the projection of the observation matrix over the eigenvector V.

$$\begin{aligned} M_{RTD} &= \underline{M}_{RTD(i)} + w_{RTD(i)} V_{RTD(i)} \\ T_{RTD} &= \underline{T}_{RTD(i)} + w_{RTD(i)} V_{RTD(i)} \end{aligned} \tag{13}$$

5.5 *Some Experimental Results*

There exists many interesting and popular applications for simulation of facial aging. These applications can be found in a lot of different disciplines like medicine, forensic, anthropology, biometrics, art, entertainment, and computer graphics. The following results illustrated in Fig. 17 are generated automatically by applying the rejuvenation simulation model on 3D images and are applicable to many of these applications. As can been from Fig. 17, the backward trajectory model of an individual's appearance in their younger age is simulated and the outputs are obtained in 3 dimensional environment.

6 Conclusion

In this chapter, 2D and 3D Facial Rejuvenation Models [4, 5, 23] have been considered. These models, for the first time, digitally rejuvenates an adult face appearance down to its childhood, based on statistical measurements.

First, the Geometrical Model has been applied on the face. Face components and face contour are non-linearly modified using a robust Geometrical Model. In order to adjust the texture to the target age, the weighted sum calculation has been applied on the geometrical results and the related Reconstructive Face Template.

Effects of non-facial factors that can influence the perception of an individual's age, such as hair style, facial hair, and makeup are not taken into account. Actually, it can be said, all the actions affecting the natural process of the facial growth such as heavy makeup and cosmetic surgeries are challenges for reconstructing a child face image from an adult face image. Their effects can influence the accuracy of the model.

Besides biometric security and healthcare biometrics, rejuvenation techniques, or de-aging can be considered in other applications dealing with special effects, as it is the case in movies. This challenge can certainly be extended to other innovative applications.

In the next chapter Facial Ageing Model will be presented. This model predicts adult face appearance in the future by considering the natural ageing process.

References

1. A. Bastanfard, O. Bastanfard, H. Takahashi, M. Nakajima, Toward anthropometrics simulation of face rejuvenation and skin cosmetic. Comp. Anim. Virtual Worlds **15**(3–4), 347–352 (2004)
2. G. Guo, Digital anti-aging in face images, in *2011 International Conference on Computer Vision* (2011), pp. 2510–2515
3. M. Minear, D.C. Park, A lifespan database of adult facial stimuli. Behav. Res. Methods Instrum. Comput. **36**(4), 630–633 (2004)
4. E. Farazdaghi, A. Nait-Ali, Backward face ageing model (B-FAM) for digital face image rejuvenation. IET Biometri. (2017)
5. E. Farazdaghi, F.M.Z. Heravi, L. Chatelain, A. Nait-Ali, Reverse facial ageing model for youthful appearance restoration from adult face images, in *2016 6th European Workshop on Visual Information Processing (EUVIP)* (2016), pp. 1–6
6. P. Sharma, A. Arora, A. Valiathan, Age changes of jaws and soft tissue profile. Sci. World J. **2014**, e301501 (2014)
7. S. Premkumar, *Textbook of Craniofacial Growth* (JP Medical Ltd., 2011)
8. R.K. Harley, G.A. Lawrence, L.D. Sanford, R. Burnett, *Visual Impairment in the Schools*, 3rd edn. (Charles C Thomas Publisher, 2000)
9. R.S. Snell, M.A. Lemp, *Clinical Anatomy of the Eye*, 2nd edn. (wiley). [Online] Available: http://www.wiley.com/WileyCDA/WileyTitle/productCd-063204344X.html. Accessed 14 Feb 2017
10. J.D. Subtelny, A longitudinal study of soft tissue facial structures and their profile characteristics, defined in relation to underlying skeletal structures. Am. J. Orthod. **45**(7), 481–507 (1959)

11. S.R. Coleman, R. Grover, The anatomy of the aging face: volume loss and changes in 3-dimensional topography. Aesthet. Surg. J. **26**(1_Supplement), S4–S9 (2006)
12. J.A. Salzmann, Handbook of facial growth. Am. J. Orthod. **71**(3), 340–341 (1977)
13. D.F. Huelke, An overview of anatomical considerations of infants and children in the adult world of automobile safety design. Annu. Proc. Assoc. Adv. Automot. Med. **42**, 93–113 (1998)
14. C.J. Lux, C. Conradt, D. Burden, G. Komposch, Three-dimensional analysis of maxillary and mandibular growth increments. Cleft Palate Craniofac. J. **41**(3), 304–314 (2004)
15. A.D. Dixon, D.A.N. Hoyte, O. Ronning, *Fundamentals of Craniofacial Growth* (CRC Press, 1997)
16. J. Hall, J. Allanson, K. Gripp, A. Slavotinek, *Handbook of Physical Measurements* (Oxford University Press, 2006)
17. L.G. Farkas (ed.), *Anthropometry of the Head and Face*, 2nd sub edn. (Raven Pr, New York, 1994)
18. C. Solomon, T. Breckon, Morphological processing, in *Fundamentals of Digital Image Processing* (Wiley, 2010)
19. F.Y. Shih, *Image Processing and Mathematical Morphology: Fundamentals and Applications* (CRC Press, 2009)
20. A. Takeuchi, K. Nagao, Communicative facial displays as a new conversational modality, in *Proceedings of the INTERSECT'93 and CHI'93 Conference on Human Factors in Computing Systems* (ACM 1993), pp. 187–193
21. M. Chan, P. Delmas, G. Gimel'farb, P. Leclercq, Comparative study of 3d face acquisition techniques, in *Proceedings of International Conference on Computer Analysis of Images and Patterns (CAIP'05)*. LNCS, vol. 3691 (Versaille, France, 2005), pp. 740–747
22. D.H. Enlow, R.E. Moyers, W.W. Merow, *Handbook of Facial Growth* (Saunders, 1975)
23. F.M.Z. Heravi, M. Eslahi, E. Farazdaghi, A. Nait-Ali, A Morphable Model to simulate rejuvenation trajectory of 3D face images: preliminary results, in *2016 International Conference on Bio-Engineering Smart Technology BioSMART 2016* (2017), pp. 1–4
24. D.I. Perrett et al., Effects of sexual dimorphism on facial attractiveness. Nature **394**(6696), 884–887 (1998)
25. E. Farazdaghi, F.M.Z. Heravi, L. Chatelain, A. Nait-Ali, Reverse facial ageing model for youthful appearance restoration from adult face images, in *European Workshop on Visual Information Processing* (EUVIP, Marseille, France, 2016)
26. L. Sirovich, M. Kirby, Low-dimensional procedure for characterization of human faces. J. Opticalsoc. Am **4**, 519–524 (1987)
27. M. Chan, P. Delmas, G. Gimel'farb, P. Leclercq, Comparative study of 3d face acquisition techniques, in *Proceedings of International Conference on Computer Analysis of Images and Patterns (CAIP'05)*. LNCS, vol. 3691 (Versaille, France, 2005), pp. 740–747
28. J. Hall, J. Allanson, K. Gripp, A. Slavotinek, *Handbook of Physical Measurements* (Oxford University Press, 2006)

Facial Ageing Modeling

Farnaz Majid Zadeh Heravi, Elham Farazdaghi and Amine Nait-Ali

Abstract In this chapter, facial biometrics is discussed from the ageing point of view. Both geometrical and textural aspects are discussed. This challenging biometrics can be useful in numerous applications, including security, biomedical engineering, plastic surgery, and forensics.

1 Introduction

In this chapter Facial ageing models are presented in order to predict the facial appearance of a young adult in its future. These models are based on the effects of the generic factors that are involved in natural ageing process.

The purpose of this chapter is not only presenting the models to predict the facial ageing in adult image faces, but to build a model as a foundation for the novel and practical forthcoming models. As a matter of fact, in this chapter the influence of the natural ageing trajectory on the different aspects of the face is modeled, and in the next chapter the effects of the different external factors on this process will be integrated into this base model.

Unlike rejuvenation models (saw in the previous chapter), in which the focus is essentially on geometrical aspects of the face, in ageing models the geometrical variations are subtle and alterations are mostly in facial texture.

2 Structural Layers of the Facial Ageing

Proposing a reliable model for facial ageing requires an understanding of both youthful and elder face anatomy and ageing stigmata on the face. It is substantial to know

F. M. Z. Heravi · E. Farazdaghi · A. Nait-Ali (✉)
Université Paris-Est, LiSSi, UPEC, 94400 Vitry-sur-Seine, France
e-mail: naitali@u-pec.fr

A. Nait-Ali (ed.), *Biometrics under Biomedical Considerations*,
Series in BioEngineering, https://doi.org/10.1007/978-981-13-1144-4_4

which parts of the face are affected by ageing and how these changes will appear on the face.

The outcomes of the morphological studies indicate the face does not age homogeneously, rather, its ageing is associated with many dynamic components. It can be said that the facial ageing is associated with changes in all structural layers of the face, namely, skeleton, muscle, fat, and skin [1].

In the following, there is a short statement of the ageing effects on each mentioned layer.

2.1 Facial Skeleton Alterations with Ageing

Facial skeleton foundation undergo changes through the life. This alterations are either differential growth, which accrues during the childhood to adulthood and enables the infant skull to assume its adult form, or resorption of the certain areas of facial skeleton, that arise with ageing [2].

Facial skeleton resorption by ageing is heterogeneous. All different bones are not affected in the same level. Therefore, for a better understanding of the skeleton resorption impact on the face age, first the facial skeleton is divided to some sub-regions, then the receding level and the orientation in each area is studied. Figure 1 shows these different sub-regions of facial skeleton.

Periorbital area: The periorbital area consists of the forehead, the eye socket and rims, the eyelids and surrounding areas, including the eyebrows. This area seems to be the first area to show the signs of ageing.

Studies indicate that the orbital aperture undergo resorption by ageing. However, changes occur at different rates, are site specific and uneven. Many studies have been done on the effects of ageing on this region. Here some of the outcomes are cited; there is no significant changes in orbit angle in ageing [3]. Orbital aperture width and area has significant increase with increasing the age [4]. Superior and inferior orbital rim receded with the age [5] (see Fig. 2a). One of the causes of increasing vertical eyelid length, lower eyelid curve distortion and hollow eyes is the lake of the rim support.

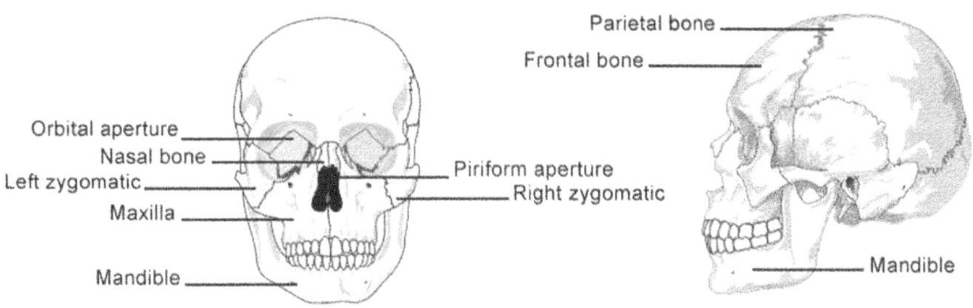

Fig. 1 Illustration of the primary bones of the face

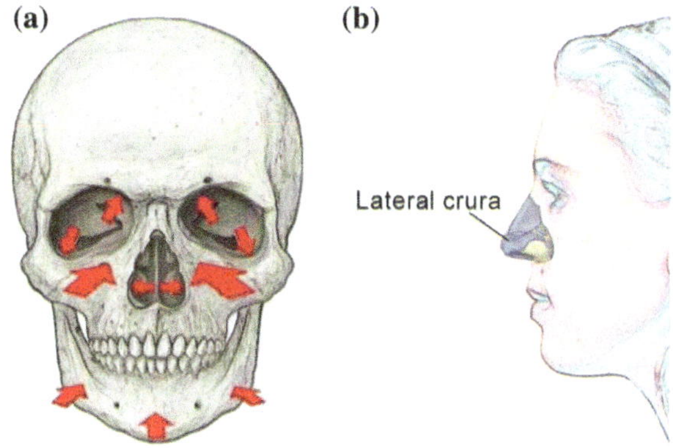

Fig. 2 **a** Indicating the areas of the facial skeleton susceptible to resorption with ageing. The magnitude of the arrows represents the amount of the resorption (Figure is adapted from [2, Fig. 5]). **b** Displaying the lateral crura on the nose which descends in the result of ageing

Midface: The midface skeleton consists of the maxilla bone and left/right zygomatic bone [6] (see Fig. 1). The grade of bone resorption in the midface is also not uniform. The maxilla is more susceptible to have age-related alterations than the zygomatic bone. Studies show maxillary angle decreased about 10° in old individuals (>60 years) comparing to young people (<20 years) [2]. This maxillary retrusion might cause the prominence of the nasolabial fold [7].

Nasal area: Similar to the situation of the orbital aperture, as the edges of the nasal bones recede with age, piriform aperture enlarges with manifestation of ageing. The posterior displacement of the bone rim at the lower pyriform aperture is greatest. This area is very important for supporting the lateral crura (Fig. 2b) and tip of the nose. Therefore, one of the characteristics of the aged face is the long and dropped nose [2, 8]. Moreover, studies indicate special bone loss in the lower part of the pyriform aperture with ageing which contributes to deepening of the nasolabial fold with age [9].

Lower face and mandible: There is no significant age-related changes in the mandible and ramus width. Mandibular body height and length and ramus height have significant decrease with age. Also, Mandibular angle increase with increasing the age. These alterations can be the cause of the appearance change in the lower third of the aged individuals' face, such as sagging chin and jowl. Missing teeth can escalate ageing signs in this region.

2.2 Facial Muscles Transformation Through Ageing

Facial muscles' size and architecture are transformed in ageing process. Most of these alterations emerge from the loss of skeletal muscle mass and strength that happen as a result of ageing [10, Chap. 1].

Numerous studies prove the role of superficial muscular aponeurotic system (SMAS), in facial ageing. SMAS is a mid-level muscular layer of the face that separates the deep facial fat from the superficial facial fat [11, 12]. Analyses of MRI

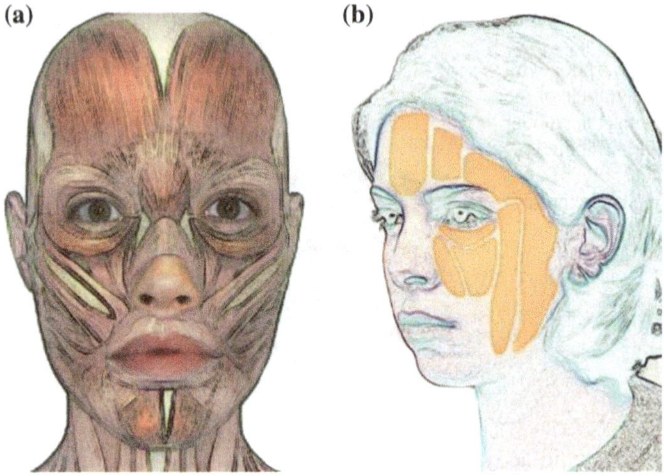

Fig. 3 **a** Illustration of the facial muscles. **b** Facial fat distribution on the face (Figure is adapted from [14, Fig. 5])

(Magnetic Resonance Imaging) images from individuals in different ages indicate that the ageing muscles in the midface shorten and straighten, as they seem in spasm [9]. It can be said that the attenuation in these muscles can cause the sagging appearance to the face. Changes in these muscles can also affect the nasolabial lines known as laugh lines.

Moreover, orbicularis muscles lose their strength by ageing. It causes descent and deflation of the upper eyelids, as well as laxity in the lower ones [13]. It makes the brows, particularly the outer corners, descend over time.

Changes in the muscle encircles the mouth including lisps (Orbicularis Oris muscle) can cause wrinkles around the lips and drooping mouth corners. Figure 3a shows the facial muscles.

2.3 Facial Fat Compartment and Its Changes with Ageing

Pattern of the fat deposition on the face undergoes specific alterations due to the ageing process. For instance, fat loss, deformation, and descending are the facial ageing signs.

Facial fat is composed of different parts such as nasolabial, jowl, malar, forehead, lateral temporal-cheek, periorbital fat and etc. [14]. Different fat compartments in the face are altered at different rates. Facial fat redistribution, accumulation and atrophy by ageing is even different from one person to another. Some of these facial fat disposition changes that affect the facial ageing are cited below:

- Forehead fat loss can cause deep wrinkles in forehead and drooping of the eyebrows downward.
- Orbital fat expansion occurs with age and displace the soft tissue of the lower eyelid and it causes double convexity deformity and under-eye bags [12].
- Malar and jowl fat descend that can lead to nasolabial lines, square facial contour and double chin [11, 15]. Figure 3b illustrates the fat distribution on the face.

2.4 *Facial Skin Ageing*

Skin is the most superficial and complex structural layer of the face that its appearance is the primary indicator of the age. To better express, facial ageing processes has strong connection with phenotypic changes in cutaneous cells [16].

Skin consists of three layers: "epidermis", "dermis" and "subcutaneous tissue" [17] (see Fig. 4). The epidermis is the most-outer layer of the skin that represents the appearance of texture and color. It consists of keratinocytes and melanocytes. The dermis is the deeper layer, forming the main volume of the skin. It is made up of fibres, cells and substances and is responsible to support the blood vessels and nerves that are included in it. The fibres available in the dermal layer are mostly collagen and elastin. About 80% of adult skin dry weight consists of collagen [16]. Collagen fibres give a high tensile strength to the skin and protect it against overstretching. Elastin constitutes about 5% of the dermis. It gives elasticity and resilience to the skin. As the skin quality deteriorates with age, the dermis also undergoes morphological, physical and chemical changes with ageing.

Overall, many of the skin's functions decline with age; epidermal turnover, sebum function, immune function, wound healing, vascular reactivity, sweat production, and

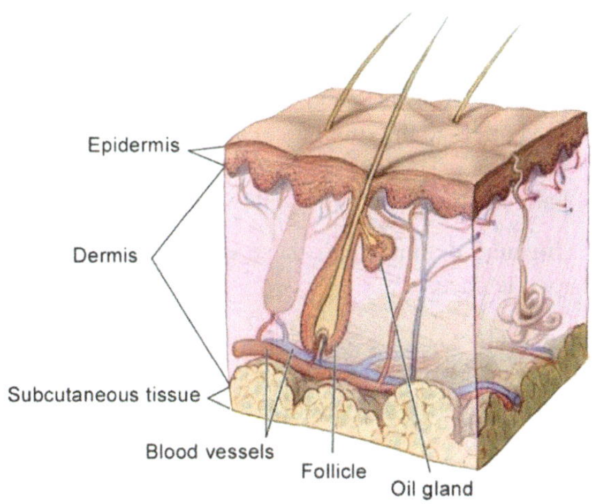

Fig. 4 Anatomy of the skin and its different layers consist of "epidermis", "dermis" and "subcutaneous tissue"

vitamin D production and etc. [16]. The skin changes in function and structure and becomes thinner with age [18]. Collagen atrophy is a main factor in skin ageing which leads to increase in laxity and wrinkles.

Similar to the collagen, elastin is a fiber-forming functional protein. Elastin fibres and collagen fibrils are closely linked together, so they are prevented from over-stretching and can rebound after a short-term stretching. Degenerative changes in dermal elastic fibres can also be speeded up with ageing.

"Transepidermal water" content of the skin decrease with age due to the skin barriers function alterations [19, 20]. It can lead the skin to dryness and deep facial creases and wrinkles.

In addition to all of the above, some pigmentary changes occur on the face with ageing. The number of melanocytes, a mature melanin-forming cell in the skin, decreases by ageing. The development of new melanocytes also declines. As a result, the skin of the elderly is more marked by pallor. Hence, an aged face skin is wrinkled, saggy, and pale.

To brief, changes in the balance of all above mentioned facial layers cause the hallmark of ageing, based on which a Predictive Facial Ageing Model is built without considering each layer separately. These signs are summed up as follows [21] (note that, '*t*' is for texture-related changes and '*g*' is for geometrical changes):

- Forehead wrinkles (t)
- Lateral Canthal lines or crow's feet wrinkles (t)
- Glabellar frown lines which are the vertical lines in between brows (t)
- Brow droop (g)
- Full bags (Under eye bags) (t)
- Increase in vertical lower-eyelid length (g)
- Wrinkles beneath eye (t)
- Nasolabial folds (t)
- Nose elongation and tip movement (g)
- Marionette lines that run from the corner of each side of the mouth to the corner of the chin (t)
- Upper lip wrinkles (t)
- Ptosis of lower face and sagging chin (g)
- Jowl refers to the small focal accumulation of fat in the lower cheek overlying the jaw bone (g)
- Paler and turned grey or yellow skin (t)

Figure 5 demonstrates these symptoms in the different parts of the skin and on the whole face.

3 2D Ageing Model Development

Same as for the facial growth, age-related facial changes can be assorted as geometrical and textural. However, inverse of the growth process, in ageing trajectory,

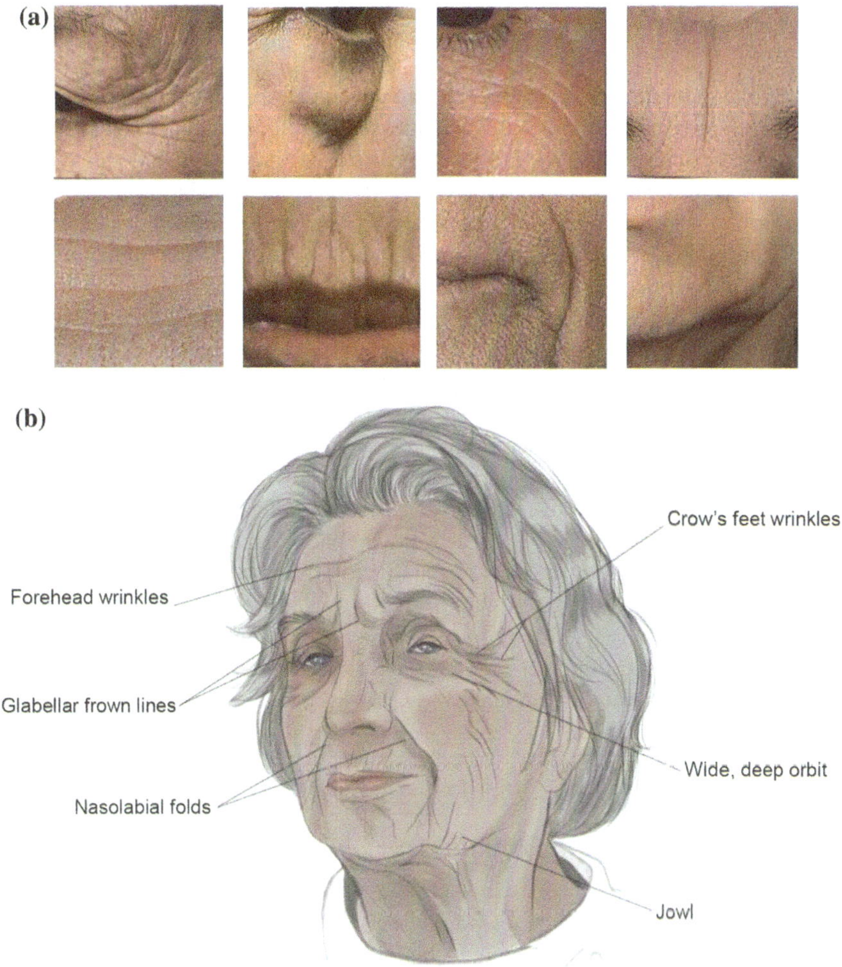

Fig. 5 Signs of facial ageing, **a** the highest grade of the ageing manifestation in different parts of the Caucasian face from [21], **b** illustration of the ageing marks in whole face

geometry undergoes tiny changes and most of the transformations are textural. To proof, from above-mentioned face ageing hallmarks, assorted to the geometrical (marked with '*g*') or textural (marked with '*t*'), most of them are texture-related changes. Therefore, face transformation from one age to another is performed in two geometry and texture parts.

3.1 Geometrical Model

To apply the geometrical face changes, even delicate, craniofacial measurements in different age groups are required. These measures include: eye size, inner Canthal distance, nasal height, nasal width, distance between nose and lips, and mouth width. Face shape is also defined by 20 facial feature points and nasal bridge.

Since the face has reached its adult shape and the maximum size by entering to the adulthood, there is no detailed accomplished geometrical measurements for the adulthood, as have been done for the childhood. To extract these measures in different age groups, the facial images from the FFAC album of the FAMD dataset are used. To remind, there are 6 afferent sub age groups in adulthood named Cluster 21–30, Cluster 31–40, Cluster 41–50, Cluster 51–60, Cluster 61–70, and Cluster 71–80, arbitrarily defined for this study. The age clusters are defined in a way that the ageing differences between two continuous clusters are distinctive.

All the mentioned measures are achieved by calculating the average number of size and distances obtained from all the face images in the 6 age clusters. Note that, to this purpose genders are not separated. Figure 6a illustrates the changing process of components' size and distribution from 20s to 80s. Moreover, the size of the 20 facial radiuses in different age clusters from 20s to 80s, extracted by calculating the average of facial radiuses in each of 6 age clusters, are shown in Fig. 6b.

It can be seen that differences between the two age clusters are very tiny, however even these minor alterations can make the face look older and these tiny variations' impact could not be denied.

3.2 Geometrical Transformations

As in Chap. "Facial Rejuvenation Modeling", the geometrical transformation is applied in two parts by employing the facial feature points:

Components' size and their distribution: as the facial components have the maximum of their scale and find their right location on the face, some slight modifications such as distance between eyebrows and eyes, nasal height and lips size should be applied as previously explained as the signs of an aged face. These modifications are each performed according to the scale factor which is defined between two input and output age clusters illustrated in Fig. 6a.

For these changes based on Eq. 1:

$$Si = \gamma Sj + d \quad (i > j) \tag{1}$$

however for forward face age modeling, the input age is less than the output age (i>j) and there is no pattern for γ, it can be less or more than one.

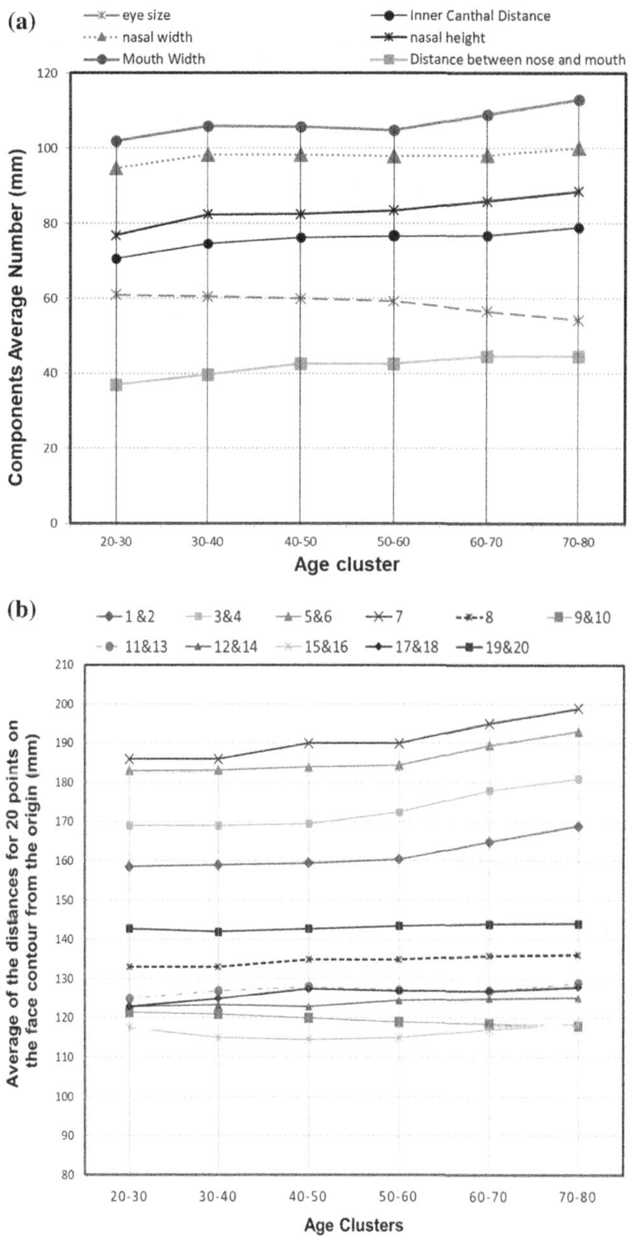

Fig. 6 **a** Facial components size and distances—both genders, six age clusters—extracted by calculated average of size and distances from all facial images in an age cluster. **b** Average distance of each point on the face contour from the origin–both genders, six age clusters—extracted from calculated average of radiuses of all facial images in an age cluster

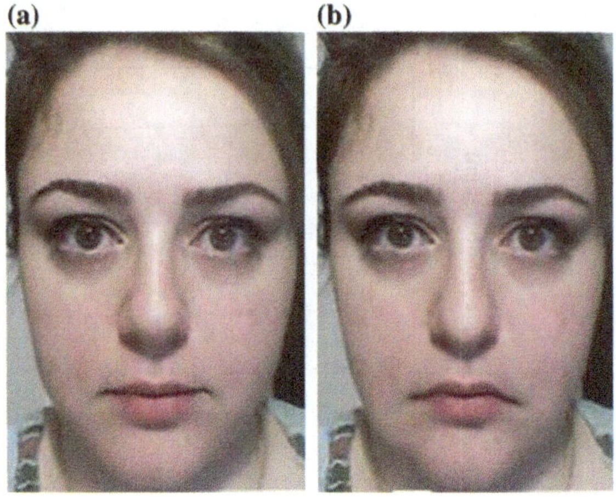

Fig. 7 Illustration of geometrical change in nose, eyebrows, lips and jowl; it can be noticed that even with these minor alterations the age of the face is changed. **a** Input adult face image. **b** Geometrically changed face

Face contour: as before, the nasal bridge is considered as the origin of the face and facial shape and its scale are defined by position of 20 points on the face contour and their distance from the origin.

Studies indicate in the process of the facial ageing, face size does not have a significant change especially in 2D frontal view. Thus, the probable changes may be related to the face shape, distinctively in its lower third part. Because one of the age variation effects associated with shape is the jowl alteration that can happen as a result of fat loss and muscle reduction.

To modify the frontal facial shape, some facial feature points on the chin are employed along with the origin. Recurrently, holding the angle to apply the changes is important. The same Eqs. 2–6 from Chap. "Facial Rejuvenation Modeling" can be employed to model the facial shape. However, for forward model, $j<i$ and β can be less, more or equal one.

The result of shape alterations is shown in Fig. 7b. It can be seen that even with these minor changes the perceived age of the face can be different.

To depict these alterations, geometrical transformation should be operated on the image domain. In fact, these alterations can be performed by moving the content from the coordinates in the image not yet changed to the dislocated coordinates. It is the image warping that has already been elaborated in Chap. "Iris Recognition in Cases of Eye Pathology".

3.3 *Textural Model*

As previously mentioned, most of the alterations in the ageing trajectory, are textural variations that come from either skin's direct change. Or they are alterations associated with facial skeleton, muscles, and fat in ageing process that can affect facial geometry, but are reflected on the texture. For instance, as the result of some bone resorption there is a concavity and darkness under eyes that can affect the face profile view, but in 2D frontal view it should be visualised by textural modifications. Therefore, Textural Model has an important role in the forward ageing model.

3.4 Predictive Face Templates Construction

Since the ageing process is under the influence of different factors like genetic, health and lifestyle, age-related effects and ageing rate is different from one person to another. Nevertheless, the common age-related aspects of the faces in one specific age range cannot be ignored. As an example to prove, a person in their 60s, in the best conditions of genetic, lifestyle, health and intensive skin cares never can look like as her 20s. This is the attribute of facial ageing that is employed to propose an ageing model. However, it can sometimes happen in reverse, that is to say, a 20 years old person can look like an individual in their 30s or 40s in the special conditions of health and lifestyle that will be fully discussed in the next chapter.

As cited before, mathematical average of the faces categorised in an age cluster and same gender can show all the common characteristics in the facial features that they all share. Moreover, a Face Template illustrates the common differences between age cluster's members. The more these two age clusters are far from each other, the better these distinctions are determined.

To build the Face Templates for F-FAM, the second album of the FAMD database consists of 900 adult facial images in 6 different adult age clusters (Cluster 21–30, Cluster 31–40, Cluster 41–50, Cluster 51–60, Cluster 61–70, and Cluster 71–80) and two genders have been used.

Once more, similar to what had been done for growth process and what was fully explained in Chap. "Iris Recognition in Cases of Eye Pathology" for Face Template construction, the average of facial images for each age cluster is calculated, using the extracted feature points, image warping, and Eq. 7.

After having a deeper look at the calculated Face Templates (Fig. 8a) especially for the older age groups, it is clear that the effects of the high-frequencies, creases and wrinkles, which are representative of the ageing marks in facial images, have been diminished. It is obvious for everyone that texture average calculation, from a set of images, defaces the high frequencies and details. It gives a smooth texture result since it eliminates very high and very low frequencies. Therefore, for a better representation of the facial deep wrinkles and high-frequencies on the Face Templates, their influence should be amplified. To amplify the high-frequencies and emphasise the fine details, the high-pass filter has been applied on all the face images in an age cluster before averaging.

$F_{\text{Sharpened}}$ is an enhanced facial image by increasing the contrast of the given facial image F around the thresholds, considering that the added noise in the identical sections of the facial image shouldn't be a lot [22, pp. 60–62, 23, pp. 133–136].

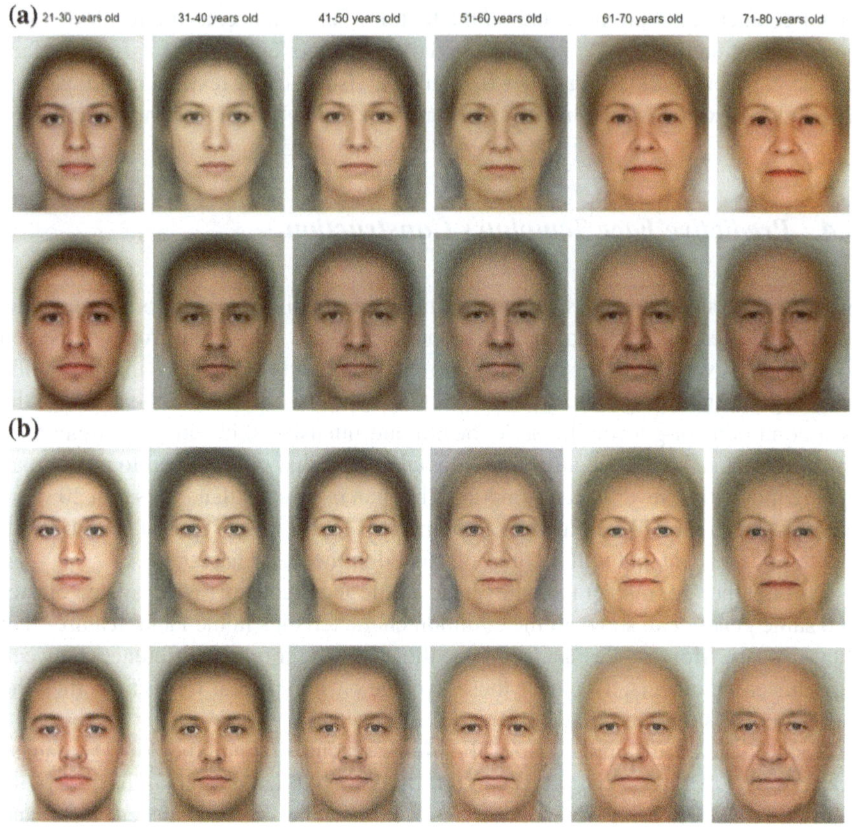

Fig. 8 **a** Illustration of the failures in creating the Predictive Face Templates, before filtering. **b** Predictive Face Templates, built after high pass filtering, for both genders and 6 different age clusters

$$F_{Sharpened} = F + \lambda[F - F * H], \quad \lambda > 0 \tag{2}$$

where λ is an ascending factor, H is the kernel applied to enhance the facial image and F * H is the smooth version of the facial image F.

According to the definition of 2D convolution the smooth version of F, FS, is:

$$F_S = F(x, y) * H(x, y) = \int_{\tau_1=-\infty}^{+\infty} \int_{\tau_2=-\infty}^{+\infty} F(\tau_1, \tau_2).H(x - \tau_1, y - \tau_2)d\tau_1 d\tau_2 \tag{3}$$

Therefore;

$$F_{Sharpened} = F + \lambda[F - F_S], \quad \lambda > 0 \tag{4}$$

Now, again, the average of all the facial images in the age cluster should be calculated, as regards in Eq. 7, $f = F_{\text{Sharpened}}$.

As a result of sharpening, some parts of the face such as nasolabial folds, lower eyelids and the lines at the corner of mouth are more recognizable in the new version of the Face Templates. These Face Templates are named *Predictive Face Templates*.

Predictive Face Templates related to the different age clusters are shown in Fig. 8b. As it can be seen, ageing progression in different decades, from 20s to 80s, is thoroughly displayed by these Face Templates. Indeed, changes in the size and form of the face components and muscle reduction are completely noticeable.

The more accurate the database classification is, the more distinctive analogies are. As a result, the higher quality Face Templates are achieved. To explain, if the database is classified according to different ethnics, each ethnicity will have its dedicated Face Template that illustrate the precise and subtle differences and characteristics special to that specific ethnicity. Consequently, the results of studies on the age-related facial characteristics will be more reliable, and the simulated faces will also be more realistic. Different ethnicities differ in the facial components size and form, face shape, and texture ageing process in both time of ageing signs appearance and the face regions that signs emerge (ethnic related factors that affect the ageing will be discussed in the next chapter).

3.5 Application of Textural Model

As the most tangible changes with ageing are related to the texture, the output result of the Geometrical Model is not really close to the target age and should be manipulated by the Textural Model. To do this, geometrically modified facial image should be combined with the Predictive Face Template belonging to the desired age cluster. For this objective, the two images should be fused, using the weighted sum, and image warping as explained in Chap. "Facial Rejuvenation Modeling".

The facial shape does not generally changed during adulthood, except in the conditions that the shape of the face or its components transformed due to a special disease or gain/lose weight. Therefore, to apply the fusion, instead of warping the geometrically changed face image and associated Predictive Face Template to the average coordinates, the Predictive Face Template is warped to the geometrically changed face coordinates. Eventually, these two face images are fused.

Making decision for share of each face image in the final result is very important. Since, texture has the more share of variations in ageing process, assigning the higher percentage to the Face Template, should normally deliver the result look more closely to the target age. Most often, a percentage between 60 and 70 for the Face Template, give relatively acceptable result. For the weighted sum Eq. 8 is used.

To achieve the percentage range between 60 and 70, a survey, similar to what was done in B-FAM, is performed. We asked people to choose whichever percentage that has the closer associated result to the target age. Obviously, to the extent that the

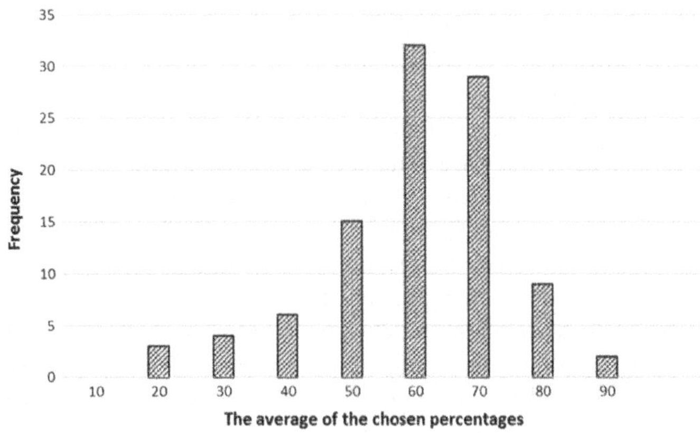

Fig. 9 Histogram of the percentages of the highest rated results, for the share of the geometrically changed faces

person's face characteristics does not change completely and is identifiable. Figure 9 demonstrates the average of the 100 responses for 20 sets of results.

Moreover, for applying the Textural Model, some details should be considered to have the better results, as they have been applied for B-FAM. For instance, the original eye colour should be preserved, as it is very important for identification and having the natural and realistic results. To do this, as it has been done in Chap. "Facial Rejuvenation Modeling", a binary mask covers the surface of the eyes area in the Predictive Face Template to prevent the masked area from being included in the average calculation. The eyes region is selected using double thresholding with pixel values and morphology-based opening and closing.

3.6 *Forward Facial Ageing Model (F-FAM)*

Based on the above-mentioned phases, the F-FAM is proposed for the natural face ageing trajectory:

$$F_{(i)} = \left[w_1 T_{p(i)} + w_2 \psi\left(f_{(j)}\right)\right] \quad \{i > j \; i - j = 10k, k \in N \; 0.6 < w_1 < 0.7 \; w_1 + w_2 = 1\} \quad (5)$$

where, f(j) is the input face in age j. F(i) is the simulated face in the age of i, Tp(i) is the Predictive Face Template in the same age range, and ψ is the transformation linear function applied on the face f(j) consists of all geometrical alteration by keeping the details related to the input facial image f(j) intact. i is the target age and j is the input face age. In F-FAM $i>j$ and i − j = 10k, k ∈ N since the age clusters are spanning 10 years between 20 and 80. The bigger i − j the more differences become visible

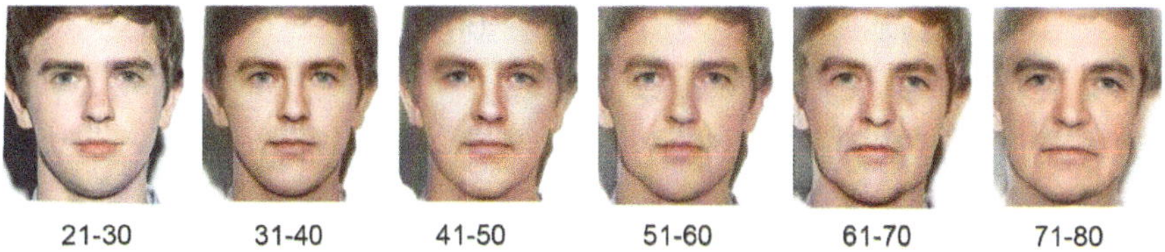

Fig. 10 Illustration of the ageing results on one subject in different age clusters

between input and simulated face. w1 and w2 are the weights given to the Predictive Face Template and the transformed input faces.

3.7 Experimental Results

In order to test the model, the F-FAM is applied on 20 young adult face images form PAL database [24], which are frontal and almost without any facial expressions.

According to the presented model, the input is a normal adult (20–30 years old) face image (Fig. 11b) and the result is how it might look in the future, more specifically in the 41–50 age range. To show the results of the F-FAM, we intentionally use young adult faces, around 20 years. Since young adult faces have the minimum signs of ageing, thus, the distinguishing of the simulated ageing signs on the output facial images compared to the input is easier.

For the results, the target age 40–50 is chosen with the purpose to add the effects of different lifestyle factors in the future. Note that, if the simulated faces are in the range of 60s or 70s, adding the effects of the different external factors will be very difficult. In these age ranges, face have already highest grade of the ageing marks, so the distinguishing between the ageing signs caused by the natural ageing and behavioural ageing is very difficult. Some of the results are shown in Fig. 11c.

However, F-FAM can generate faces in all specified age clusters (Cluster 21–30, Cluster 31–40, Cluster 41–50, Cluster 51–60, Cluster 61–70, and Cluster 71–80). Figure 10 illustrates the results of the ageing progress simulation on one young adult face image.

4 3D Ageing Model Development

Following the previous section, in which the facial ageing modeling in 2D was discussed, the goal of this section is to extend the work to third dimension. In the research of Heravi et al. [25], the Ageing Model is proposed, a 3D morphable model of ageing trajectory simulation in which a child's face appearance travels to its adulthood by morphing to show its growth path.

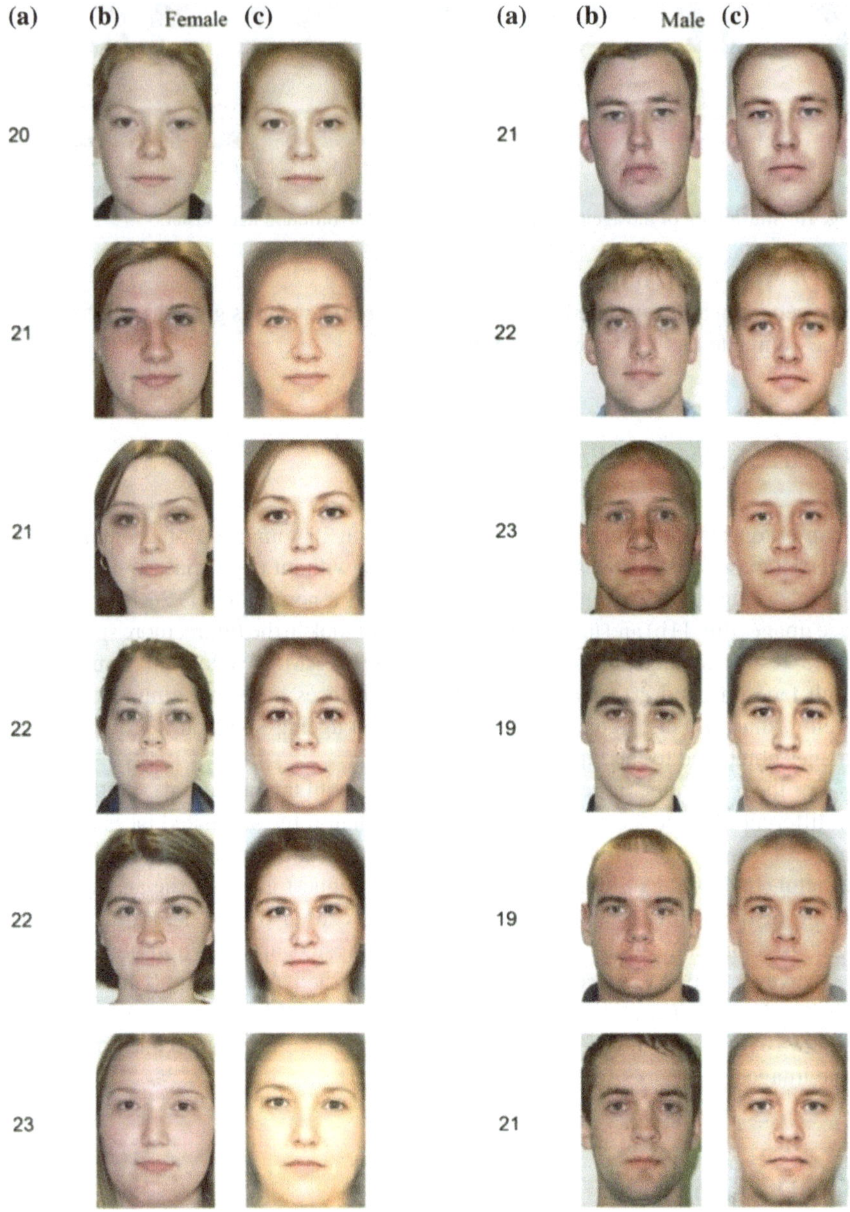

Fig. 11 Illustration of the results of applying F-FAM. **a** Input image age. **b** Input image. **c** Outputs of applying the model

Dealing with the aging process of an individual has been generally overlooked until recently. As the face matures, it changes some of its properties in geometrical aspect (e.g., shape of the cranium) and acquires new traits in textural aspect (e.g., wrinkles). These changes are the basis of information about the aging of the face.

So far we become familiar with the phenomenon of "aging of face" and gerontology, the comprehensive study of aging. Now we want to analyze the technical aspects of ageing effects on the 3D face images and visualize it in 3 dimension.

In order to discuss the 3D model, it is useful to carefully define a few key terms. Throughout this discussion the term 3D face image $F(i)$ will be used to refer to an entity that has a 3D surface geometry $F_M(i)$ and texture $F_T(i)$. The shape and mesh of an object refer to the set of points in object space that comprise the object's surface. The term model will be used to refer to any complete description of the shape and texture of an object. The so called model is mathematically explained as follows:

$$F(i) = (F_M(i); F_T(i)), \quad R(i) = (R_M(i); R_T(i)) \tag{6}$$

As a person matures the face "ages". Essentially as any individual's age beyond about 24 years the face begins to gradually match the shape less and less. As studies show, the most significant changes in human face geometry mesh takes place in the first 16–18 years of an individual's living and after puberty the changes take place mostly on the textural part of the human face. This of course is not an overnight phenomenon but as a rule is gradual occurring over a period of years. This "face aging" is actually a function of several physiologic changes that occur with time. Based on that, the Ageing Training Database (ATD) is designated to ageing development containing the 3D faces between 18 and 75 years old.

$$M_{ATD} = \underline{M}_{ATD(i)} + w_{ATD(i)} V_{ATD(i)}$$

$$T_{ATD} = \underline{T}_{ATD(i)} + w_{ATD(i)} V_{ATD(i)} \tag{7}$$

where M and T are representatives of shape (mesh) and perception (texture) of the face. Using PCA, the model is proposed in which w is the projection of the observation matrix over the eigenvector V.

4.1 3D Ageing Model (AM)

The goal of the 3D ageing face model is to represent the appearance of faces in a mathematical way. Using PCA [26–28], the faces of females and males are parameterized as triangular meshes of the facial mask separately. One most challenging part of this topic is to keep personal characteristics in every age while simulating. The model proposed in Majidzadeh et al. [25] reconstructs an aged face from the input child face image by performing Principal Component Analysis (PCA) and extracting eigenvalues of the texture map and surface mesh of 3D faces on each age category.

$$F_{AM}(i) = \left\{ T_{AM/input}(j) + wT_R(i); \underline{R} + \sum \beta V_{AM} \right\} \quad (8)$$

where $F_{AM}(i)$ is the target aged output of the 3D face at the given age, $\underline{R}$ is the Reference 3D image which consists of R_M and R_T, the reference Mesh and Texture respectively. Coefficient β is specified such that the personal characteristics similarity measurement between current age and target age can be kept as much as possible. In other words, to synthesize the different measures using the geometrical and textural models, it is necessary to calculate varying model coefficients. The approximation of the coefficients is not the same for shape and perception since most of the changes in the polygon mesh of the face occurs until the age of 18 [29], while the changes in texture take place when facial muscles begin to lose their elasticity and the wrinkles appear.

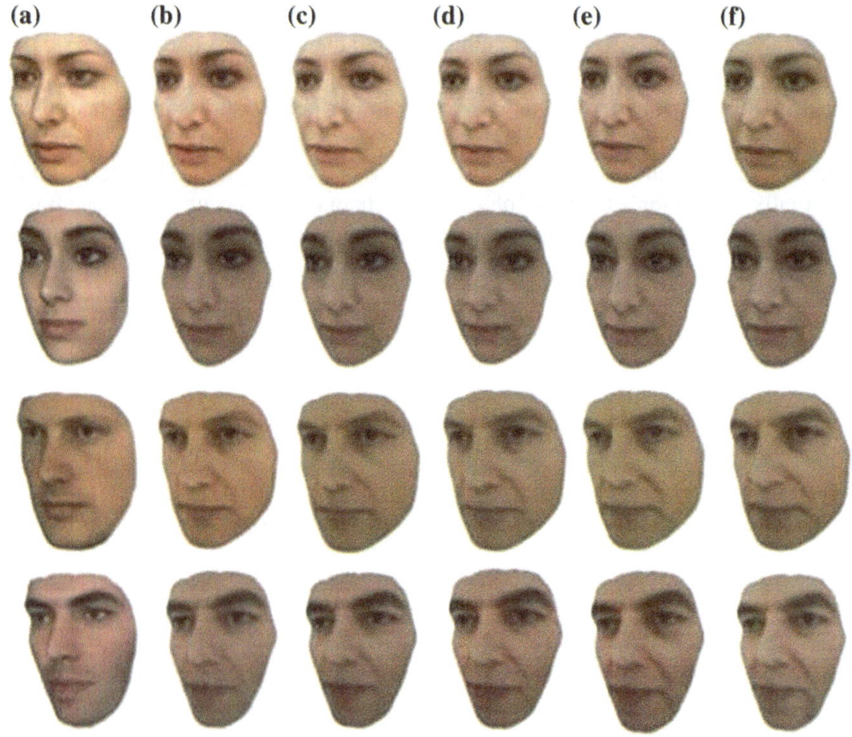

Fig. 12 Simulation of 3D facial aging trajectory in six different ages, **a** current age, **b** 40 years old, **c** 45 years old, **d** 50 years old, **e** 55 years old and **f** 60 years old

4.2 *Some Experimental Results*

The following results illustrated in Fig. 12 are generated automatically by applying the ageing simulation model [25] on 3D images and are applicable to many of applications. As can be seen from Fig. 12, the forward trajectory model of an individual's appearance in their older age is simulated and the outputs are obtained in 3 dimensional environment.

5 Conclusion

In this chapter, we described an approach for constructing the Facial Ageing Models, for both 2D and 3D [25]. The associated models are presented to predict an adult face appearance in its old ages based on the natural facial ageing trajectory. Although the geometrical variations are delicate and most of the changes are associated with texture, their impact is not denied. Hence, the geometrical aspects of the face are adjusted to the target age by applying a Geometrical Model. Then, Predictive Face Templates have been employed to adapt the face to the target age using the weighted sum and image warping. Note that, in the proposed models, the effects of fine wrinkles are not integrated, nevertheless, the results seem in the target age cluster. It is proving that the manifestation of ageing is not only related to the wrinkles, rather it depends on the changes in different layers of the face.

This Predictive Facial Ageing Model, in addition to covering the need of a model for the natural ageing process, can be a foundation for the more specialized face ageing models such as Behavioural Facial Ageing Model which will be discussed in the next chapter.

References

1. R. Fitzgerald et al., Update on facial aging. Aesthetic Surg. J. **30**(1_Supplement), 11S–24S (2010)
2. B. Mendelson, C.-H. Wong, Changes in the facial skeleton with aging: implications and clinical applications in facial rejuvenation. Aesthetic Plast. Surg. **36**(4), 753–760 (2012)
3. J.E. Pessa, An algorithm of facial aging: verification of Lambros's theory by three-dimensional stereolithography, with reference to the pathogenesis of midfacial aging, scleral show, and the lateral suborbital trough deformity. Plast. Reconstr. Surg. **106**(2), 479–488 (discussion 489-490) (2000)
4. R.B. Shaw Jr. et al., Aging of the facial skeleton: aesthetic implications and rejuvenation strategies. Plast. Reconstr. Surg. **127**(1), 374–383 (2011)
5. D.M. Kahn, R.B. Shaw, Aging of the bony orbit: a three-dimensional computed tomographic study. Aesthetic Surg. J. **28**(3), 258–264 (2008)
6. B. Jahan-Parwar, Facial bone anatomy: overview, mandible, maxilla. http://emedicine.medscape.com/article/835401-overview, June 2013

7. J.E. Pessa et al., Relative maxillary retrusion as a natural consequence of aging: combining skeletal and soft-tissue changes into an integrated model of midfacial aging. Plast. Reconstr. Surg. **102**(1), 205–212 (1998)
8. R.B. Shaw, D.M. Kahn, Aging of the midface bony elements: a three-dimensional computed tomographic study. Plast. Reconstr. Surg. **119**(2), 675–681 (discussion 682-683) (2007)
9. J.E. Pessa, D.E. Slice, K.R. Hanz, T.H. Broadbent, R.J. Rohrich, Aging and the shape of the mandible. Plast. Reconstr. Surg. **121**(1), 196–200 (2008)
10. S. Eremia (ed.), *Office-Based Cosmetic Procedures and Techniques*, 1st edn. (Cambridge University Press, Cambridge; New York, 2010)
11. S.R. Coleman, R. Grover, The anatomy of the aging face: volume loss and changes in 3-dimensional topography. Aesthetic Surg. J. **26**(1_Supplement), S4–S9 (2006)
12. A.E. Wulc, P. Sharma, C.N. Czyz, C.N. Czyz, The anatomic basis of midfacial aging, in *Midfacial Rejuvenation*, ed. by M.E. Hartstein, A.E. Wulc, D.E. Holck (Springer, New York, 2012), pp. 15–28
13. H.H. Publications, *The Aging Eye: When to Worry About Eye, Eyelid Problems* (Harvard Health) [Online]. Available: http://www.health.harvard.edu/healthbeat/the-aging-eye-when-to-worry-about-eye-eyelid-problems
14. J.H. Liu, F.E. Bartun, The aging face: rhytidectomy and adjunctive procedures. Sel. Readings Plast. Surg. **6**(C3) (2001)
15. J.M. Stuzin, Restoring facial shape in face lifting: the role of skeletal support in facial analysis and midface soft-tissue repositioning. Plast. Reconstr. Surg. **119**(1), 362–376 (discussion 377-378) (2007)
16. E.C. Naylor, R.E.B. Watson, M.J. Sherratt, Molecular aspects of skin ageing. Maturitas **69**(3), 249–256 (2011)
17. J. Calleja-Agius, Y. Muscat-Baron, M.P. Brincat, Skin ageing. Menopause Int. **13**(2), 60–64 (2007)
18. C. Castelo-Branco, M. Duran, J. González-Merlo, Skin collagen changes related to age and hormone replacement therapy. Maturitas **15**(2), 113–119 (1992)
19. J.M. Waller, H.I. Maibach, Age and skin structure and function, a quantitative approach (II): protein, glycosaminoglycan, water, and lipid content and structure. Skin Res. Technol. **12**(3), 145–154 (2006)
20. S. Luebberding, N. Krueger, M. Kerscher, Age-related changes in skin barrier function—quantitative evaluation of 150 female subjects. Int. J. Cosmet. Sci. **35**(2), 183–190 (2013)
21. R. Bazin, E. Doublet, *Skin Aging Atlas*, vol. 1 (Éd. Med'com, Paris, 2007)
22. R. Klette, *Concise Computer Vision: An Introduction into Theory and Algorithms* (Springer Science & Business Media, 2014)
23. R.J. Schalkoff, *Digital Image Processing and Computer Vision* (Wiley, Australia Limited, 1989)
24. M. Minear, D.C. Park, A lifespan database of adult facial stimuli. Behav. Res. Meth. Instrum. Comput. **36**(4), 630–633 (2004)
25. F. Majidzadeh, Heravi, A. Nait-Ali, *A 3D Dynamic Shape Model to Simulate Rejuvenation & Ageing Trajectory of 3D Face Images* (2017) pp. 0–4
26. L. Sirovich, M. Kirby, Low-dimensional procedure for characterization of human faces. J. Optical Soc. Am. **4**, 519–524 (1987)
27. M. Kirby, L. Sirovich, Application of the KL procedure for the characterization of human faces. IEEE Trans Pattern Anal. Mach. Intell. **12**(1), 103–108 (1990)
28. M. Turk, A. Pentland, Eigenfaces for recognition. J. Cognitive Neurosci. **3**(1), 71–86 (1991)
29. J. Hall, J. Allanson, K. Gripp, A. Slavotinek, *Handbook of Physical Measurements* (Oxford University Press, 2006)

Lifestyle Facial Ageing Models

Elham Farazdaghi, Farnaz Majid Zadeh Heravi and Amine Nait-Ali

Abstract This chapter is an extension of the previous chapter whereby digital face ageing is explained by taking into account lifestyle factors. The objective is to be able to predict the biometric facial appearance by including some prior and behavioral information like addictions of some drugs, alcohol, and smoking.

1 Introduction

Although, the facial similarities between people from the same age group are irrefutable, yet ageing pattern in each person is unique and associated with the various intrinsic and extrinsic determinants [1–3].

Setting the effects of lifestyle habits up to a facial ageing model can lead to more realistic results. The main aim of the Behavioural Face Ageing Models is to provide the perspective of the face in the future in case of having risky lifestyles. This means that ageing process may be very sensitive to some specific factors such as addiction to: some drugs, alcohol, smoking, etc. Excluding genetic factors, other factors may affect the ageing. For instance, it can be due to a stressful lifestyle, or even due to an excessive sun exposure. Therefore, when dealing with ageing models, it may be interesting to integrate these factors as model parameters in order to increase the accuracy in predicting the face appearance. Basically, we are currently working on these ageing models, for education purpose. Such tools can be used to make people (especially young people), more sensitive to some risky behaviors which may help them to positively control their lifestyle. Obviously, the idea is to dal with biometrics models, constructed from measurements and observations. Such models, can be exploited in other fields, including plastic surgery, clinical investigations, forensics, etc.

E. Farazdaghi · F. M. Z. Heravi · A. Nait-Ali (✉)
Université Paris-Est, LiSSi, UPEC, 94400 Vitry-sur-Seine, France
e-mail: naitali@u-pec.fr

A. Nait-Ali (ed.), *Biometrics under Biomedical Considerations*,
Series in BioEngineering, https://doi.org/10.1007/978-981-13-1144-4_5

This chapter is organised as below: First, facial ageing influential determinants are discussed in Sect. 2. 2D behavioural facial ageing models Sect. 3. In Sect. 4 3D behavioural facial ageing models are defined. Chapter is concluded in Sect. 5.

2 Facial Ageing Influential Determinants

There are two ways by which facial ageing is affected: internally and externally. Intrinsic ageing of the face inevitably occurs as a consequence of physiological changes over the time and individual has almost no control over them. Intrinsically aged face is typically smooth and uncontrollable. Contrariwise, extrinsic factors, in different levels, are controllable and varied.

In the following, first, these factors will briefly be discussed, then some models are proposed according to the most important and influential ones.

2.1 Intrinsic Determinants

Intrinsic factors are defined by individuals genetic. Mainly, these factors cannot be easily refined. Intrinsic determinants may influence things such as the colour of the skin, skin thickness, amount of collagen, elastin production and etc. Some of these factors are reviewed in the following.

2.1.1 Ethnicity

Ethnic differences in the face can explain genetically disparities seen in facial ageing progress. Some of the facial age related factors that can be differed from one ethnic to another are the amount, the composition (or colour) and production of the melanin [4], skin thickness, the amount of the collagen fibrils, natural moisturizing and etc. [5]. Since deep explanation of all of these factors is not related to this study, only some of them will be discussed.

First, let us come up with the definition of the ethnic. In this study the term ethnic is not defined as its common literal meaning which is the wide groups of populations with common culture and language. Rather, from now on in this discussion, ethnicity defines one specific population in terms of genetic similarities.

There are 3 major ethnic groups in the world according to Ethnographic division from Meyers Konversations-Lexikon of 1885–90:

- Caucasian: (Europeans, people of the Middle East and India)
- Mongolian or Asian (northern Mongolian, Chinese and Indo-Chinese, Japanese and Korean, Tibetan, Malayan, Polynesian, Maori, Micronesian, Eskimo, American Indian)

Fig. 1 Illustration of Fitzpatrick skin type classification

- Negroid (African, African Americans, African Caribbean, Australian Aborigine)

This classification is used in many age studies.

Another classification that can be very useful for the ageing studies is Fitzpatrick skin type or photo-type [6–8]. It depends on the amount of the melanin pigment in the skin. This classification shows the strong correlation between the skin colour with its ability to respond to UV light, that is one of the most important extrinsic factor in ageing skin, with tanning or burning reaction [4]. There are 6 different categories according to this classification, I–VI, from light to dark (see Fig. 1).

- Type I—Pale white; blond or red hair; blue eyes; freckles. Always burns, never tans.
- Type II—White; fair; blond or red hair; blue, green or hazel eyes. Usually burns, tans minimally.
- Type III—Cream white; fair with any hair or eye color. Sometimes mild burn, usually tans.
- Type IV—Moderate brown; typical Mediterranean skin tone. Burns minimally, always tans well.

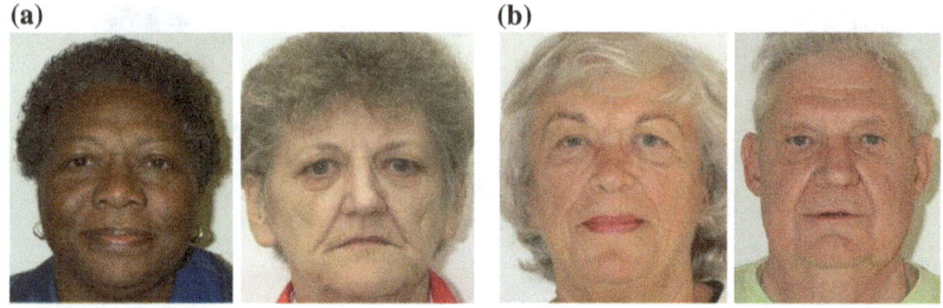

Fig. 2 **a** 65 years old African and Caucasian women facial image to observe the difference between facial ageing marks. **b** 70 years old female and male facial images to compare the signs of ageing that consist of fine and deep wrinkles. Images are from PAL database [16]

- Type V—Dark brown; Middle Eastern skin types. Very rarely burns, tans very easily.
- Type VI—Deeply pigmented dark brown to darkest brown; African descendant. Never burns, always tans [9].

Naturally, the darker the skin phenotype, the greater the skin protection against UV irradiation. The most lightly pigmented skin types have approximately half as much epidermal melanin as the most darkly pigmented skin types [10]. Therefore, there is greater probable dermal damage in the lighter ethnic groups. Accordingly, Caucasians showing earlier onset of photo-ageing compared to the other groups [11].

Moreover, the dermis, which is the thick layer of living tissue below the epidermis that forms the true skin, of black skin contains many more fiber composed of collagen fibrils and glycoproteins. People with darker skin are overall thought to have smoother and firmer skin than ones with lighter skin in the same age.

Increasing evidence has demonstrated ethnic variability in the skin's natural moisturizing. Skin moisturizing properties can be partitioned to water loss, water content of the outermost layer of the skin and its sensitivity to exogenously factors such as dry climates and colder winter [12]. These factors have very important role in the skin ageing process. Most studies about water loss support some differences between Black, Caucasian, and Asian skin. These studies show water loss in African-American skin is significantly higher compared with Caucasian skin. These studies indicate no significant differences could be detected in water loss among Japanese (Mongolian) and German (Caucasian) women in the same age group [12].

Assess for sensitivity in the black skin can be difficult and determination of redness of the skin cannot be very accurate. Yet, studies have shown the lows skin reactivity is in Black skin by using some objective techniques to evaluate blood flow have been employed to solve this problem.

In general, it can be said that the clinical manifestations of ageing in lighter skin types are more severe and mostly occur decades earlier than those of age-matched dark counterparts (see Fig. 2a).

2.1.2 Gender

There are many gender differences observed in human skin that can have a strong influence on gender-linked differences in facial ageing [13, pp. 113–114].

Studies indicate men's and women's skins differ in structure, biochemistry and functionality. Male skin is, in average, thicker than female skin. It contains more amount of collagen and has a tighter, firmer appearance. Sebum (a light yellow, oily substance that keeps the skin and hair moisturised) production is more in male's skin. Therefore, their skin is oilier than female skin, as a result its pH is lower and they rarely have drier skin compared to women. In men water loss is higher than women.

Some of the studies show gender-dependent differences in the degree of facial wrinkles. In general men have more deep wrinkles than women. Also, forehead wrinkles are increased by age in men compared with women (see Fig. 2b) [14]. At the end, male wrinkles occur later but, when they do appear, they are more slightly.

2.1.3 Hormonal Changes

The human skin is the target as well as the producer of hormones [15]. These hormones have an important duty in the development and physiological functionality of the skin tissues.

Hormones are evidently involved in intrinsic ageing that is accompanied by reduced secretion of the pituitary, adrenal glands, and the gonads. Thus, with ageing, the level of hormones, such as estrogen, testosterone, dehydroepiandrosterone (DHEA) sulfate and growth hormone are decreased [17].

Estrogen includes a group of hormones which are chemically alike: Estrone, Estradiol (the most abundant in women of reproductive age) and Estriol. They are also referred to as primary female sex hormones. Overall, estrogen is produced in the ovaries, however, adrenal glands and fat tissues also make some estrogen, which is why men will have estrogen but at lower levels than women. One of the highest levels of estrogen receptors are seen on the facial skin, therefore it is a crucial factor in the facial ageing. A number of studies have shown that estrogens have many important beneficial and protective roles in skin physiology [18] such as improved collagen content and quality, increased skin thickness and also improved blood supply to the skin [19].

Testosterone is a hormone supplied by the testicles and is liable for the proper development of male sexual characteristics. Females also produce testosterone, but usually in much smaller amounts. In women, testosterone is produced in the ovary, adrenal gland and the peripheral tissues from the various precursors produced in the ovaries and adrenal gland. Testosterone keeps the skin from sagging and symptoms of its loss can be dry and thin skin with poor elasticity [20].

Dehydroepiandrosterone sulfate (DHEA) is a weak male hormone produced by the adrenal glands in both men and women. It is a precursor hormone which means it has little biological effect in itself; however, it is extremely important because it makes estrogen, progesterone, and testosterone in both men and women [20] so its level has powerful effects on the skin.

Growth hormone (GH) is a small protein that is made by the pituitary gland and secreted into the bloodstream [21, pp. 136–147]. GH is essential for skin-cell repair and the prevention of skin sagging. The level of this hormone decrease as a result of ageing and indication of its drop off can be the sagging skin, loss of skin's thickness and firmness especially in cheeks and chin [22, Chaps. 2, 3, 10].

In females, serum levels of 17β-estradiol, DHEA, progesterone, growth hormone (GH), are significantly decreased with increasing age. In males, serum levels of GH and IGF-I decrease significantly, whereas it can decrease in late age in a part of the population. The effect of these decreases on the skin has been poorly documented, although more data are available for estrogen than for other hormones [17]. Less estrogen production that naturally occurs with age makes our skin thinner and less elastic, which leads to more wrinkling and sagging. As estrogen dips, less collagen and elastin are produced.

2.1.4 Anatomical Variation

Some anatomical variations have been reported in the histological structure of various sites of the skin as well as age related differences. It means different skin's regions such as forehead, upper eyelid, nasolabial area, perioral area, nose, cheek, chin, neck and volar forearm may have different level of biological parameters such as "transepidermal water loss", skin hydration, skin surface temperature, skin pH and sebum [23, pp. 24–35]. Moreover, these measures differ from young to aged people.

Further, there are large differences in skin thickness in different body sites, ranging from 0.5 mm on the eyelids to more than 6 mm on the bottom of the feet [24]. Also, in areas of the body with high blood flow (e.g. lip, finger, nasal tip and forehead), blood flow decreased with age compared to areas with baseline low blood flow, in which no difference was detected [25].

2.2 Extrinsic Determinants

Over the course of life, appearance and structure of the face will change not only because of chronological ageing and effects of intrinsic factors, but also due to several external factors that are caused by outside influences. These determinants can be: gravity, geographical area, working environment, high levels of pollution, temperature, and illnesses.

Moreover, lifestyle influence is a certain and definite factor that have impacts on health and facial ageing. Lifestyle habits includes: nutrition, sleep, exercise, drug abusing, and exposure to UV light (photo-ageing), smoking, and alcohol abuse.

Addressing all the cited factors is beyond the scope of this study, but two of the most effectual determinants, drug abusing and exposure to UV light (photo-ageing) will be briefly discussed. Then, as a main part of the work, a Predictive Behavioural Face Ageing Model will be proposed, based on each lifestyle behaviour to illustrate merely a part of their devastating impact on the face age.

3 2D Behavioural Facial Ageing Models

3.1 *Primal Facial Ageing Model*

To propose a facial ageing model based on any specific extrinsic factor that can affect the face age, first, a fundamental face ageing model should be constructed. Then, effects of each lifestyle factor will be added to this model. This fundamental facial ageing model is the model that make the person's face look older up to 80 years old in the natural conditions (F-FAM) as it was completely discussed in the previous chapter.

To remind, since in the adulthood most of the changes are texture related, minor geometrical changes are required. These geometrical alterations in the face components and shape, can be done by employing the formerly extracted geometrical feature points and image warping. To manipulate the textural aspect of the face, geometrically modified faces should be fused with the Predictive Face Templates belonging to the desired age decade.

Thus, Forward Facial Ageing Model (F-FAM), which is constructed based on natural ageing process, can be used as the fundamental for any supplementary model based on any extrinsic factors. This model, from now on for Behavioural Models, is called *Primal Facial Ageing Model*. Derived from the Eq. 4 in Chapter "Facial Ageing Modeling":

$$F_{Primal(i)} = w_1 T_{p(i)} + w_2 \psi\left(f_{(j)}\right) \quad \{i > j \; 0.6 < w_1 < 0.7 \; w_1 + w_2 = 1\} \qquad (1)$$

$F_{Primal(i)}$ is the face which is aged to the age i by considering the natural ageing pattern and it can be the base of any behavioural facial ageing model. $j > 16$, it means that the input is a young adult face image.

3.2 *Effectiveness Factor on Apparent Age (Δ)*

The studies indicate each external or behavioural factor causes a person look years older, as we will see in the following sections. To construct the Behavioural Facial Ageing Model, an *Effectiveness Factor* must be calculated to show the average of the years that people look older than ordinary people by being exposed to each of these factors. In fact, this Effectiveness Factor is achieved by calculating the difference between the perceived or apparent age (estimated age of the individual), and the actual or chronological age (the actual number of years that individual has lived):

$$\Delta = \frac{1}{n} \sum_{m=1}^{n} (Apparent\, age - Actual\, age)_m \tag{2}$$

Δ is the Effectiveness Factor, n is the number of facial images in the dataset.

In essence, each of the lifestyle behaviour determinants has its specific Effectiveness Factor that can be statistically obtained assuming that there is a facial database of the people that are exposed to that specific factor or an existing population to performing the clinical examination.

It is worth mentioning that, undoubtedly, if someone's lifestyle contains of more than one behavioural factors, calculating the sum of these Effectiveness Factor's number to obtain the future perceived age is senseless. The solution could be defining an impact percent for each of these factors and by considering the degree of its impact on the facial ageing. Of course, it requires extensive and meticulous clinical studies since it is very difficult to define a borderline for each behaviour factor impact. For example, the people who are addicted to drugs are more likely to have an alcohol abuse, or people who drink are more likely to smoke, therefore, distinguishing between the ageing effects of each behaviour is very difficult. Moreover, the number of the years that person is exposed to the factor is very important and can be taken into account.

3.3 *Behavioural Facial Ageing Model*

The statistical analysis shows that each of the negative lifestyle habits make the individuals look a few (Δ) years older for their age. Therefore, to build a facial ageing model for each lifestyle behaviour factor, a model is constructed on the basis of the Primal Facial Ageing Model. Except that instead of using the Predictive Face Template for the same age group ($Tp(i)$), the one corresponding to the age group $(i+\Delta)$ is used. It means that the facial ageing is intensified Δ years by this special lifestyle habit and as a result the person leading this lifestyle looks Δ years older.

Aside from geometrical alterations related to the natural ageing pattern that are applied via transformation function ($\psi(f(j))$), age related geometrical changes caused by the behavioural factor under discussion are applied in the face transformation function $\delta\ (f(j))$. For example changes in the face shape, its fatness and thinness, or changes in facial components' angles such as eyes, eyebrows and etc.

$$F_{Behavioral(i)} = w_1 T_{p(i+\Delta)} + w_2\left(\psi\left(f_{(j)}\right) + \delta\left(f_{(j)}\right)\right) \tag{3}$$

where $F_{Behavioral(i)}$ is the affected face by one of the behavioural determinants in the age of i, $T_{p(i+\Delta)}$ is the Predictive Face Template that must be used instead of $T_{p(i)}$, so Δ is the Effectiveness Factor which shows the number of years that an affected face looks older than a normal face.

3.4 Methamphetamine Addicts' Facial Ageing Model

Millions of people are trapped in a cycle of drug addiction and the number of addicts is increasing. Unfortunately, most of the people use drugs for the first time when they are teenagers [26]. Although, a lot have been discussed about drug abuse and its effects, in the accounts of drug addiction, rarely there are attempts to show the drug tolls on the face. We believe that one of the most successful ways to raise drug awareness is to show substances' catastrophic effects on a user's face over the time. Thus, we hope showing these negative devastating effects on the face, will play this important role in preventing and even advance stopping the addiction to the fatal substances.

Any drugs can affect users in different ways when they are continuously abused. All substances such as heroin, cocaine, crack, hallucinogen (PCP and LSD), amphetamines, marijuana and etc. can cause negative mental effects as well as serious physical tolls on the addicted person. Some of these physical effects that can touch the person's face age are: flushing and a rash of red bumps all over the skin, drop in appetite and dangerous malnutrition and as a result weight loss, significant dental diseases and grinding of teeth [27], diminished skin collagen and an appearance of premature ageing, wrinkles, redness, and skin elasticity loss.

As regards to methamphetamine, it is often seen as one of the most visibly destructive drugs that cause facial wasting. This issue will particularly be discussed in this study and a Predictive Behavioural Model is proposed due to its abuse.

3.4.1 How Does Methamphetamine Affect Its User's Face?

According to the National Institute on Drug Abuse [28], one of the most striking effects of methamphetamine is the change in the physical appearance and consequently the face of meth users. To have a realistic face ageing model, knowing these effects is necessary. A few of them include: the appearance of acne due to the dry skin, extreme itchiness due to the sensory hallucination of having bugs crawling under the skin and repetitive skin picking which causes ulcers, sores that take longer to heal, lackluster skin, loss of skin elasticity, facial musculature destruction, the pale skin colour, grayish rough texture of the skin with the coriaceous texture that causes a lot of wrinkles, abnormally quick loss of teeth as a result of dried out salivary glands and changes in salivary Ph. [27, 29], loss of appetite and as a result remarkable weight loss, and many other side effects that can make the user appear years, or even decades older than they actually are [30].

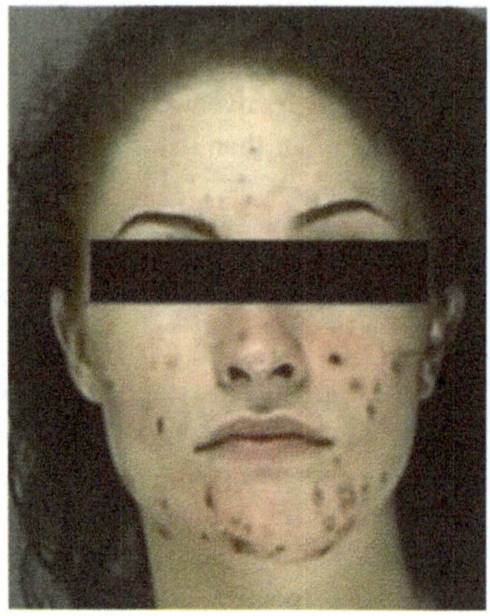

Fig. 3 A sample of the images shown to the respondents for statistical calculation of Δm

3.4.2 Methamphetamine's Effectiveness Factor (Δ_m)

For calculating Methamphetamine's Effectiveness Factor (Δ_m), a database composed of 23 methamphetamine addicts face images are used to study the effects of meth abusing and observe the difference between an addicted face and a normal face age [31]. A statistical analysis was performed using a user study to achieve Δ_m. We asked people to estimate the age of the shown face images. By having the real and the estimated age of the given faces, Δ_m is calculated. A sample of displayed images for Δ calculation is in the Fig. 3. The face was shown to the respondents to estimate the age of the person. For this shown face, for instance, the average of estimated age by the respondents is 29.5 and the real age is 24; thus, the average difference between perceived age and the apparent age is 5.5 years.

Of course, by increasing the number of the face images in the database, a more accurate number for Δ_m will be attained. Moreover, the number of the years that person has used the methamphetamine is surely an important factor that can be add to the model in the future.

Table 1 shows the average and standard deviation of the results coming from the web survey. The total number of the responses used is 100. Respondents are separated by their gender since studies have indicated that visual perception is different from women to men for future possible studies in the future. Total number of Δm for the methamphetamine addicts is equal to 5.2 ($\approx$5 years).

3.4.3 Model Construction

The statistical analysis on the methamphetamine addicts face database proves that in average, methamphetamine users look $\Delta m \approx 5$ years older for their age. Therefore, to build a facial ageing model for the individual with this lifestyle behaviour, a model

Table 1 Calculated average and standard deviation of delta in methamphetamine addicts' faces—separated by respondents' gender

Respondents		Δm
Gender	Number	
Female	71	5.5 ± 3.3
Male	29	4.7 ± 2.5
Both male and female	100	5.2 ± 3.1

is constructed on the basis of Behavioural Face Ageing Model. Accordingly, in this case, the Predictive Face Template corresponding to the age group $i+\Delta_m=i+5$ is used. It means that the person's age will be perceived 5 years more than her/his actual age if she/he abuses methamphetamine.

In addition, some geometrical modifications are applied in the face transformation function ($\delta m\ (f(j))$), for example face becomes thinner, eyes become tighter and the mouth and eyes turn a bit downward. All of these transformations are found out from meth addicts' facial images. To have more precise and accurate scale numbers for these geometrical alterations a bigger database is needed to compare all the different facial components before and after using meth.

Thus, based on Behavioural Facial Ageing Model, to model the facial ageing in meth addicts:

$$F_{Meth(i)} = w_1 T_{p(i+\Delta_m)} + w_2\left(\psi\left(f_{(j)}\right) + \delta_m\left(f_{(j)}\right)\right) \quad (4)$$

where $F_{Meth}(i)$ is a meth addict face in the age of i, $T_p(i+\Delta m)$ is the Predictive Face Template that must be used and Δ_m is the coefficient that shows the difference between a methamphetamine addict and a normal face age.

$$F_{Meth(i)} = w_1 T_{c(i+5)} + w_2\left(\psi\left(f_{(j)}\right) + \delta_m\left(f_{(j)}\right)\right) \quad (5)$$

In the end some colour adjustments such as saturation modification may be needed to give a pale look to the face.

3.4.4 Some Experimental Results

The Methamphetamine Addicts' Face Ageing Model is applied on the same database that have been employed for Forward Facial Ageing Model testing (Chapter "Facial Ageing Modeling"). In this model, the input is a normal adult face image and the result is how it might look in one of the mentioned age decades (41–50 years) if the person gets addicted to the methamphetamine. Some of the results are shown in Fig. 4.

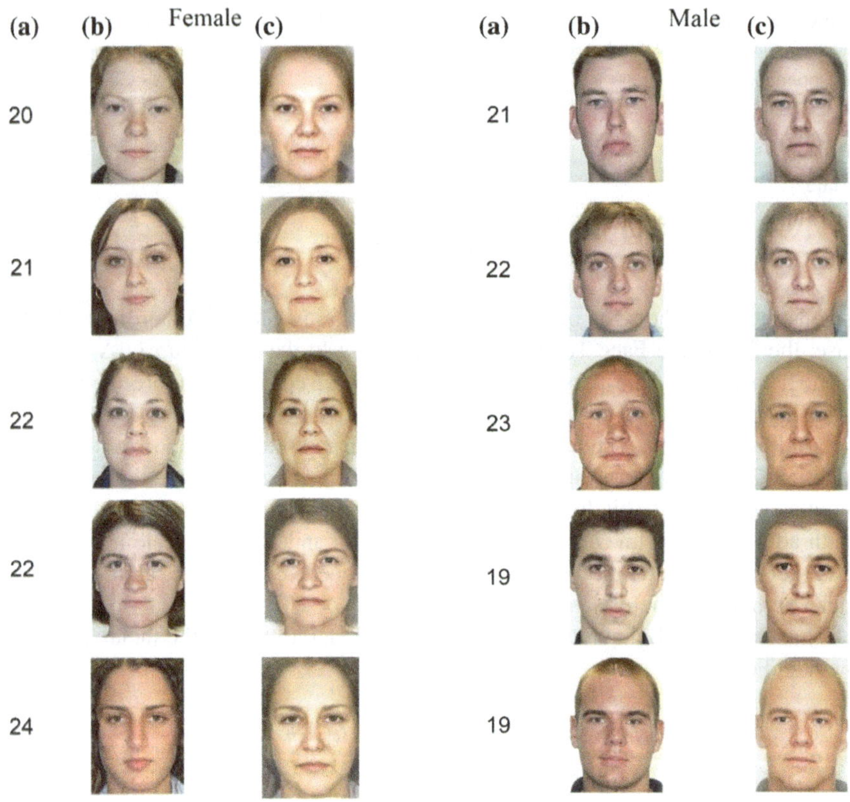

Fig. 4 Illustration of the results of applying Methamphetamine Addicts' Facial Ageing Model. **a** Input age **b** input image **c** outputs of applying the proposed model. The output age cluster is 41–50

3.4.5 Sores Modeling

Even though the results are significantly acceptable, integration of some sores that are normally in the meth addicts' face due to the extreme itchiness, may give a more realistic appearance to the results and exhibit the terrible effects of the meth using. However, this improvement does not necessarily affect the perceived age.

To do that, different masks are created for the forehead and cheeks. Some samples of real sores from addict faces in the database are stored. Then, random pixel locations are selected on the masks, depending on the size of the sores. It has to be checked if sores can be fitted inside the mask without interfering with the borders or not. At last, a random sore from the database will be used in the specific chosen location. The number of the sores for each area can be defined. To illustrate these sores on the simulated faces, image fusion using Discrete Wavelet Transform (DWT) has been used. Some of the results of sores integration are shown in Fig. 5.

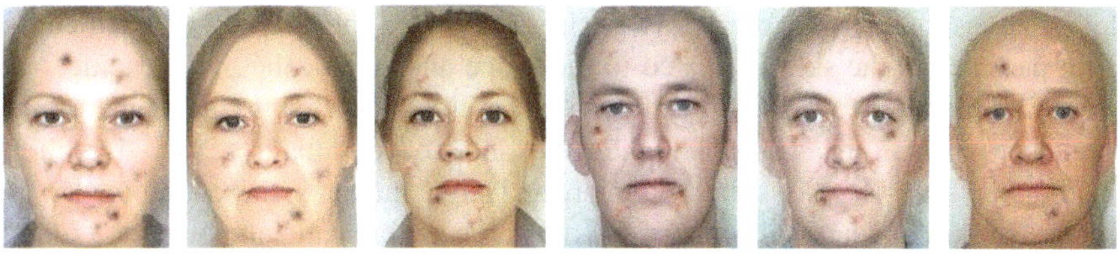

Fig. 5 Illustration of the results of applying Methamphetamine Addicts' Facial Ageing Model with sores modeling to have more realistic results. The results are in the age range 40–50

3.5 Sun Damage Facial Ageing Model

Ultraviolet radiation can be referred as the most important environmental factor influencing the facial ageing, since studies have divided the skin ageing into two basic parts: intrinsic ageing and photo-ageing [32]. The term 'photo-ageing' was used for the first time in 1986 by Kligman [33] to represent the significant effects of UV light exposure on the skin ageing. The importance of photo-ageing lies on the enormous number of individuals, that intentionally (or unintentionally), are cumulatively exposed to the sun and scarify their health and face appearance. Therefore, a visual representation of its irreparable effects on the face appearance can push tan seeking people to stop extensive exposures to the sun.

3.5.1 Melanin, Skin Colour and Its Relation to the Sun Exposure

The natural colour of the skin has an impact on how the skin reacts to the sun and environment. In fact, this reaction is noteworthy for ageing studies since this is the intersection point between an intrinsic factor and an extrinsic factor.

Human skin colour can vary from very white that always burns to almost black that never burns. This range of colors comes from the amount and type of melanin in the skin. Melanin, the skin's brown pigment, is produced at the base of the epidermis by special cells called melanocytes [34]. These cells have photosensitive receptors that detect ultraviolet radiation from the sun and other sources. In response, they produce melanin. In general, people regardless of skin colour, have the same number of the melanocytes; however, in darker skins melanocytes are more active and produce about four times as much as melanin as those in people with lighter skin. The less pigment the lighter the skin. There are two types of melanin manufactured by the skin: eumelanin and pheomelanin. People with fair skin and red hair produce more pheomelanin, and brown and black people produce more eumelanin. In general, the more eumelanin in the skin, the darker the skin. Thus, the people who make more pheomelanin than eumelanin have lighter skin [35].

Eumelanin absorbs most of the UV radiation that is why it is considered as natural sunscreen. In other words, darkened pigmentation of the skin protects it. Pheomelanin, in contrast, acts as a photosensitizer, making the skin more sensitive to sunlight;

this is necessary, for example, to synthesise vitamin D. The size and shape of eumelanin particles in the skin is related to the shape and size of the melanosomes, the organelles in melanocytes that synthesise melanin. In darker skin, the melanosomes are nearly twice as big as those in lighter skin and are individually scattered throughout the skin layers. Darker skin make more and larger melanosomes, compared to the Caucasian skin in which the smaller and more concentrated melanosomes are found. Alaluf et al. [10] demonstrated that Caucasian skin is characterised by a low number of melanocytes, small melanosomes and light pigments, while the black skin is characterised by the presence of a higher number of melanocytes, larger melanosomes, a higher quantity of melanin and more eumelanins. Therefore, in general, Caucasian skins are very sensitive to the UV light and show the signs of photo-ageing earlier than the darker skins.

3.5.2 How Does Sun Exposure Affect the Face

"Photo-ageing causes premature ageing of the skin through cumulative exposure to ultraviolet radiation (UV) from the sun and artificial UV sources" [32, pp. 1–1].

UV rays are categorised into three types based on wavelength: UVC (100–290 nm) is largely blocked by the ozone layer and has almost no influence on skin. UVB (290–320 nm) penetrates only into the epidermis and cause sunburns. UVA (320–400 nm) penetrates into the dermis and it may be responsible for most of the long-term skin damages associated with photo-ageing, immune dysfunction, and some skin cancers [36] (see Fig. 6).

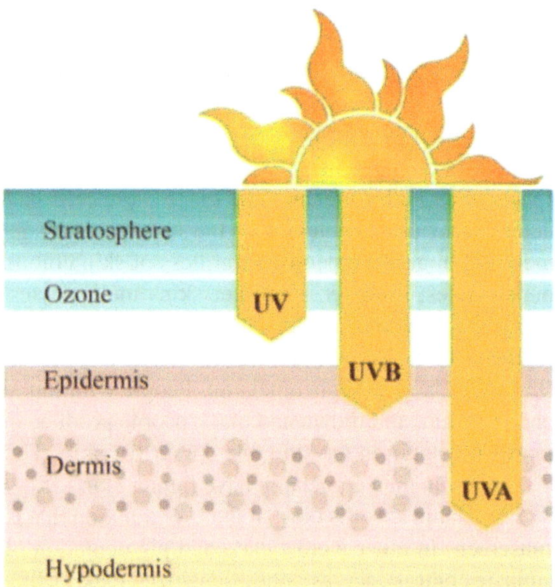

Fig. 6 Representation of different UV rays' penetration in the skin

The face is continuously exposed to sunlight, so after each year, the skin will accumulate damage until the time the visible signs of ageing appear. Continuous UV exposure can lead to other changes such as pigmentation disorder that is the most visible and most specific effects of photo-ageing, loss of skin elasticity, and degradation of skin texture described as photo-induced damages [37, 38].

Many studies have been done on solar exposure and facial ageing signs. For instance one noteworthy study compared photo-ageing and chronological ageing by observing two different groups of Caucasian women: sun seeking and sun phobic [39]. This study shows that skin pigmentations are the most linked parameter to photo-ageing but are not very related to the age. Whatever the age is, a prevalence of pigmentation disorders occurs in individuals who regularly are exposed to the sunlight with significant differences with those who are not frequently exposed to the sun.

Concerning wrinkles and the skin texture quality, remarkable differences between the two groups appear after the age of 50. Photo-ageing is basically determined by "coarse and fine wrinkles" compared with chronological ageing that can be described with fine wrinkles [40].

In general, photo-aged women look older than women who protect themselves from the sun, especially when they are younger [39, 41]. Thus, it can be said that photo-aged skin is a good example for the term "prematurely aged skin".

Figure 7 can be a good example of the UV impact on the face apparent age. Figure 7a shows identical twins. This image demonstrates two people with the same DNA could look so different according to their profile. Though, the woman B has the profile of smoking and weight loss, the main factor for looking a decade older than her sister can be her cumulatively exposure to the sun as she has admitted her strong interest in sunbathing. This sun exposure, in addition to causing dark skin and age spots, reduces skin's elasticity and increases wrinkles [42].

A fortiori, Fig. 7b is the most compelling argument for photo-ageing by showing only one face without any time differences. A 69 years old truck driver, Bill McElligott, who had spent 28 years driving on the job, has sideway sun exposure. The UVA rays transmitted through the window of his truck, have strongly damaged the left side of his face. As a result, the left side of the face looks around 20 years older than the right side [43].

3.5.3 Sun Damage Effectiveness Factor (Δ_{SD})

The measurement of the sun exposure's effect in facial ageing is very difficult, especially without the required medical and dermatological background, correct definition of the clinical factors influencing facial ageing, and only based on image data. Moreover, separating out the related factors involved in both chronological ageing and photo-ageing is a very complex process. Many worthy studies have been done in this field by dermatologist. Of course, not using their results and starting from the scratch without the facilities and dedicated knowledge is not wise. Therefore,

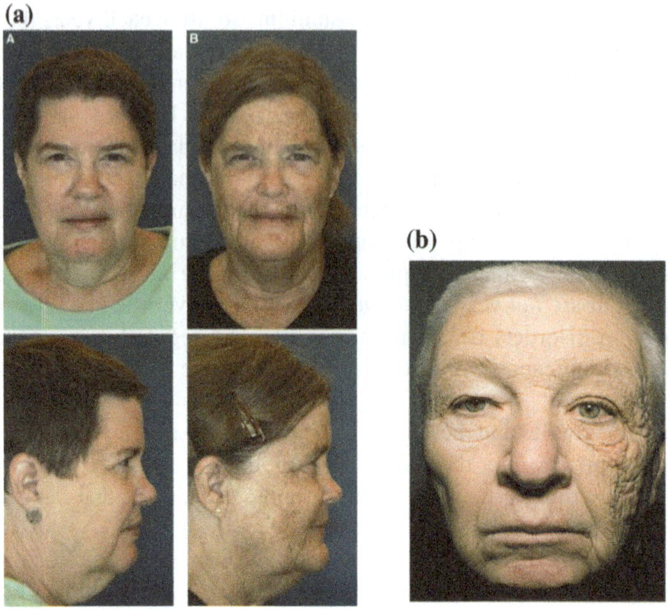

Fig. 7 **a** Identical twins—Age 61—Twin B has smoked for 16 years of her life, sunbathes, and weighs 15 lb less. Since her 20 s, she has spent as much time as she could in the sun. Twin A, on the contrary, has had as little exposure to sun as possible [42]. **b** A 69-year-old man that had driven a truck for 28 years and the left side of his face had been exposed to the UVA rays transmitted through window glasses [43]

for the sun damage Effectiveness Factor (Δ_{SD}) elicitation, the result of some strong dermatology studies on the effect of sun exposure on facial ageing are used [39].

Flament et al. [39] have provided some elements for measuring the effect of UV on the skin. Their criteria and results have been used as a main approach for sun damage Effectiveness Factor (delta) extraction. The study has been performed on 298 Caucasian women (more sensitive to the sun), aged from 30 to 78 years. They were divided into two groups: sun-seeking (S-S) and sun-phobic (S-P). Study placed in a town with more than 110 days of sunshine every year. The women had different skin types (dry, oily, and combination) and were balanced across photo-types I–IV according to Fitzpatrick classification. Then, they were regrouped into 10-year age clusters.

Afterwards, by considering the importance of UV exposure for different items: residence location, occupation, passive UV exposure, active UV exposure, and photo-protection patterns, a score between 0 (none) and 3 (very) was given for each 10-year cluster after performing a clinical examination and after asking the sun behaviour history.

Sun behavioural score history (SBSH) is the sum of the scores for all the items. In fact, "SBSH is a key descriptor of the UV exposure level of each panelist" which is linked with age and photo-type. Therefore, the panel descriptor and labeling of S-S

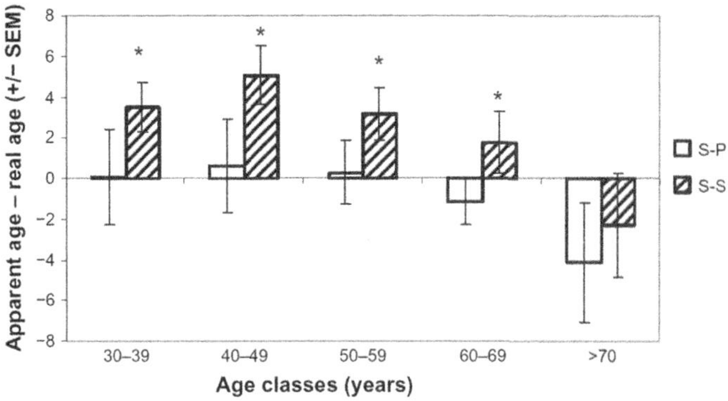

Fig. 8 Illustration of the difference between apparent and chronological age for the S-S and S-P groups (original figure is from [39]). "*" shows significant difference between bars. A positive difference means that the person looks older than her age. Abbreviations: S-P—sun-phobic; S-S—sun-seeking; SEM—standard error of the mean

and S-P was made with the following thresholds: 25 for the cluster aged 30–39 years, 34 for the 40–49 years, 43 for the 50–59 years, 51 for the 60–69 years, and 60 for the cluster aged 70–78 years.

After that, some clinical signs like pigmentation, elastosis, vascular disorders, fine lines and wrinkles were described. Beside, five grades of photo damage from 0 to 4 were defined: 0, none; 1, minor; 2, moderate; 3, important; and 4, major. The clinical evaluation was done by using two set of photographic scales; the original skin ageing atlas [44] and the clinical scales specific for the population that have been more affected by UV exposure.

By merging the two sets of clinical measures, a recording of facial ageing was obtained that covers four major axes: wrinkles and relief texture, lack of firmness, pigmentary disorders, and vascular alterations.

The process was ended by taking the average score of all the panelists and obtaining a clinical evaluation of each sign for each volunteer.

Finally, a survey was added to the research in order to record the perceived apparent age by asking the question: "what age do you think this woman is?" from 30 people after showing them the full-face picture of the study volunteers. The effects of ageing and UV exposure were characterised by a sum of the clinical criteria divided into four clinical factors (wrinkles, sagging, pigmentary disorders, and vascular alterations).

The study of the effects of UV exposure on the appearance is summarised in Fig. 8. It can be seen that with the exclusion of the oldest cluster, S-S volunteers looked older than their chronological age [39].

Concerning wrinkles and skin texture quality, noteworthy differences between the two groups appear after age 50 as it can be seen in Fig. 9.

A sum was done of all signs most affected by UV exposure by using the quantification of each volunteer which was then compared with the sum of all clinical signs established for facial ageing.

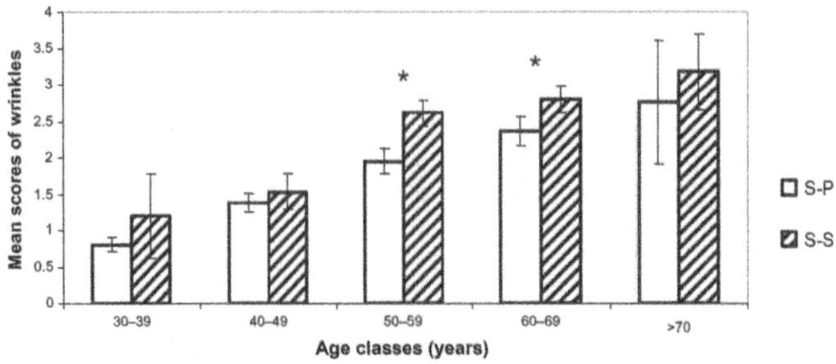

Fig. 9 Comparison of wrinkles and relief texture. "*" shows statistically significant difference. Abbreviations: S-P—sun-phobic; S-S—sun-seeking. Original figure is from [39]

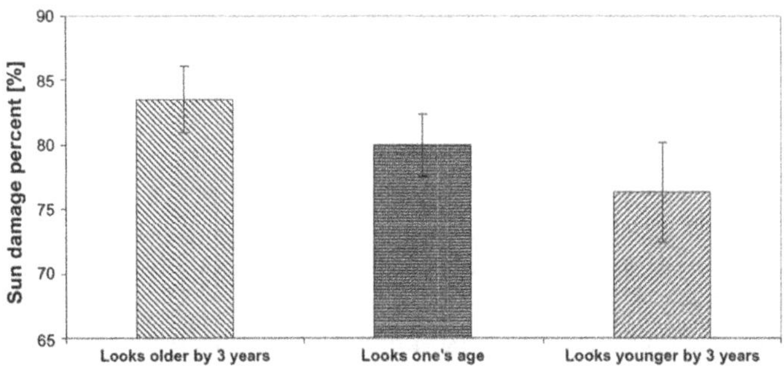

Fig. 10 Percentage of sun damage shows that how old a woman looks. Original figure is from [39]

A new ratio, Sun Damage Percentage (SDP), was determined that represents the percentage between specific photo-ageing signs and clinical signs. By computing SDP, the effect of sun exposure on the face can be estimated. On average SDP is $80.3 \pm 4.82\%$.

Eventually, it has been derived that "at a threshold of 80% for the SDP, women have similar apparent age to actual age. If SDP increases (82%), then apparent age becomes higher than real age, and this woman looks older. Conversely, a decrease of SDP (78%) means that the woman looks younger" (Fig. 10).

On the basics of all steps mentioned above we can extract the Sun Damage Effectiveness Factor (Δ_{SD}) as below:

$$\Delta_{SD} = \{0, SDP = 80 \pm 3, SDP = 80 \pm 2\} \qquad (6)$$

where Δ_{SD} the Sun Damage Effectiveness Factor and SDP is the Sun Damage Percentage. For the average SDP of 80%, Δ_{SD} is zero which means the apparent age and the real age are similar. If SDP increases by 2%, then apparent age becomes 3 years higher than the real age which means the person looks older. Contrariwise, a decrease of SDP by 2% means that the woman looks 3 years younger.

3.5.4 Model Construction

As discussed above, people who are constantly exposed to the sun, look $\Delta_{SD}(\approx 3)$ years older for their age. To construct the face ageing model for sun damage, in Eq. 3 in Chapter "Facial Rejuvenation Modeling", $i+\Delta = i+\Delta_{SD} = i+3$. Thus, another time the Coalesced Face Template corresponding to the target age should be used. Therefore we have:

$$F_{SD}(i) = w_1 T_c(i+3) + w_2(\psi(f(j)) + \delta_{SD}(f(j))),$$
$$\{i > j,\ 0.6 < W_2 < 0.7,\ w_1 + w_2 = 1\} \quad (7)$$

where, $F_{SD}(i)$ is the simulated face have been constantly exposed to the sun in age i and $T_c(i+3)$ is the Coalesced Face Template that must be used. In this study $\delta_{SD}(f(j)) = \varepsilon$ and its impact can be ignored. However we do not omit it from the model's equation as it the geometrical detailed alterations related to the Sun Damage Facial Ageing Model and it can be a part of future model improvements.

Since Δ_{SD} is close to the Δ_m, therefore, both models employ the same Face Template. As a result, the simulated faces for the same target age cluster (41–50 years old) can be very close to the meth addict's results. The differences between these two results are mainly the difference in the transformation functions $\delta_{SD}(f(j))$ and $\delta_m(f(j))$. For instance, the changes on the face geometry like face thinness and eyes angles may not be applied. However, some modifications on the texture can be added to improve the results: the appearance of the deep wrinkles and the colour of the face. For example, this time instead of a pale face, the face will be tanned.

3.5.5 Wrinkle Modeling

Facial wrinkles are not important only in terms of face age but they can also provide information about person's lifestyles and habits. For instance, facial wrinkles can represent the facial expressions or indicate some habits such as smoking [45] or sun exposure. As discussed before, there is a significant correlation between photo-ageing and the number and severity of the wrinkles, due to the damage of the elastin in skin. Therefore, to consecrate the results of Sun Damage Facial Ageing Model, some wrinkles are interpolated.

To do this, first a face image with wrinkles from the age range of $i + \Delta_{SD}$ is chosen as the wrinkle source. Then, the target face (simulated face by the Eq. 7 in Chapter "Facial Rejuvenation Modeling") and the wrinkled face will share the same size. Then after, facial features are extracted for both faces and are employed to warp the wrinkled face coordinates to the target face coordinates. Finally, the image fusion is used to fuse the wrinkles to the target face. For image fusion the discrete wavelet transform (DWT) is used. To fuse $I1(x, y)$ and $I2(x, y)$ with wavelet transform W using fusion rule Ø we have [46]:

$$I(x, y) = W^{-1}(\varnothing(W(I_1(x, y)), W(I_2(x, y)))) \quad (8)$$

where W^{-1} is the inverse discrete wavelet transform (IDWT) [46]. Thus, by substituting $F_{SD}(i)$ and wrinkled face image in this equation we have;

$$F_{SDW}(i) = W^{-1}(\varnothing(W(F_{SD}(i)(x, y)), W(R(x, y)))) \tag{9}$$

where FSDW(i) is the simulated face after wrinkle interpolation in age i, $R(x, y)$ is the wrinkled mask image which is used as the wrinkle source.

Finally, some colour enhancement such as saturation is applied on the final image to make it more realistic as a tanned face.

3.5.6 Some Primary Results

The Sun Damage Facial Ageing Model is applied on the same database that had been employed for the F-FAM and Methamphetamine Addicts' Facial Ageing Model. In this model, the input is a normal adult face image and the result is how it might look in the age decade 41–50 years if the person was exposed to the sun continuously. Some of the results are shown in Fig. 11e.

Figure 11 shows the input image and results of applying different models to compare.

3.5.7 Sun Spots Modeling

One of the most important symptoms of direct and chronic exposure to the ultraviolet rays is sun spots, also known as solar lentigines, and in some extreme cases mottled pigmentations appear on the face. While, sun damage in irreversible and cumulative, most of the times the signs of damage is invisible until the skin damage reaches a critical stage.

Ultraviolet photography is a way to show the sun damage which is hidden in normal photos. In this part of the work, we benefit from UV photos to integrate the undercover pigmentations into our model to achieve the more legalistic results.

Ideal situation for using this technique is to capture photos from each subject with both conventional camera and UV camera at the same time and with the same pose. The facial image captured using normal camera can be used as the input in the model. The output of the model, then, can be fused with the sunspots extracted from the face image captured by UV camera (named UV facial image) using wavelet transform. The results show how the person's face will look like in the not too distance future in case of repeating sun exposure without protections.

Since the UV camera is not available for us, the results are achieved with available samples of the face portraits captured using UV camera. First a UV facial image is selected as the sun spot source (Fig. 12).

Then, this UV facial image is mapped to the target face. Then after, a mask is created from the UV face image by erasing the eyes, nose and lips from the face and extracting the sunspots using wavelet transform. Finally, the image fusion using DWT is applied on two images. Some primary results are shown in Fig. 13.

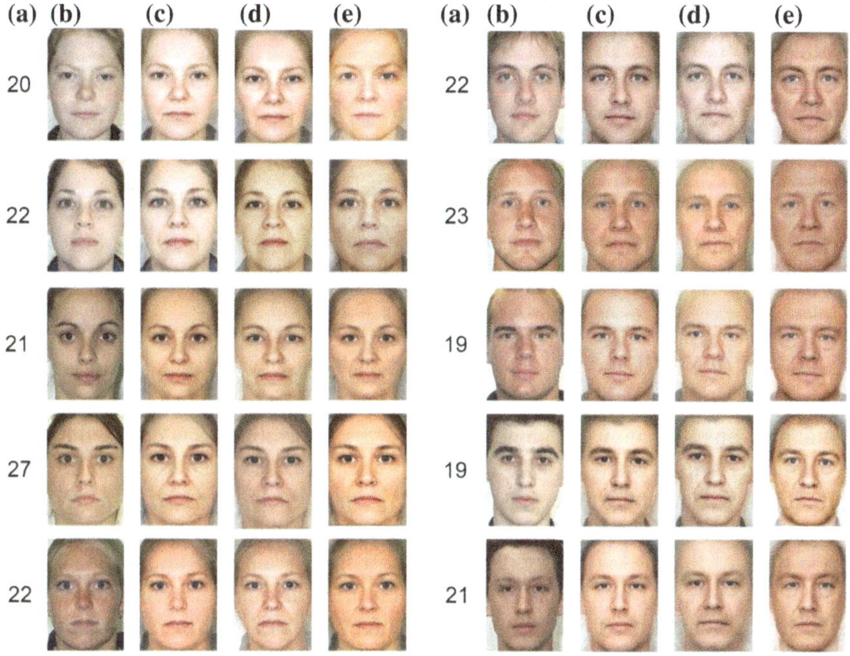

Fig. 11 Illustration of the results of applying different models to compare. **a** Input age **b** input image **c** outputs of applying F-FAM (natural ageing) **d** outputs of applying the Methamphetamine Addicts' Facial Ageing Model without considering the effect of wrinkles **e** results of applying the Sun Damage Facial Ageing Model considering the effect of wrinkles. Note that output age range for all the results is 40–50

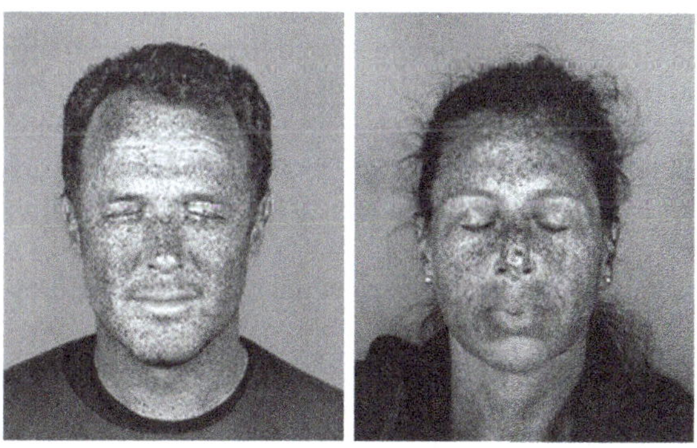

Fig. 12 UV face image samples. Photos by Cara Phillips [47]

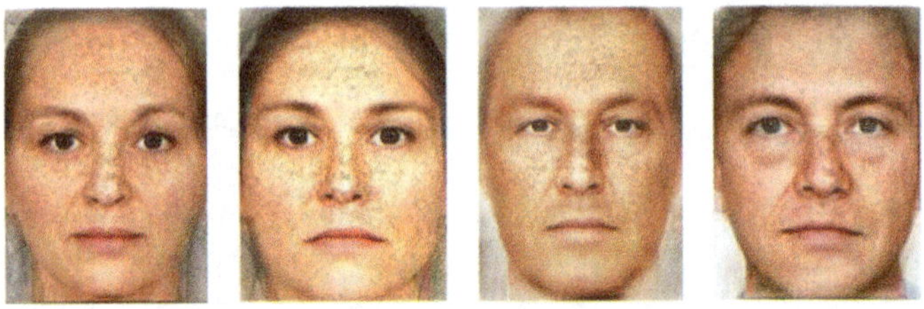

Fig. 13 Illustration of the results of applying Sun Damage Facial Ageing Model with the sun spots modeling to have more realistic results. The results are in the age range between 40 and 50

4 3D Behavioural Facial Ageing Models

In the previous section, we became familiar with all the changes occur on the facial appearance by ageing as well as considering some behavioral factors which can have significant impact on the process of ageing. Aging of the human face is the result of both superficial textural wrinkling of the skin and changes in the three-dimensional space (3D) of the face structure. The skin, soft tissues, and facial craniofacial are individually affected by the aging process, but they also act in dynamic harmony to determine the trajectory progression of the face throughout life. The major traits resulting in facial aging include gravity, craniofacial alteration, fat loss, hormonal imbalance, environmental impacts such as sun exposure and smoking. Other environmental factors that are claimed to affect facial appearance include mental stress, diet and nutrition, exercise habits, drug abuse, and disease [48].

We started with a brief discussion of the nature of three-dimension environment and how to apply it on facial visualization. Subsequently, the visualization techniques of the facial features in three-dimension is presented. In practice it is usually easy to define the appropriate technique, but compared to two-dimension space, it requires some extra work, requiring the knowledge of the geometry involved, to construct the three-dimension model. Although the ageing section may be familiar to the readers as many research has been done on this subject so far, the age reversal or regression calls rejuvenation is a newly absorbing topic which is approached thoroughly in the following chapter.

Being able to modify three-dimension facial image requires precise three-dimension models including such mouthfuls as "Geometrical Mesh and Textural Model".

In recent decades, lifestyle as an important factor of health has been of great interest to the researchers. According to WHO, 60% of related aspects to individual health and quality of life are related to lifestyle [49]. Unfortunately, millions of people follow an unhealthy lifestyle and bad habits. Consequently, they encounter illness, disability and in the worst case death. Those bad habits and lifestyle choices can do huge damage to mind and body and specifically on how good, or not, the facial skin looks in no time. Face perception is powerful channel to human social interaction [50]. According to neuroscience, it is widely accepted that face perception is perhaps the most highly advanced human visual skill [51]. The relationship of lifestyle and health should be highly considered.

We improve on previous proposed models in previous chapters by offering a fusion of two models of ageing and rejuvenation with facial shape variations due to lifestyle and behavioral factors using Principal Component Analysis (PCA). This model can simulate the requested target age of the 3D input either at older or younger age and can change the face shape of an individual. By extracting 3D Face information and manipulating them, wide variety of applications for education, healthcare, security and entertainment purpose could be expected.

Specific facial elements such as the eyes, mouth and nose have significant importance as they deliver most of the audiovisual information expressed by humans [52]. As one of the most complex facial traits, the human facial shape topology, is deferred due to many factors such as genetics, ethnicity, age, gender, and health. Male and female faces differ in their shape. Mature features in adult human faces reflect the masculinization or feminization of their characteristics that occurs at puberty. Larger jawbones, more prominent cheekbones and thinner cheeks and lips are all features of male faces that differentiate them from female faces [53, 54]. It is obvious that normal face shape differences are controlled by one's genes. Craniofacial development is a complex process determined by genetics influence facial morphology [55, 56] although there are multiple attributes regarding one's lifestyle and behavioral traits which contribute to facial shape variations and bring about for example a fat or slim face appearance.

Accordingly, facial modeling and simulation can not only be helpful in forensics in the aim of finding the missing children but also in health and educational purpose to make children aware of the bad habits and lifestyle impacts by showing them the simulation of their face due to consuming too much processed and fast food and/or spending countless hours in front of the television being sedentary and as a result assisting in controlling the degeneration of health among many people today.

Our aim in this chapter is to discuss on the model approach which is able to take the 3D image of a particular person and to simulate a rejuvenation and ageing trajectory specific to that particular individual and as well as visualizing the perceptible changes occur in facial appearance and shape changes due to the lifestyle and behavioral traits.

The importance of lifestyle and health and its impact on facial appearance brought about the results of the 3D dynamic shape model proposed in Heravi et al. [57] in which the 3D Face Time Machine Matrix (FT2M) presented as a fusion of two models of ageing and rejuvenation with facial shape variations due to lifestyle and behavioral factors. We improve on previous proposed models in the previous chapters by offering a fusion of two models of ageing and rejuvenation with facial shape variations due to lifestyle and behavioral factors using Principal Component Analysis (PCA).

This model can simulate the requested target age of the 3D input either at older or younger age and can change the face shape of an individual. what makes this model exclusive, is the face shape variance along the horizontal axis displaying the probable facial shape changes from relative thin to relative fat regarding the person's behavior and lifestyle which can affect multiple aspects of the health including the aging process and how physical health changes with age. Thus, this 3D model

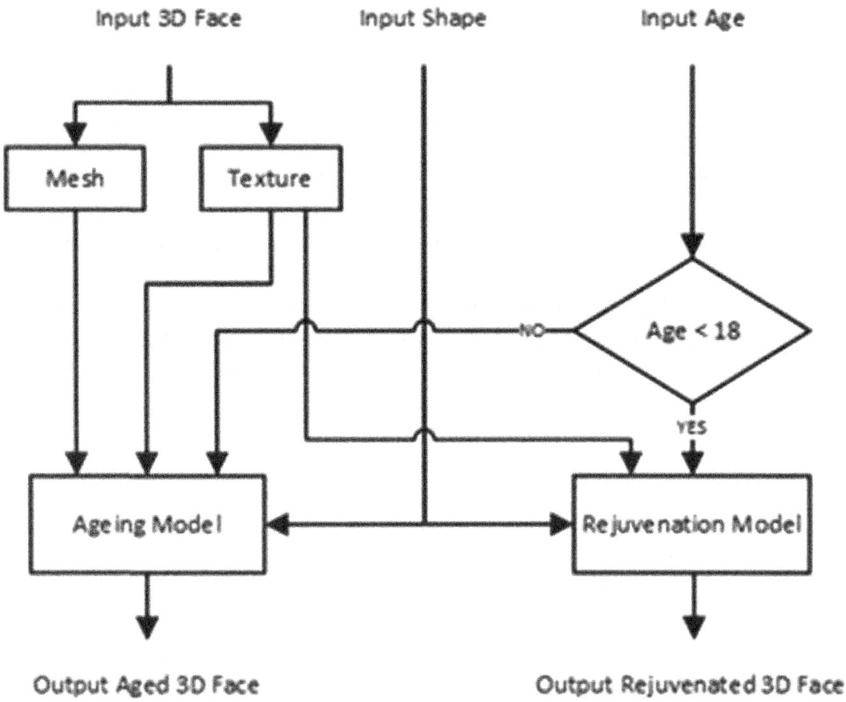

Fig. 14 The 3D Dynamic Shape Model to simulate Rejuvenation and Ageing flow diagram

permits transforming an image of the face of a person into a corresponding image of that person as he would appear in any target age and perceptible shape either a number of years later, or earlier, thinner, or heavier respectively. By extracting 3D Face information and manipulating them, wide variety of applications for education, healthcare, security and entertainment purpose could be expected.

3D database featuring face models are not common and defective; however, 2D datasets are more readily available. We therefore decided to reconstruct the 3D images from multiple 2D images [58] using a face-fitting method, and then extract the 3D information regarding the geometry and texture of the 3D face to construct the "Geometrical Mesh and Textural Model".

Heravi et al. [57] proposed an approach to model 3D face images such that the Mesh and Texture of the input 3D face are extracted and based on the input age >18 or <18, the Ageing Training database (ATD) or Rejuvenation Training Database (RTD) is considered to apply Ageing Model or Rejuvenation Model, respectively. This approach is illustrated in Fig. 14.

4.1 Face Time Machine Matrix Model

The goal of a face model is to represent the appearance of faces in a mathematical way. Using PCA [58–60], the faces of females and males are parameterized as triangular meshes of the facial mask separately. One most challenging part of this topic is to keep personal characteristics in every age while simulating. Our model reconstruct an aged/rejuvenated face from the input face image by performing Principal Component Analysis (PCA) on each age category.

The appropriate coefficients are designated by a special similarity measurement between current age and target age to keep personal characteristics as much as possible. The system is able to generate aged/rejuvenated faces using the original face shape as an input and considering the probable shape changes (slim or fat face) and age-related features such as spots, wrinkles, and sagging while keeping personal characteristics throughout all ages.

The model starts out with the recognition that there are a finite number of typical faces in different age groups that has already been explained thoroughly in previous chapters. In each age group, typical face categories exist as average female face, fat female face, thin female face, average male face, fat male face and thin male face. In other words, in each of the age categories, there are certain characteristics which change in almost all faces of a given type, e.g., the types of wrinkles which appear, the loss of facial elasticity, etc.

In order to study the perceptible facial changes along age and shape axis, the experiment has been done on females and males database. According to the presented

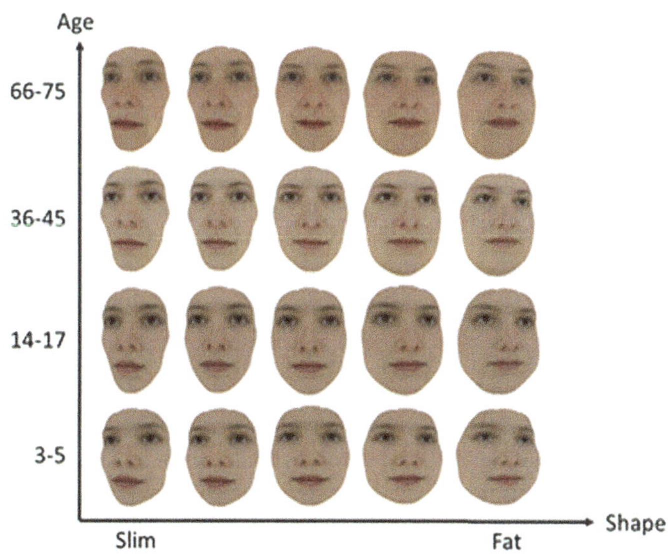

Fig. 15 Simulation of 3D face time machine matrix in Age and Shape axis

model, various outputs in different target ages and shapes can be obtained based on the choice of the user. One can be able to simulate his face in younger, older, slim or fat version. Many interesting applications can be proposed using this simulation method. These applications can be found in a lot of different disciplines like medicine, forensic, anthropology, biometrics, art, entertainment, and computer graphics. The following results illustrated in Fig. 15 are generated automatically by applying the Face Time Machine Matrix (FT2 M) model on 3D images and are applicable to many of these applications. As can be seen in Fig. 15, the trajectory of ageing evolution and regression in 4 different age groups of 3–5, 14–17, 36–45 and 66–75 years old along the vertical axis is illustrated and the facial shape variance along the horizontal axis from skinny to fat face shape is represented.

5 Conclusion

In this chapter, Behavioural Facial Ageing Models are proposed. To start, various intrinsic and extrinsic determinants that affect the facial ageing process are briefly discussed. Among extrinsic factors, those lifestyle behaviours that have more effects on the face age and that avoiding them, can help to prevent premature ageing, are selected to add their effect to the ageing model.

In the proposed approach, for each lifestyle behaviour, one Effectiveness Factor (Δ) is calculated which is the difference between perceived age and actual age of the people who have that lifestyle habit. Therefore, in the ageing model Predictive Face Templates are replaced with Face Templates which are Δ years older than the target age. The geometry and colour of the face image are modeled according to the effects of each lifestyle behaviour.

The facial ageing model due to the methamphetamine abuse is proposed. $\Delta m \approx 5$ is achieved using a survey and meth users database. Furthermore, Sun Damage Facial Ageing Model is presented. Its Effectiveness Factor is achieved according to the outcomes of a dermatology study on sun seeking people. Since one of the effects of UV radiation is manifestation of the fine and deep wrinkles, to give more realistic look to the resultant, the effect of the wrinkles is modeled. Since the serious source of the visible signs of ageing are directly linked to the environmental and behavioural factors, Behavioural Facial Ageing Models can create a revolution in the facial ageing studies. This study provides preliminary results on behavioural ageing. Of course there is a lot of space for improvement and many interesting and useful applications in biometrics and medical fields can be imagined for the future.

In the 3D proposed model, the dynamic 3D model of Face Time Machine Matrix (FT2 M) has been presented [57]. This model simulates the trajectory of the ageing in either forward or backward path in the designed age time machine matrix in which an input 3D face image travels to its target age. It is also capable to represent the probable changes in one's facial shape regarding the lifestyle or behavioral pattern alterations by using the original face shape as an input. The geometrical and textural model of the 3D face image are modifying simultaneously. We demonstrated the potential of our approach in designing 3D face ageing and rejuvenation simulation system robust to lifestyle variation. The extension of mesh and texture modeling from 2D to 3D domain gives additional capability of compensating for pose and lighting variations.

References

1. M.A. Farage, K.W. Miller, P. Elsner, H.I. Maibach, Intrinsic and extrinsic factors in skin ageing: a review. Int. J. Cosmet. Sci. **30**(2), 87–95 (2008)
2. H. Rexbye, I. Petersen, M. Johansens, L. Klitkou, B. Jeune, K. Christensen, Influence of environmental factors on facial ageing. Age Ageing **35**(2), 110–115 (2006)
3. K. Sveikata, I. Balciuniene, J. Tutkuviene, Factors influencing face aging. Literature review. Stomatologija **13**(4), 113–116 (2011)
4. A.V. Rawlings, Ethnic skin types: are there differences in skin structure and function? Int. J. Cosmet. Sci. **28**(2), 79–93 (2006)
5. N.A. Vashi, M. Buainain De Castro Maymone, R.V. Kundu, Aging differences in ethnic skin. J. Clin. Aesthetic Dermatol. **9**(1), 31–38 (2016)
6. S. Sachdeva, Fitzpatrick skin typing: applications in dermatology. Indian J. Dermatol. Venereol. Leprol. **75**(1), 93 (2009)
7. M.A. Pathak, In memory of Thomas Bernhard Fitzpatrick. J. Invest. Dermatol. **122**(2), xx–xxi (2004)
8. "Fitzpatrick scale," *Wikipedia, the free encyclopedia*, 01 Sept 2016
9. T.B. Fitzpatrick, Soleil et peau [Sun and skin]. J. de Médecine Esthétique **2**, 33–34 (1975)
10. S. Alaluf, D. Atkins, K. Barrett, M. Blount, N. Carter, A. Heath, Ethnic variation in melanin content and composition in photoexposed and photoprotected human skin. Pigment Cell Res. **15**(2), 112–118 (2002)
11. D.A. Weigand, C. Haygood, J.R. Gaylor, Cell layers and density of Negro and Caucasian stratum corneum. J. Invest. Dermatol. **62**(6), 563–568 (1974)
12. D.C. Wan, V.W. Wong, M.T. Longaker, G.P. Yang, F.-C. Wei, Moisturizing different racial skin types. J. Clin. Aesthetic Dermatol. **7**(6), 25–32 (2014)
13. Z.D. Draelos, *Cosmetics and Dermatologic Problems and Solutions*, 3rd edn. (CRC Press, 2011)
14. K. Tsukahara, M. Hotta, O. Osanai, H. Kawada, T. Kitahara, Y. Takema, Gender-dependent differences in degree of facial wrinkles. Skin Res. Technol. **19**(1), e65–71 (2013)
15. C.C. Zouboulis, Human skin: an independent peripheral endocrine organ. Horm. Res. **54**(5–6), 230–242 (2000)
16. M. Minear, D.C. Park, A lifespan database of adult facial stimuli. Behav. Res. Methods Instrum. Comput. **36**(4), 630–633 (2004)
17. T.J. Phillips, Z. Demircay, M. Sahu, Hormonal effects on skin aging. Clin. Geriatr. Med. **17**(4), 661–672, vi (2001)
18. M.J. Thornton, The biological actions of estrogens on skin. Exp. Dermatol. **11**(6), 487–502 (2002)
19. M.P. Brincat, Hormone replacement therapy and the skin. Maturitas **35**(2), 107–117 (2000)
20. P. Smith, A comprehensive look at hormones and the effects of hormone replacement, in *14th Annual International Congress on Anti-Aging Medicine*, Orlando, FL, 2005
21. M. Sulindro, M. Lam, *How To Stay Young and Live Longer*, 2nd edn. (Academy of Anti Aging Research, Pasadena, CA, 2001)
22. R. Klatz, C. Kahn, *Grow Young With Hgh: The Amazing Medically Proven Plan to: Lose Fat, Build Muscle, Reverse the Effects of Aging, Strengthen the Immune System, Improve Sexual Performance*, 1st edn. (Harpercollins, New York, 1997)

23. M.A. Farage, K.W. Miller, H.I. Maibach, *Textbook of Aging Skin* (Springer Science & Business Media, 2009)
24. M. Gray, Preventing and managing perineal dermatitis: a shared goal for wound and continence care. J. Wound Ostomy Continence Nurs. **31**(1 Suppl), S2–9; quiz S10–12 (2004)
25. J.M. Waller, H.I. Maibach, Age and skin structure and function, a quantitative approach (I): blood flow, pH, thickness, and ultrasound echogenicity. Skin Res. Technol. **11**(4), 221–235 (2005)
26. N.I. on D. Abuse, *Nationwide Trends*. [Online] Available: https://www.drugabuse.gov/publications/drugfacts/nationwide-trends. Accessed 16 Jan 2017
27. H. Shekarchizadeh, M.R. Kahmi, S.Z. Mohebbi, H. Ekhtiari, J.I. Virtanen, Oral health of drug abusers: a review of health effects and care. Iran J. Public Health **42**(9), 929–940 (2013)
28. www.drugabuse.gov
29. M. Cretzmeyer, J. Walker, J.A. Hall, S. Arndt, Methamphetamine use and dental disease: results of a pilot study. J. Dent. Child. (Chic) **74**(2), 85–92 (2007)
30. A. Mandal, Opioid side effects. News Med. Life Sci. (Oct 2013)
31. www.facesofmeth.us
32. R. Yin, Q. Chen, M.R. Hamblin, *Skin Photoaging* (Morgan & Claypool Publishers, 2015)
33. L.H. Kligman, Photoaging. Manifestations, prevention, and treatment. Dermatol. Clin. **4**(3), 517–528 (1986)
34. D. Gawkrodger, M.R. Ardern-Jones, *Dermatology E-Book: An Illustrated Colour Text* (Elsevier Health Sciences, 2016)
35. N.G. Jablonski, *Living Color: The Biological and Social Meaning of Skin Color*, 1st edn. (University of California Press, 2012)
36. B.A. Gilchrest, A review of skin ageing and its medical therapy. Br. J. Dermatol. **135**(6), 867–875 (1996)
37. L. Baumann, Skin ageing and its treatment. J. Pathol. **211**(2), 241–251 (2007)
38. J. Uitto, Understanding premature skin aging. N. Engl. J. Med. **337**(20), 1463–1465 (1997)
39. F. Flament, R. Bazin, S. Laquieze, V. Rubert, E. Simonpietri, B. Piot, Effect of the sun on visible clinical signs of aging in Caucasian skin. Clin. Cosmet. Investig. Dermatol. **6**, 221–232 (2013)
40. C. Griffiths, The clinical identification and quantification of photodamage. Br. J. Dermatol. **127**(S41), 37–42 (1992)
41. F. Flament et al., Solar exposure(s) and facial clinical signs of aging in Chinese women: impacts upon age perception. Clin. Cosmet. Investig. Dermatol. **8**, 75–84 (2015)
42. B. Guyuron, D.J. Rowe, A.B. Weinfeld, Y. Eshraghi, A. Fathi, S. Iamphongsai, Factors contributing to the facial aging of identical twins. Plast. Reconstr. Surg. **123**(4), 1321–1331 (2009)
43. J.R.S. Gordon, J.C. Brieva, Unilateral dermatoheliosis. N. Engl. J. Med. **366**(16), e25 (2012)
44. R.D. Bazin Eric, *Skin Aging Atlas t.1; Caucasian Type*, MED'COM edn. (MEDCOM, Paris, 2007)
45. H.C. Okada, B. Alleyne, K. Varghai, K. Kinder, B. Guyuron, Facial changes caused by smoking: a comparison between smoking and nonsmoking identical twins. Plast. Reconstr. Surg. **132**(5), 1085–1092 (2013)
46. N. Nahvi, D. Mittal, Medical image fusion using discrete wavelet transform. Int. J. Eng. Res. Appl. **4**(9), 165–170
47. "Ultraviolet Beauties," *Cara Phillips*. [Online] Available: http://www.cara-phillips.com/ultraviolet-beauties/. Accessed 12 Sept 2017
48. S.R. Coleman, R. Grover, The anatomy of the aging face: volume loss. Aesth. Surg. J. **L** (2006)
49. E. Ziglio, C. Currie, V.B. Rasmussen, The WHO cross-national study of health behavior in school aged children from 35 countries: findings from 2001–2002. J. School Health **74**(6): 204–206 [PubMed] (2004)
50. A. Takeuchi, K. Nagao, Communicative facial displays as a new conversational modality, in *Proceedings of the INTERSECT'93 and CHI'93 Conference on Human Factors in Computing Systems*, ACM (1993), pp. 187–193

51. J. Haxby, E. Hoffman, M. Gobbini, The distributed human neural system for face perception. Trends Cogn. Sci. **4**(6), 223–233 (2000)
52. M. Chan, P. Delmas, G. Gimel'farb, P. Leclercq, Comparative study of 3d face acquisition techniques, in *Proceedings of the International Conference on Computer Analysis of Images and Patterns (CAIP'05)*, Versaille, France, LNCS 3691, pp. 740–747, 2005
53. D.M. Enlow, *Handbook of Facial Growth*, 2nd edn. (Saunders, Philadelphia, PA, 1982)
54. A.C. Little, B.C. Jones, L.M. DeBruine, Facial attractiveness: evolutionary based research. Phil. Trans. R. Soc. B **366**, 1638–1659 (2011)
55. F. Liu, F. van der Lijn, C. Schurmann et al., A genome-wide association study identifies five loci influencing facial morphology in Europeans. PLoS Genet. **8**, e1002932 (2012). |Article|PubMed|CAS
56. L. Paternoster, A.I. Zhurov, A.M. Toma et al. Genome-wide association study of three-dimensional facial morphology identifies a variant in PAX3 associated with nasion position. Am. J. Hum. Genet. **90**, 478–485 (2012). |Article|PubMed|ISI|CAS
57. F.M.Z. Heravi, A. Nait-Ali, *A 3D Dynamic Shape Model to Simulate Rejuvenation & Ageing Trajectory of 3D Face Images*, pp. 0–4, 2017
58. V. Blanz, T. Vetter, A morphable model for the synthesis of 3D faces, in *Proceedings of the 26th Annual Conference on Computer Graphics Interaction Technology—SIGGRAPH'99*, pp. 187–194, 1999
59. L. Sirovich, M. Kirby, Low-dimensional procedure for characterization of human faces. J. Opticalsoc. Am. **4**, 519–524 (1987)
60. M. Kirby, L. Sirovich, Application of the KL procedure for the characterization of human faces. IEEE Trans. Pattern Anal. Mach. Intell. **12**(1), 103–108 (1990)

3D Face Analysis for Healthcare

Guillaume Stern, Zehua Fu and Mohsen Ardabilian

Abstract This chapter is entitled "3D Face Analysis for Healthcare". It describes 3D face acquisition aspects and protocols. It discusses computer-aided three dimensional face biometrics in health for surgery, prosthesis, ophthalmology, dermatology and cosmetics purposes.

1 Introduction

Due to the improvement of hardware and software, digital three-dimensional models have been increasingly used in the last decade for face biometrics. Generally, security comes to mind, especially with two-dimensional face imaging, but health applications has also greatly benefited from the development of three-dimensional face imaging. Surgery support or diagnostic assistance are examples of applications. They will be further detailed in this chapter.

Three-dimensional models deal with different types of data including texture and depth. Texture is related to the color of the material captured by the camera whereas depth is related to the shape of the material. Texture and depth are stored pixel-by-pixel after acquisition and processing of the digital images.

Needs and objectives: Computer-aided three-dimensional face biometrics can be used in several medical areas such as diagnostics, surgeries, production of prostheses or patient identification. It allows fast, accurate and reproducible measurements of the face. It provides a tool for physicians to improve their diagnostics. It also gives surgeons and patients a better overview of the outcome of an operation by simulation. This chapter is complementary to the previous one dedicated to (face biometrics: from security to biomedical applications). It is organized as follows: in the next section, 3D acquisition aspects and protocols will be considered. Afterwards, we will focus on computer-aided three dimensional face biometrics in health, particularly in surgery,

G. Stern · Z. Fu · M. Ardabilian (✉)
Laboratoire LIRIS - UMR 5205 CNRS,
Université de Lyon - Ecole Centrale de Lyon, Lyon, France
e-mail: Mohsen.Ardabilian@ec-lyon.fr

A. Nait-Ali (ed.), *Biometrics under Biomedical Considerations*,
Series in BioEngineering, https://doi.org/10.1007/978-981-13-1144-4_6

prosthesis, ophthalmology, dermatology and cosmetics. Finally, a small section is dedicated to patient identification and ethical questions.

2 Acquisition and Processing

To perform a three-dimensional face analysis, data has to be acquired and processed. The overview of this process is illustrated in Fig. 1.

The data acquisition can be done based on different techniques such as stereo cameras, optical scanners, CT scans (Computed Tomography) or CBCT scans (Cone Beam Computed Tomography). Stereo cameras are made of two (or more) cameras watching the same scene from different positions (see Fig. 2). Optical scanners comprise a camera and a light source illuminating the object to be captured (see Fig. 3). CT-scans are based on X-rays captured with different angles and combined in order to produce slices of a three-dimensional model (see Fig. 4). Depending on the technique used, the acquisition is considered as active or passive. The acquisition is active if a source of energy is used. For instance, a light beam or a X-ray beam is sent to the object and the sensor measures the transmitted or the reflected beam. The acquisition is passive if the acquisition is only based on measurements of the object with a sensor [1–3].

In case of stereo vision, the data processing starts with determining the transformation between the three-dimensional real world and the corresponding two-dimensional captured images. The pinhole camera model assumes that the transformations are projections fully defined by extrinsic (translation and rotation between

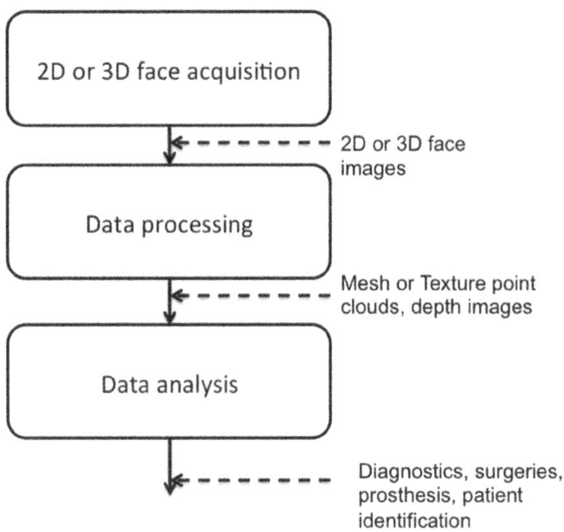

Fig. 1 General process for a three-dimensional face analysis

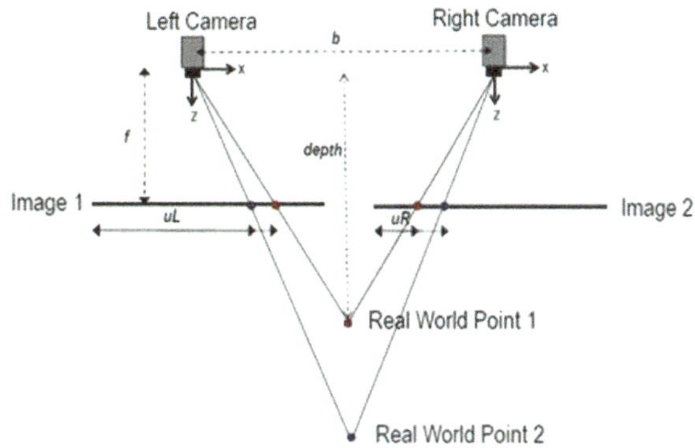

Fig. 2 Principle of stereo cameras [4]

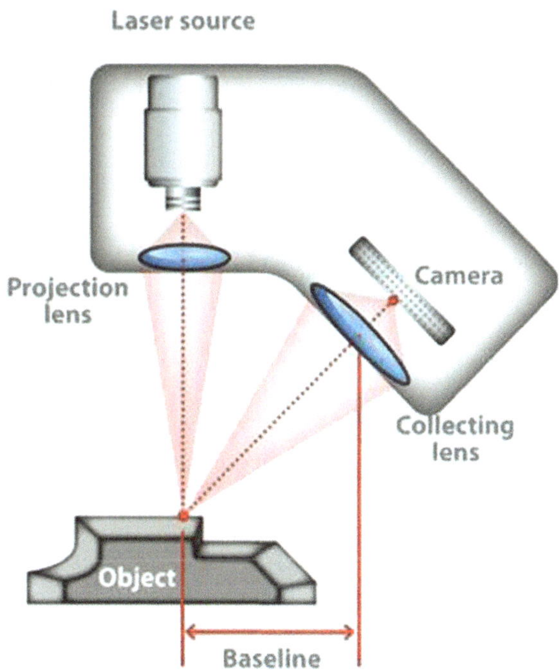

Fig. 3 Optical scanning principle [5]

stereo rig) and intrinsic (focal length, principal point) cameras parameters (see Fig. 5). A projection can be represented by the following equation:

$$x = K[t]X$$

with X the 3D-coordinates of a world point, t the translation vector and R the rotation matrix between the world coordinate system and the camera coordinate system, K the matrix containing the principal point p and the focal length f, and x the 2D-coordinates of the corresponding image point.

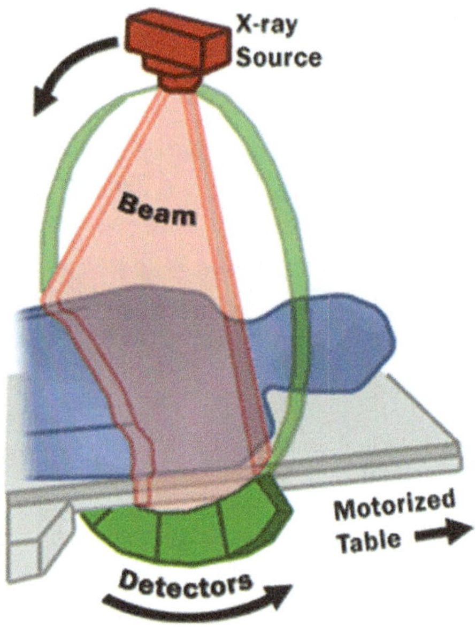

Fig. 4 CT (computerized tomography) scan principle [6]

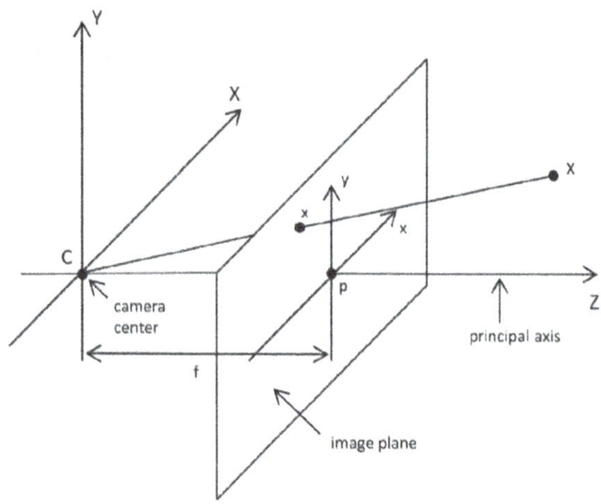

Fig. 5 Pinhole camera model

In addition to focal length and principal point, distortion parameters can be taken into account as intrinsic parameters. They correct for radial and tangential deformations caused by the camera lens and the optical system.

To determine the projections between the world and the images, the extrinsic and the intrinsic cameras parameters have to be estimated. This operation is called camera calibration and is done in an off line stage [7].

As soon as both stereo cameras are calibrated, the images can be rectified. Image rectification consists of cancelling the rotations and the scales of the images, so that only the translation between the images remains. This is done in order to simplify

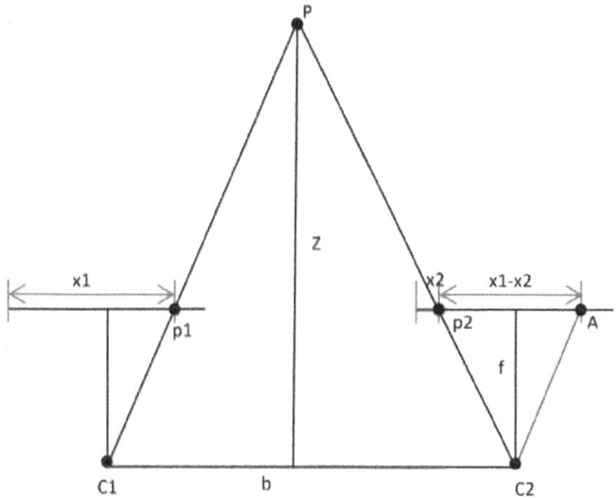

Fig. 6 Stereo matching after image rectification. C1 and C2 are two cameras watching at the same 3D point P. p1 and p2 are the corresponding 2D image points

the stereo matching and to keep the only transformation that generates a disparity between stereo cameras.

In an online stage, stereo matching corresponding pairs, e.g. p1 and p2, in the left and in the right rectified images are identified (see Fig. 6 and [8, 9]). The distance between two corresponding points is called disparity. The disparity calculated for all points of the scene is called disparity map where the disparity is encoded as intensity in grayscale. The disparity is useful to reconstruct three-dimensional models as it is inversely proportional to the depth. Indeed, the following relationship holds:

$$Z = \frac{f \cdot b}{x_1 - x_2}$$

with Z the depth, f the focal length, b the baseline between cameras calculated from the calibration, and $x_1 - x_2$ the disparity. So far, an overview of stereo vision has been presented. As mentioned above, besides stereo cameras, there are other solutions to reconstruct three-dimensional models such as optical scanners (see Fig. 2).

In case of optical scanners, one of the cameras is replaced by a light beam source and combined with a light beam. The light source can be a laser but not necessarily. The reconstruction is active but the process is similar. First, the camera has to be calibrated, then stereo matching can be applied between the known illumination pattern and the pattern projected on the three-dimensional object. Figure 7 gives an overview of projected patterns. The advantage of an active system is that it speeds up the stereo matching stage.

At the end of the processing step, different ways of representing the data are possible, for instance mesh or textured point clouds, or depth images (see Fig. 8). A point cloud is a set of three-dimensional vectors in a given coordinate system. Thanks to surface reconstruction, it can be converted to a mesh which is a set of triangles that are connected by their common edges. A depth image is a grayscale image that indicates for each pixel the distance to the surface.

Fig. 7 **a** Binary-coded patterns projection for 3D acquisition, **b** n-ary-coded patterns for 3D acquisition (extracted from [1])

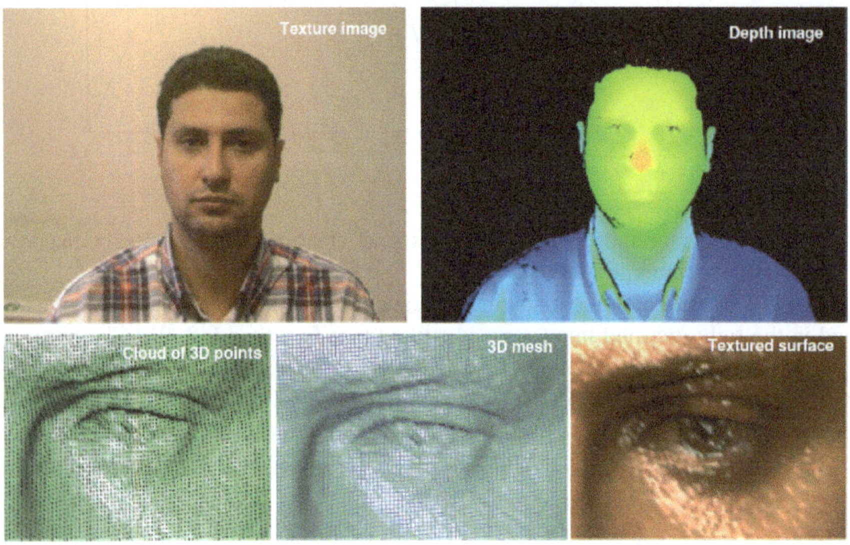

Fig. 8 Examples of different representations (from [1]). From left to right, and from top to bottom: texture image, depth image, cloud of 3D points, 3D mesh and textured shape

Several commercial or open source solutions exist for acquisition and processing, for instance Artec, 3dMD, the DermaTOP from Eotech, or by universities, for instance MorphAnalyzer from the University of Aberystwyth or the Antera 3D from the Department of Physics of Trinity College, Dublin.

Finally, the physicians can analyze the reconstructed three-dimensional faces and deliver diagnostics, perform surgeries or produce prostheses.

3 Advantages of Computer-Aided Three-Dimensional Face Biometrics in Health

Computer-aided three-dimensional face biometrics are usually faster, more accurate and more reproducible than existing techniques. It automates the data acquisition and processing, which reduces the time to get the three-dimensional models and eliminates human failures. It improves the visual representation of the face for medical doctors and patients, helping for better diagnostics and surgeries.

In the following, several use cases of three-dimensional face biometrics will be dealt with. The same criteria as in the introduction will be used and completed, namely the objectives and needs, details about acquisition and processing, and the advantages of computer-aided three-dimensional face biometrics with respect to existing techniques.

3.1 Surgery

Multi 2D and 3D face biometrics is widely used in plastic surgery, especially for reconstructive and cosmetic surgery. Reconstructive surgery aims at restoring the shape and the function of a missing body part whereas cosmetic surgery aims at improving the appearance of an existing body part.

In [10], the authors explain that 3D models have significantly improved craniomaxillofacial surgery in the last decade but still suffer from a lack of normalized datasets. They propose to compute average 3D skulls from CT images with 21 anthropometric landmarks. This database could then be used to estimate missing anatomy and evaluate surgery results.

In [11], a similar idea is tested. A patient's face is captured by two stereo cameras in order to create a 3D model and an hologram. As a result, the virtual representation of the 3D face should provide support before, during and after a plastic surgery.

In [12], the authors also use computer-assisted 3D models of the face in the field of surgery. As an input, they have computed tomographic and magnetic resonance imaging stereolithographic data. They process it to simulate the skin behavior during a surgery, which can be of prime importance for residents and young doctors.

A last example in the surgery field is related to masculinization of the male face [13]. Some men are dissatisfied with their faces and want to improve their attractiveness, for example having a balanced and harmonious soft tissue facial envelope or a beautiful smile aesthetics. Thanks to three-dimensional face biometrics, the mathematical proportions of attractive men could be determined. The positions and the orientations of the upper lip, the chin or the jaws were particularly studied. The authors used clinical photographs, rotational tomograph, lateral cephalogram and cone beam CT-scan images. They did a two-dimensional planning with the Facewizz software in order to plan and simulate the osteotomies. They assessed a three-dimensional planning with the ProPlan software and the Geomagic Sculpt software in order study the symmetry of the skull. Finally, they state that the level of surgical refinement is much better with computer-aided models than with previous techniques.

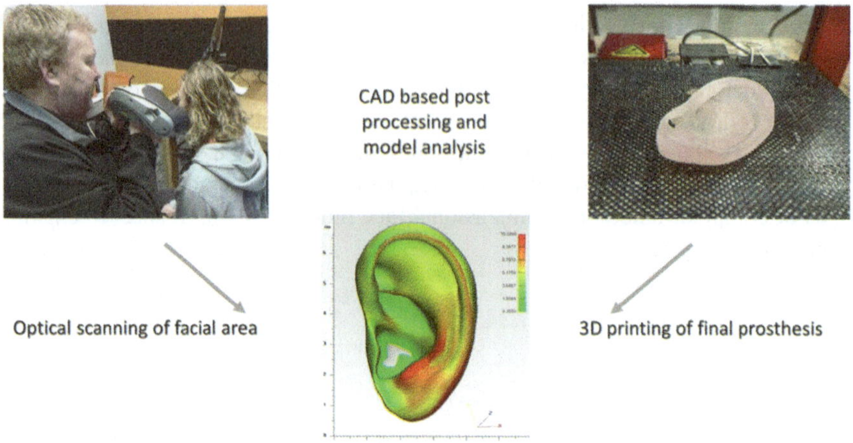

Fig. 9 Prosthesis design and production

3.2 *Prosthesis*

In some cases, a prosthesis is an alternative to a surgery. However, designing and producing a prosthesis is currently a long, costly, invasive and subjective process. 3D face biometrics could play a major role to improve the prosthesis production [14].

The idea consists of scanning the surface topography of the nose and the ear of a volunteer's face, making corresponding 3D models and using them to design prostheses (see Fig. 5). The optical scanning system is the Artec Spider, which is laser free in order to avoid laser exposure into the eyes. Its resolution is between 50 and 100 μm, which is satisfying for modeling the anatomy. The full scanning lasts approximately 5 min. The optical scanner comes with a software that transforms the images into a point cloud and finally to a contour map. In addition, the 3-Matic software package was used to improve the surface topography by smoothing rough surfaces or removing features leading to print failures.

Such a process is faster, more precise and more flexible than existing techniques. It is also non-invasive, optimized and low-cost. Last but not least, it enables the creation of a database that could be used as a reference for patients whose initial anatomy is unknown (Fig. 9).

3.3 *Facial Asymmetry Assessment*

Computer-aided face biometrics is also widely used for facial asymmetry assessment. Asymmetry helps evaluating dynamic changes of the face before and after an operation or a treatment, thus, it should be determined with accuracy. The usage of the three-dimensional information is highly appreciated.

In [15], the authors propose a technique to automate facial asymmetry assessment. They perform an experiment where they scan the faces of 55 adults. The scanning system is a 3dMDface from the company 3dMD and consists of three cameras. Two cameras are positioned on the sides so that three-dimensional images can be produced with stereo-photogrammetry techniques. Another software called 3dMDpatient from the same company refines the scanned images, eliminating outliers and cutting off neck and shoulders. A third software called 3dMDvultus quantifies the facial asymmetry by registration between an image and its mirror image. The level of asymmetry is finally calculated from RMS errors between the two registered images. The full process is summarized in a block diagram in Fig. 6.

Such a method is fast and effective and presents the advantage of being independent of landmarks (Fig. 10).

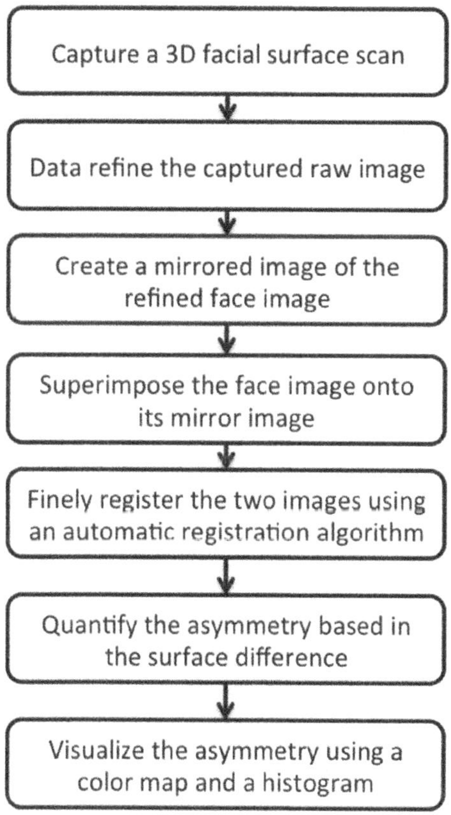

Fig. 10 Process of facial assessment

3.4 *Facial Assessment for Health or Disease*

Another application of facial assessment is related to the detection of healthy or sick patients where again, three-dimensional information is helpful (Fig. 11).

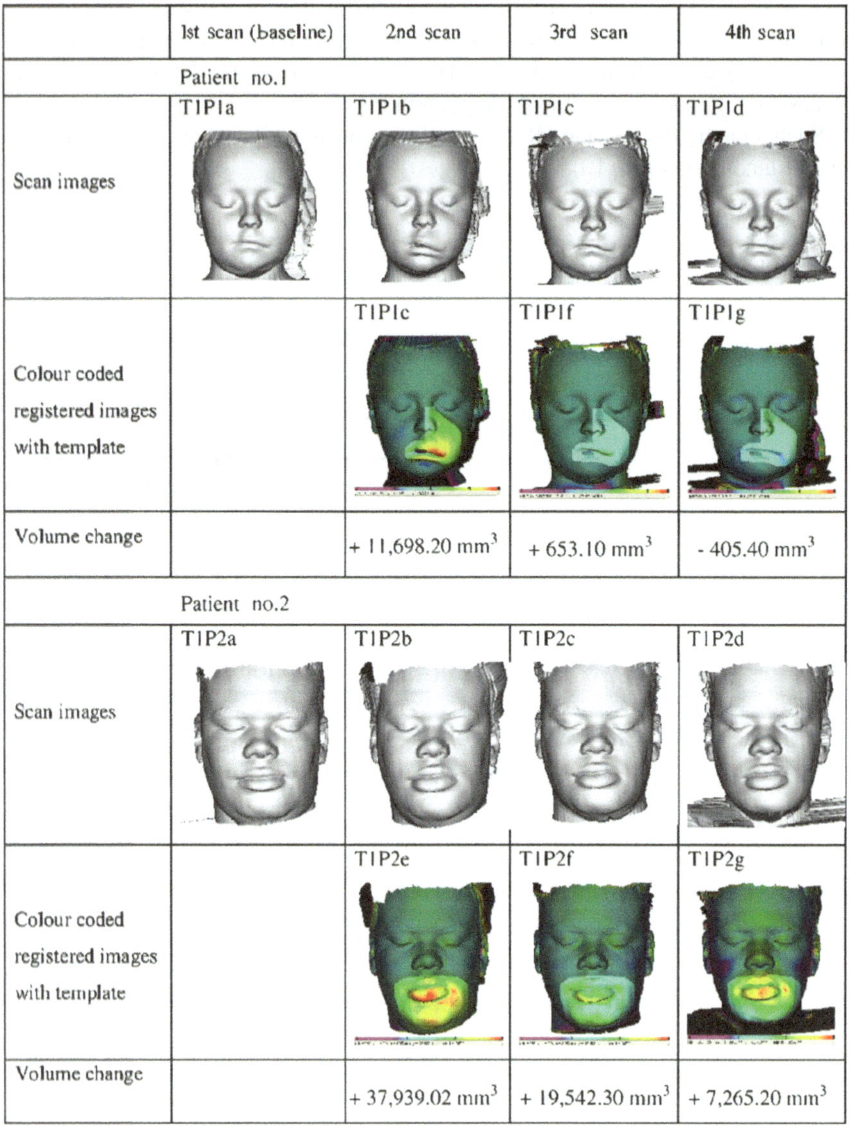

Fig. 11 Examples of detection and quantification of the Crohn's disease at different stages

In [16], the authors combine multiple facial cues measured from two-dimensional and three-dimensional images to estimate whether a person is healthy or not. They take into account several criteria, namely men's masculinity, facial symmetry, facial adiposity, skin condition (texture and color) and face expression. They capture three-dimensional images of 68 Caucasian women and 50 Caucasian men with a camera from the company 3dMD. The scanned images are sent to the software MorphAnalyzer, which computes 49 landmarks, and aligned in orientation, rotation and scale using superimposition. Finally, a Principal Component Analysis is run on the three-dimensional head models in order to extract a small amount of parameters. In parallel, basic information such as age, gender, ethnicity, height or weight was recorded manually in order to compute evaluation criteria such as body/mass index (BMI). The authors show that face shape and skin color independently contribute to perceived health. They also compare results between two-dimensional and three-dimensional face images. It turns out that three-dimensional images give more accurate information about health than two-dimensional ones.

In [17], the idea is similar but focused on a particular disease, namely Crohn's disease. Crohn's disease affects the gastrointestinal tract and presents peri-oral swellings resulting in severe facial disfigurement. The authors develop a system based on three-dimensional face biometrics to quantify the extent and progression of Crohn's disease. Thirteen patients are scanned with an in-house laser scanner. A CCD camera captures a series of facial profiles within a rotation range of 220° I steps of 5°. The acquisition time lasts 7 s. Images are registered using the algorithm of Iterative Closest Point (ICP) based on landmarks such as nasal bridge, ear lobes or forehead areas. Finally, a three-dimensional software called Cloud UCL is used to calculate volumes. In addition, the scans are performed five times in order to estimate the measurement reproducibility. As a result, it turns out that three-dimensional facial biometrics is a fast, accurate, robust, objective and economical method to identify and quantify Crohn's disease (see Fig. 7).

3.5 *Ophthalmology*

Apart from surgery and facial assessment, other medical fields can benefit from applications, in particular ophthalmology. One patent and one publication underline the need for custom-fit glasses and how computer-aided three-dimensional face biometrics could improve the current situation.

In [18], the author presents a method to design and produce custom-fit glasses. He proposes to perform facial biometrics measurements and create a three-dimensional version of the patient's head in order to carefully design corresponding eyeglasses.

In [19], the authors state that Chinese people are missing comfortable glasses because the models are generally conceived for European or American styles. They compare the heads of 62 Cantonese with the worldwide standard head (KEMAR) and notice that there are significant differences between them, especially in the areas interacting with the glasses. They conclude that the Glasses Wearing Comfortability (GWC) can be improved.

3.6 *Dermatology and Cosmetics*

Another medical field where face biometrics could be applied is dermatology, especially for the evaluation of cosmetic products.

In [20], the authors want to compare the Antera 3D from the Department of Physics of Trinity College in Dublin with other solutions such as parallel-polarized photographs captured by a high resolution digital camera or the DermaTOP from Eotech.

They select 26 healthy female Caucasian volunteers aged 45–70 years having visible crow's feet wrinkles. Wrinkles and textures are first captured by the digital camera VISIA CR provided by the Canfield Scientific Inc under parallel-polarized light. Then the images are sent to an image analysis software called Image-Pro Plus 7 from the Media Cybernetics. Wrinkles and textures are also measured by the DermaTOP and the Antera 3D systems on the temple and cheek areas. The DermaTOP system consists of a projector illuminating the skin and a camera. It makes use of fringe projection to get three-dimensional information on skin topography. The Antera 3D consists of a camera that is placed directly onto the skin and can measure different skin features without being affected by lighting conditions. A first series of measurements is done as a reference, then the volunteers apply a serum on their temple areas and a second series of measurements is performed.

Results from the Antera 3D are strongly correlated with the DermaTOP but not with the parallel-polarized photographs. In addition, the Antera 3D records significant improvements after application of the serum (see Fig. 8). Thus, the Antera 3D is a reliable tool to evaluate cosmetic products (Fig. 12).

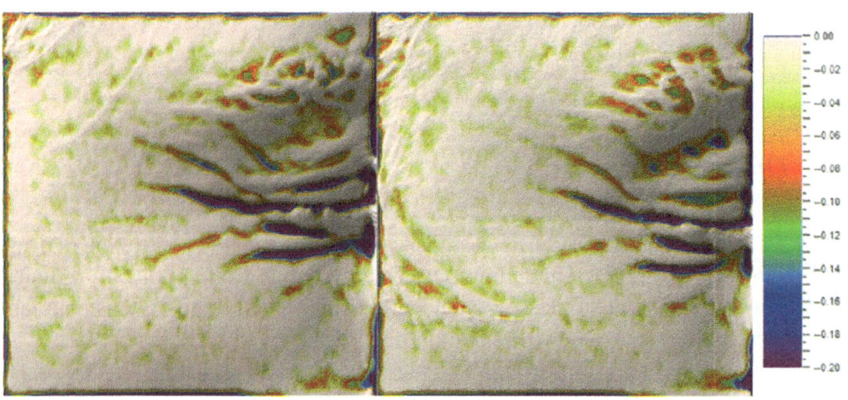

Fig. 12 Crow's feet areas measured by the Antera 3D system before (left) and after (right) application of a serum. The colors show the depth of the depression into the skin

4 Patient Identification and Ethical Questions

As mentioned in the introduction, face recognition is generally used for security and surveillance but not for health. However, some applications of face recognition exist that are directly or indirectly related to medicine.

In [21], the author reports about programs that are able to identify patients based on face recognition. This can be especially useful for elderly or young people who are lost and not able to tell their identities.

In parallel, the author underlines that health information is private and highly personal, which raises ethical issues and leads to think about the benefits and the risks of face recognition. Programs that estimate the overall health, the aging or the genetic conditions of patients could be used to anticipate the medical care, propose them good habits and improve their health situations. But on the other hand, it could be used by employers to check whether it is worth to hire or promote an employee. It could also be used by life, disability and long-term care insurances to increase premiums.

5 Conclusion

In this chapter, the need for computer-aided three-dimensional face biometrics in the medical field was described and several use cases were presented. The improvements of hardware and software in terms of quality and quantity made various applications possible, for instance in plastic surgery, prosthesis, facial assessment for asymmetry, health or disease, ophthalmology, dermatology or even patient identification.

The new tools showed interesting aspects. They are not only accurate, fast and robust, they are also cost-efficient compared to traditional methods. However, they still suffer from limitations such as the lack of available data, the lack of standardized datasets, the time to learn to use them and the ethical issues that they raise.

References

1. B.B. Amor, M. Ardabilian, L. Chen, 3D face modeling, in *3D Face Modeling, Analysis and Recognition*, 1st edn. (Wiley, New York, 2013)
2. K. Ouji, M. Ardabilian, L. Chen, F. Ghorbel, 3D deformable super-resolution for multi-camera 3D face scanning. J. Math. Imaging Vis. **47**(1–2), 124–137 (2013)
3. D. Chen, M. Ardabilian, L. Chen, A fast trilateral filter-based adaptive support weight method for stereo matching. IEEE Trans. Circuits Syst. Video Technol. (2014)
4. http://www.depthbiomechanics.co.uk/?p=102, Centre for Sports Engineering Research, Sheffield Hallam University
5. http://www.digitaleng.news/de/3d-scanning-101/, Digital Engineering
6. https://www.fda.gov/Radiation-EmittingProducts/RadiationEmittingProductsandProcedures/MedicalImaging/MedicalX-Rays/ucm115317.htm, U.S. Food and Drug Administration

7. P. Szeptycki, M. Ardabilian, L. Chen, A coarse-to-fine curvature analysis-based rotation invariant 3D face landmarking, in *Biometrics: Theory, Applications and Systems* (2009)
8. Z. Fu, M. Ardabilian, G. Stern, Stereo matching confidence learning based on multi-modal convolution neural networks, in *7th International Workshop of Representations, Analysis and Recognition of Shape and Motion from Imaging Data (RFMI)* (2017)
9. Z. Fu, M. Ardabilian, Learning confidence measures by multi-model convolutional neural networks, in *IEEE Winter Conference on Applications of Computer Vision (WACV)* (2018)
10. T.L. Teshima, V. Patel, J.G. Manprize, G. Edwards, O.M. Antonyshyn, A three-dimensional statistical average skull: application of biometric morphing in generating missing anatomy. J. Cranio-Fac. Surg. (2015)
11. T.Y. Kim, 3D plastic surgery support system based on hologram using by camera sensor/image processing, in *Proceedings of International Conference on Future Information and Communication Engineering* (2014)
12. K. Ueda, Y. Shigemura, Y. Otsuki, A. Fuse, D. Mitsuno, Three-dimensional computer-assisted two-layer elastic models of the face. J. Plast. Reconstr. Surg. (2017)
13. J. Blythe, J. Joshi-Othero, M.Y. Mommaerts, Masculinisation of the male face—'power-face' surgery. ENT Audiol. News (2017)
14. M.I. Mohammed, J. Tatineni, B. Cadd, G. Peart, I. Gibson, Applications of 3D topography scanning and multi-material additive manufacturing for facial prosthesis development and production, in *Proceedings of the 27th Annual International Solid Freeform Fabrication* (2016)
15. A. Patel, S.M.S. Islam, K. Murray, M.S. Goonewardene, Facial asymmetry assessment in adults using three-dimensional surface imaging. Prog. Orthod. (2015)
16. A.J. Henderson, I.J. Holzleitner, S.N. Talamas, D.I. Perrett, Perception of health from facial cues. Philos. Trans. R. Soc. B: Biol. Sci. (2016)
17. L. Zou, O.K. Adegun, A. Willis, F. Fortune, Facial biometrics of peri-oral changes in Crohn's disease. Lasers Med. Sci. (2014)
18. J.C.H. Kim, Methods of designing and fabricating custom-fit eyeglasses using a 3D printer. Patent US9341867B1 (2016)
19. Z. Li, X. Zhang, J. Zhang, L. Jiang, G. Yu, Analysis of glasses wearing comfortability based on 3D head shape features. Lasers Adv. Ergon. Des. (2016)
20. C. Messaraa, A. Metois, M. Walsh, S. Hurley, L. Doyle, A. Mansfield, C. O'Connor, A. Mavon, Wrinkle and roughness measurement by the Antera 3D and its application for evaluation of cosmetic products. J. Skin Res. Technol. (2018)
21. S. Mohapatra, Use of facial recognition technology for medical purposes: balancing privacy with innovation. Pepp. L. Rev. **43**, 1017 (2016)

Facial Biometrics: From Security to Biomedical Applications

Amine Nait-Ali

Abstract This chapter is entitled "Facial Biometrics: From Security to Biomedical Applications". The aim of this chapter is to bridge the gap between classical facial recognition approaches which are commonly considered for identification or verification i.e. security applications with some extended or adjusted systems related to Biomedical-engineering field.

1 Introduction

The ability of humans to recognize faces among others and the ability to deal with genders, aging and expressions, show how extraordinary the brain processes the information. Our brain is also trained to easily describe and recognize other face details such as, (e.g. dark circles, under-eyes bag, black spots, freckles, oily or dry skin etc.). Within this context, we will categorize this global human facial recognition (i.e. person identity, expressions, face health perception, etc.) in the first catégorie (i.e. category I). It summarizes somehow, some common facial recognition tasks that a normal subject can achieve in everyday life, of course, at different sensitivity, whether the perception is achieved by a young or a mature person.

As a second category (i.e. category II), one can consider that trained persons (e.g. dermatologist) can further push the facial analysis by describing skin details such as skin texture and skin diseases. In addition, some physicians are also experienced to diagnose some facial genetic disorders (e.g. Progeria, Treacher Collins, Williams Beuren, etc.).

On the other hand, we consider a third category for which some pathological persons suffer from some cognitive disorders, such as Prosopagnosia, commonly called face blindness. In fact, persons suffering from this pathology cannot recognize some familiar faces, including their own faces.

A. Nait-Ali (✉)
Université Paris-Est, LiSSi, UPEC, 94400 Vitry-sur-Seine, France
e-mail: naitali@u-pec.fr

A. Nait-Ali (ed.), *Biometrics under Biomedical Considerations*,
Series in BioEngineering, https://doi.org/10.1007/978-981-13-1144-4_7

Finally, the last category (IV) we consider in this chapter, is related to face identification, or more specifically, face reconstruction using genetic information, extracted from human DNA. This is a type of hidden biometrics showing how biology can also be connected to biometrics through artificial intelligence.

By defining these four categories, one can raise the following question: using computer vision and artificial intelligence approaches, how can face recognition technology be considered in these various categories? Actually, the main challenge is to make "computers" performing like, or even better than human beings, in some particular situations. Or, from another point of view, to use artificial intelligence in order to assist human beings in their daily tasks. This may concern, normal persons, professionals, including physicians and clinicians, or even persons suffering from some disorders.

Actually, this is what we will try to highlight in this chapter. In particular, we aim to bridge the gap between classical facial recognition approaches, commonly considered for identification or verification (i.e. security applications), and some extended or adjusted systems, as those related to biomedical engineering field.

In Sect. 2, we provide a review of some common face recognition technologies, challenging the recognition/analysis process, as described in category I. Afterwards, other challenges are extended to category II, III, and IV respectively in Sects. 3 and 4.

2 Face Recognition: Quick Review

In computer vision, face recognition systems consists in recognizing automatically individuals based their facial traits. As in many biometric systems dealing with security, recognition process can be achieved either for verification or identification. When considering the verification mode, the unrolled person should claim his/her identity each time the system is being used. Claiming the identity can be achieved by entering a code, name, or any other signal or information through any modality. Readers who are not familiar with the verification process in biometrics, can just take as reference the classical system of verification, used to access accounts in Internet. Commonly, we are used to use a login and a password. The login is needed to claim your identity (Who you are) and the password is required to verify if you are the right person. Consequently, in our case, facial biometric traits are used as a password (or a signature).

On the other hand, when considering the identification mode, the user doesn't have to claim his identity (no login word, and no code). Facial traits are directly extracted (using global or local approaches [1] and compared to each signature available in a database. If this signature is close to one of the signatures (previously unrolled), the associated identity will be considered. In security field, facial recognition has intensively been used for access control (e.g. to resources and devices) and for smart video surveillance, including on the fly face identification. However, in some uncontrolled situations, even if we consider that face recognition techniques have considerably been improved, they are still more or less sensitive to some external factors, such as: illumination, pose, occlusion, facial expressions, spoofing, etc.

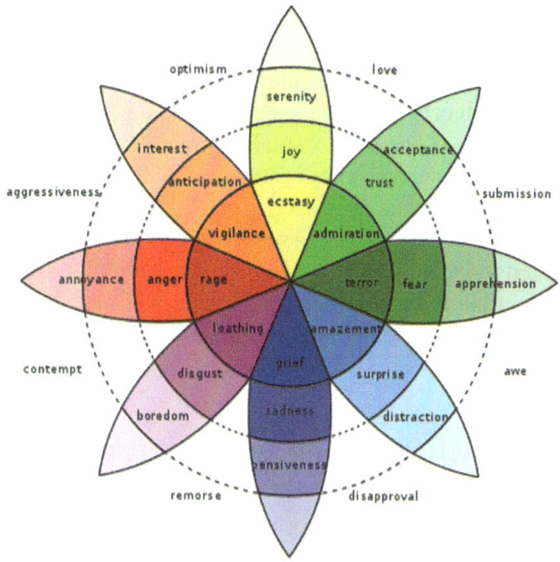

Fig. 1 Plutchik's wheel of emotions [2]

3 Face Recognition in Health-Biometrics

Face recognition technology has been extended to numerous applications dealing with face traits. If recognizing emotion expressions, age and gender estimation can also be employed for security purposes, healthcare field can also take advantage of this advanced technology. Excluding the fact that facial recognition systems may be used by physicians and patients to control the access to medical data and resources, facial recognition and analysis can be considered in Human-Computer-Interaction (HCI) systems, or in Human-Robot-Interaction (HRI) in order to assist patients in their daily tasks. We will call this field H-Biometrics, to stand for health biometrics.

As we know, intelligent robots are getting more and more involved in our daily life. Thanks to Artificial Intelligence based approaches, robots may considerably assist aged persons living alone. Through face detection, face recognition, and facial expression analysis, appropriate and adapted assistance can be delivered to these persons to decode some expressions through affective computing. Basically, one can consider some basic emotions such as: happy, excited, tender, sad, angry, and scared, as described depthly from Plutchik's wheel of emotions (Fig. 1). For better performances, face traits can be combined with other biosignals to allow an accurate identification of the emotions.

Within the same context of HCI or HRI, facial recognition in H-Biometrics can be useful for persons suffering from some neurodegenerative disease, such as alzheimer. Very often, these persons are not autonomous. They sometimes require intensive home care and heavy efforts and investments from their family members. In the future, one can imagine that HRI can be regarded as an option to assist persons concerned by this type of disease. In fact, face recognition technology associated with activities recognition, and object recognition, can be very useful and helpful to reduce some social pressures. As a possible scenario, the robot can detect, recognize and track faces and activities in real time, and provide assistance (e.g. vocal assistance) if the subject needs to locate objects (e.g. where are my glasses?, where are my keys?, etc.). The robot can also provide reminders, generate and transmit reports, call emergencies in case of uncontrolled situations, etc. Such HRI can be deployed in clinical centers to assist a group of patients suffering from Alzheimer, and other neurodegenerative disease such as agnosia (e.g. difficulty to recognize objects, sound, smells), including prosopagnosia (i.e. difficulty to recognize faces). Actually, face recognition in a clinical environment, will play an important role to connect all the other technologies through Artificial Intelligence.

4 Face Recognition and Prosopagnosia

As evoked previously, prosopagnosia (specific visual form of agnosia) is a cognitive disorder. Persons suffering from this disease have difficulties to recognize familiar faces, including their own faces. Consequently, to overcome this deficiency, they rather focus on other details surrounding the face in order to identify people they are interacting with. For example, they can recognize a person from his/her voice, hairstyle, glasses, smell, or any other details, excluding, of course, face traits. Therefore, we can imagine how difficult their life is. They continuously should make efforts to avoid critical situations which considerably impacts their social life.

It has been reported in [5] that around 2.5% of the population suffers from this disease which can be either congenital or acquired. As shown by Fig. 2, prosopagnosia is due to a damage in the Fusiform Area, called also Fusiform Face Area. It is known that this part of the brain is associated to face recognition process. Functional Medical Resonance Imaging (fMRI) can highlight an increase in blood flow when faces are shown to normal subjects. However, if objects are used as visual stimulus, other parts of the brain are activated. Persons who cannot recognize objects suffer from a visual object agnosia.

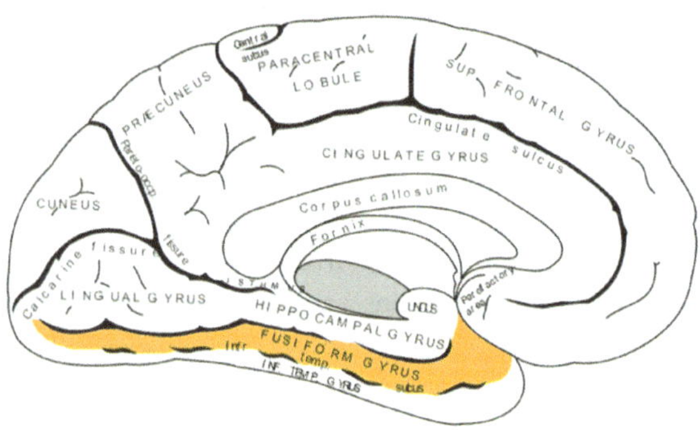

Fig. 2 Fusiform face area damage in prosopagnosia (figure [3])

As an example, persons suffering from object agnosia can basically define faces from Fig. 3 (left). However, persons suffering from prosopagnosia, can see only objects from the second Fig. 4 (right).

Based on this, we believe that Artificial Intelligence and technology can be used to assist this category of persons. From the technical point of view, many solutions can be proposed. For instance, we can suggest to use, connected smart glasses (Fig. 4) for which one can implement specific facial recognition, and object algorithms to assist patients suffering from different types of agnosia, including, of course, prosopagnosia. These kind of applications will be useful and somehow easy to implement. Either small databases can be used for small functionalities, or very big databases can be considered. In such a case, one can consider using some advanced Machine Learning algorithms, and more specifically Deep-Learning which is now commonly and frequently used in pattern recognition field. These solutions cannot solve the neurodegenerative disorders, but at least, once trained, one can expect that the subjects may see their lifestyle improved.

Fig. 3 (left) The forest has eyes painting by Bev Doolittle: people suffering from object agnosia can easily detect faces. (right) Vertumne or Vertumnus Painting (by Giuseppe Arcimboldo en 1590): people suffering from face agnosia focus on objects

Fig. 4 Smart glasses can be used in healthcare applications (figure [4])

5 Face Diseases Recognition

Beyond face recognition approaches that consist in recognizing person's identity, or person's emotions from the expressions, one can consider, instead, recognizing face pathologies. Generally speaking, the generic scheme that is commonly used in authentication or identification remains the same, except that this specific "medical-system" extracts different features, and the final decision concerns the identification or the verification of the pathologie. In fact, this is an extended scenario to consider biometrics as a general purpose tool, and not only restricted to security applications. In this section, this scenario will be highlighted through two concrete examples. In the first example, the idea is to recognize some skin pathologies from a face, and the second example consists in recognizing face pathologies due to some gene disorders.

5.1 Face Skin Diseases Recognition in Dermatology

Face diseases recognition is a recent topic considered in many studies. For instance, in [6], they propose a solution to detect visually some observable symptoms on faces. For this purpose, they used semi-supervised anomaly detection combined with machine vision algorithms. In [7], they used color analysis for hepatitis diagnosis. In particular, they proposed a KNN-based classifier to distinguish between three categories, namely: healthy, severe hepatitis with jaundice, and severe hepatitis without jaundice. In [8] they proposed some preliminary results for automating facial paralysis rehabilitation therapy, using Machine Learning techniques.

Within this context, the Biometrics Group [9] is currently working on using Machine-Learning approaches to identify skin diseases by including specific biometrics features. More specifically, Deep-learning approaches [10] are employed to detect and classify some face diseases such as: acne, Rosacea, etc. [11, 12] (Fig. 5).

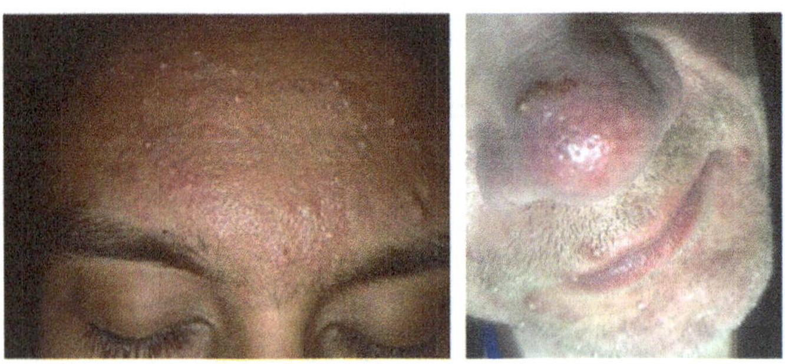

Fig. 5 (left) Acne skin diseases. (right) Rosacea skin diseases (images [11, 12])

5.2 Dysmorphic Face Recognition

Some recent researchers [13] conducted some studies to diagnose facial disorders due to genetic diseases. Based on phenotypic information extraction, and Machine Learning classification techniques, they proposed an algorithm to distinguish between some disorders such as: Progeria, Angelman, Apert, Williams-Beuren, etc.

Basically, they show that the representation of extracted features don't share the same space, and can efficiently be separated. In their database, a manual check has been achieved in order to exclude images for which the faces or eyes were not enough visible, and also those for which the clinicians couldn't visually identify the disorder. We recommend to the reader some other references related to this topic in which computer vision techniques are used to analyse dysmorphic faces [14, 15].

6 Face Modelling from DNA

As it is known, many phenotypes can be defined from genotypes. This includes, gender, hair color, eye color, etc. But, reconstructing a face appearance from genotypes is still very challenging regarding all the possible interesting applications that can be developed. For example, in forensics, one can consider the possibility of synthesizing or modelling a face from collected DNA (e.g. from a crime scene).

In this field, advanced researches [16] shown that five genes can be used to estimate some facial traits. This study was achieved using head Magnetic Resonance Imaging (MRI) (see also [17]). In some other studies, 3D face models have been generated, based on the correlation between genotypes and phenotypes, as described in the study [18], where some face categories have been considered (e.g. european, african), Fig. 6.

In [19], advanced researchers shown that multiple specific loci are involved to describe face traits such as: cranial base Intracranial width, upper facial depth, nasal protrusion, nasal width and cranial base width and recently 15 genes have been identified to shape human faces [20]. Among them, seven genes are associated to the nose. Actually, even if the reconstructed faces from genotypes are still not very accurate, it is obvious that the obtained result are interesting and can be exploited in numerous applications, including in forensics.

In another interesting research [21], the study was not limited to face phenotypes, but it has been extended to other characteristics such as the voice, eyes color, hair, age, height, etc. Actually, the main purpose of the study is to show that it is possible to identify individuals using a whole-genome sequencing. The study was presented to challenge the techniques of de-identification, commonly employed in clinical contexts. In particular, their process is described in Fig. 7, where a measure of similarity is achieved in order to compare the characteristics extracted from a DNA, with the observed phenotypes. In terms of face recognition, real faces of individuals who participated in the experience were compared to synthesized faces (Fig. 8).

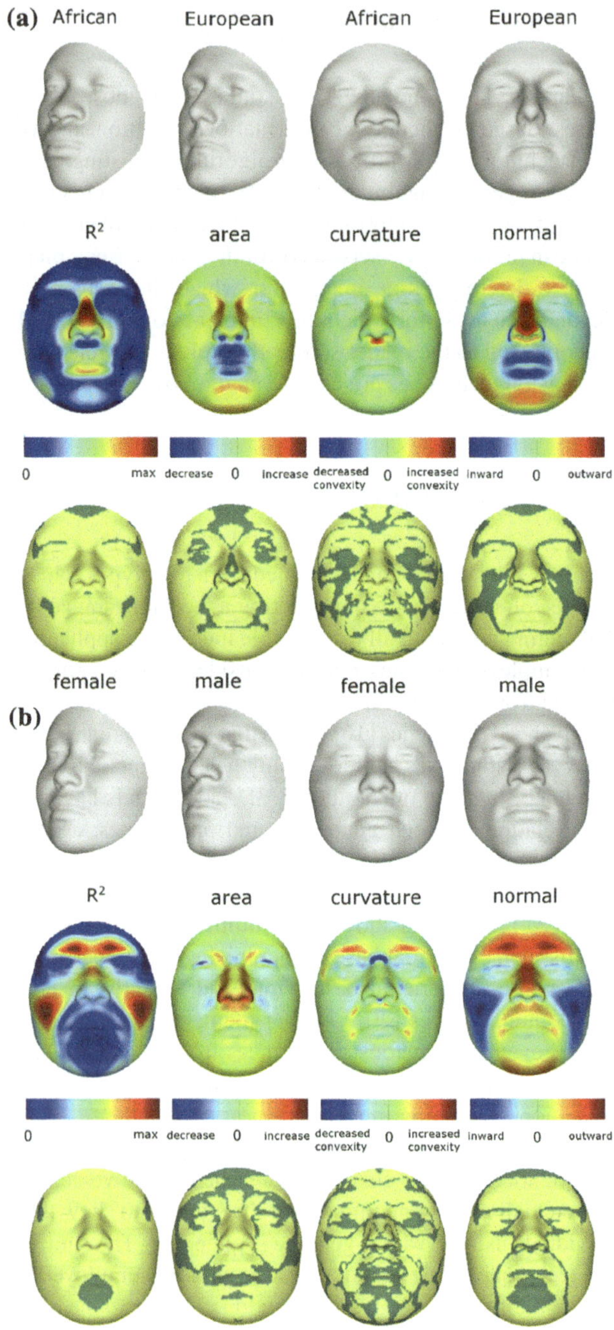

Fig. 6 Transformations and heat maps showing how face shape is affected by **a** RIP-A and **b** RIP-S according to [18]

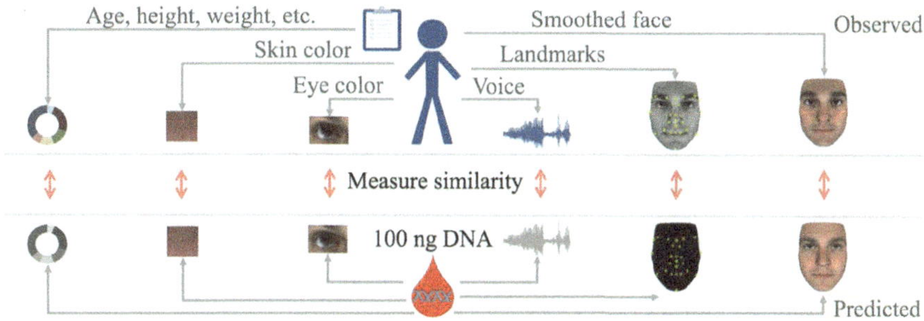

Fig. 7 Overview of the experimental approach according to the research published in [21]

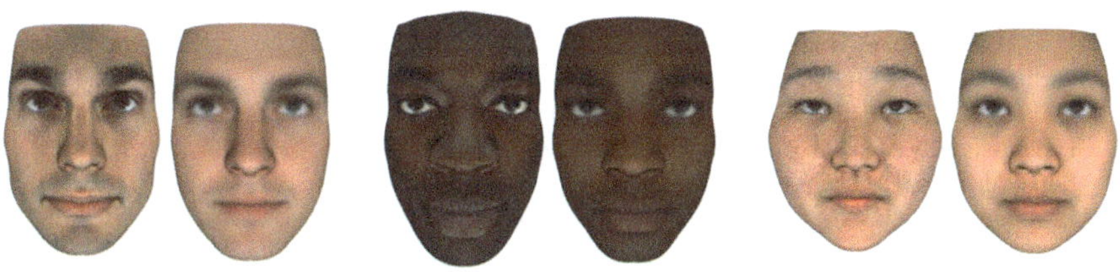

Fig. 8 Examples of real (left) and predicted (right) faces according to [21]

7 Conclusion

As conclusion of this chapter, one can mention how strongly biomedical engineering is connected to biometrics by considering face recognition, as a special case. In other words, face recognition should not be restricted to security biometrics. It is obvious that this topic can be more and more developed if it is seen as a multidisciplinary field. By doing so, we expect that many connexions can be established and many useful applications can be developed. From the technical point of view, we expect that using some Artificial Intelligence approaches (e.g. Deep learning) on big databases, may solve many issues, and may contribute to develop very challenging applications.

References

1. A. Naït-Ali, R. Fournier (eds), *Signal and Image Processing for Biometrics* (ISTE-Wiley, 2012). ISBN: 9781848213852
2. By Machine Elf 1735. By Machine Elf 1735—Own work, Public Domain, https://commons.wikimedia.org/w/index.php?curid=13285286
3. By Gray, vectorized by Mysid, coloured by was_a_bee. - File:Gray727.svg, Public Domain, https://commons.wikimedia.org/w/index.php?curid=8886915
4. By Loïc Le Meur - Flickr: Loïc Le Meur on Google Glass, CC BY 2.0, https://commons.wikimedia.org/w/index.php?curid=26050963

5. I. Kennerknecht, T. Grueter, B. Welling, S. Wentzek, J. Horst, S. Edwards, M. Grueter, First report of prevalence of non-syndromic hereditary prosopagnosia (HPA). Am. J. Med. Genet. A. **140**(1–22) (2006)
6. K. Wang, J. Luo, Detecting visually observable disease symptoms from faces. EURASIP J. Bioinform. Syst. Biol. **2016**(1), 13 (2018)
7. M. Liu, Z. Guo, Hepatitis diagnosis using facial color image, in *Proceedings of the 1st International Conference on Medical Biometrics (ICMB 2008)*, ed. by Z. David (Springer, Berlin, 2008), pp. 160–167
8. W.S.W. Samsudin, K. Sundaraj, Image processing on facial paralysis for facial rehabilitation system: a review, in *2012 IEEE International Conference on Control System Computing and Engineering (ICCSCE)*, 2012, pp. 259–263
9. Biometrics Group, http://www.amine-nait-ali.org
10. Y. Le Cun, Y. Bengio, G. Hinton, Deep learning. Nature **521**, 436–444 (2015)
11. By Roshu Bangal—Own work, CC BY-SA 4.0, https://commons.wikimedia.org/w/index.php?curid=47832556
12. User:RicHard-59—Own work, CC BY-SA 3.0, https://commons.wikimedia.org/w/index.php?curid=40915844
13. Q. Ferry, J. Steinberg, C. Webber, D.R. Fitzpatrick, C.P. Ponting, A. Zisserman, C. Nellåker, Diagnostically relevant facial gestalt information from ordinary photos. eLife **3**, e02020 (2014). https://doi.org/10.7554/elife.02020
14. P. Hammond, The use of 3D face shape modelling in dysmorphology. Arch. Dis. Child. **92**, 1120–1126 (2007)
15. H.S. Loos, D. Wieczorek, R.P. Wurtz, C. Von Der Malsburg, B. Horsthemke, Computer-based recognition of dysmorphic faces. Eur. J. Hum. Genet. **11**, 555–560 (2003)
16. F. Liu, F. van der Lijn, C. Schurmann, G. Zhu, M.M. Chakravarty, P.G. Hysi et al., A genome-wide association study identifies five loci influencing facial morphology in Europeans. PLoS Genet. **8**(9), e1002932 (2012). https://doi.org/10.1371/journal.pgen.1002932
17. L. Paternoster, A.I. Zhurov, A.M. Toma, J.P. Kemp, B.S. Pourcain, N.J. Timpson, et al., Genome-wide association study of three-dimensional facial morphology identifies a variant in PAX3 associated with nasion position. Am. J. Hum. Genet. **90**, 478–485 (2012). (pmid:22341974)
18. P. Claes, D.K. Liberton, K. Daniels, K.M. Rosana, E.E. Quillen, L.N. Pearson et al., Modeling 3D facial shape from DNA. PLoS Genet. **10**(3), e1004224 (2014). https://doi.org/10.1371/journal.pgen.1004224
19. J.R. Shaffer, E. Orlova, M.K. Lee, E.J. Leslie, Z.D. Raffensperger, C.L. Heike et al., Genome-wide association study reveals multiple loci influencing normal human facial morphology. PLoS Genet. **12**(8), e1006149 (2016). https://doi.org/10.1371/journal.pgen.1006149
20. P. Claes, J. Roosenboom, J.D. White, T. Swigut, D. Sero, J. Li, M.K. Lee, A. Zaidi, B.C. Mattern, C. Liebowitz, L. Pearson, T. González, E.J. Leslie, J.C. Carlson, E. Orlova, P. Suetens, D. Vandermeulen, E. Feingold, M.L. Marazita, J.R. Shaffer, J. Wysocka, M.D. Shriver, S.M. Weinberg, Genome-wide mapping of global-to-local genetic effects on human facial shape. Nature Genet. (2018). https://doi.org/10.1038/s41588-018-0057-4
21. C. Lippert, R. Sabatini, M. Cyrus Maher, E.Y. Kang, S. Lee, O. Arikan, A. Harley, A. Bernal, P. Garst, V. Lavrenko, K. Yocum, T. Wong, M. Zhu, W.-Y. Yang, C. Chang, T. Lu, C.W.H. Lee, B. Hicks, S. Ramakrishnan, H. Tang, C. Xie, J. Piper, S. Brewerton, Y. Turpaz, A. Telenti, R.K. Roby, F.J. Och, J. Craig Venter, Identification of individuals by trait prediction using whole-genome sequencing data. PNAS **114**(38), 10166–10171 (2017). https://doi.org/10.1073/pnas.1711125114

Eye Movement Analysis in Biometrics

Chiara Galdi and Michele Nappi

Abstract In this chapter, eye movement in biometrics is addressed. An overview of acquisition aspects and applications are discussed, and it considers using this behavioral modality for recognition purposes.

1 Introduction

Among biometric traits, it is possible to identify two main categories, denoted as physical and behavioral. The former ones describe some physiological characteristics of the human being, are usually passive in nature, and include DNA, fingerprints, face, and iris. With few exceptions, e.g. DNA, physical biometrics are strictly related to appearance characteristics and systems based on them, can be subject to spoofing attacks, i.e. when a replica of the biometric trait is presented to the system, e.g. a photo. Behavioral biometrics describe dynamic aspects of "something that a person does" and are in general more difficult to forge and replicate. They include gait, gaze, keystroke, and signature dynamics. Although behavioral biometrics can suffer from transient changes due to cognitive and/or emotional states, they can be used in conjunction with physical biometrics to further improve recognition accuracy or they can be used when physical biometric recognition cannot be applied, e.g. in a surveillance video where the suspect of a crime hides her face and eyes. The following discussion focuses on eye movement analysis. The idea at the basis of eye movement recognition, is that each person has a different way to look, due both to physical characteristics of the eye and because of the different personal background. Eye movements occur as very fast, almost instantaneous, saccades (whose duration

C. Galdi
EURECOM, 450 Route des Chappes, 06410 Biot Sophia Antipolis, France
e-mail: chiara.galdi@eurecom.fr

M. Nappi (✉)
Università degli Studi di Salerno, Via Giovanni Paolo II 132,
84084 Fisciano, SA, Italy
e-mail: mnappi@unisa.it

A. Nait-Ali (ed.), *Biometrics under Biomedical Considerations*,
Series in BioEngineering, https://doi.org/10.1007/978-981-13-1144-4_8

is usually less than 100 ms), alternating with fixation periods of about 100–600 ms (characterized by a relative stability of the eye). During saccades, we are relatively speaking blind. We see only during fixations, while the eye is holding still [1]. These eye movements usually take place in response to specific stimuli or mental processes. There exist specific devices for eye movement recording, namely the eye-trackers.

An overview on eye movement acquisition and applications is given in the remainder of the chapter, followed by a more in depth discussion on the use of eye movement analysis for biometric recognition.

1.1 Data Capture and Tracking

An eye-tracker is a device specifically designed for recording eye movements. The information retrieved includes the time, duration and location of eyes' fixations and the saccades between fixations. An eye-tracker measures changes in gaze direction with respect to the measuring system: head-mounted eye-trackers, measure eye-in-head angles (head position and eye-in-head direction are added to determine gaze direction); remote eye-trackers, measure the gaze angles directly [2]. Over the years, four categories of methods for eye-tracking have been developed: electro-oculography (EOG), scleral contact lens/search coil, photo-oculography (POG) or video-oculography (VOG), and combined pupil-corneal reflection [3].

The EOG method is based on the measurement of differences of the electrical potential detected by four electrodes placed on the skin just above, below, on the left and on the right of the eye. The electrodes detect the changes in voltage due to eyeball movements. Although this solution is very cheap, it is not very accurate. In addition, EOG is often noisier than other approaches, e.g. VOG, because the electric signals originating from facial muscle movements due to talking, smiling, frowning or eye-blinking, interfere with the eye movement signal recording [4].

The method based on scleral contact lenses and "search coil", on the other hand, is one of the most accurate, but also very invasive. The gaze direction is estimated via a mechanical (or optical) reference object mounted on a contact lens that covers both the cornea and the sclera. A stalk attached to the lens is then connected to a mechanical or optical device, such as a coil that measures the variation of an electromagnetic field.

VOG is based on image processing, a video sequence (or several photos) is analyzed to extract a set of eye characteristics during rotation and translation movements including the pupil's shape, the edge separating sclera and iris and corneal reflections caused by one or more (usually infrared) light sources.

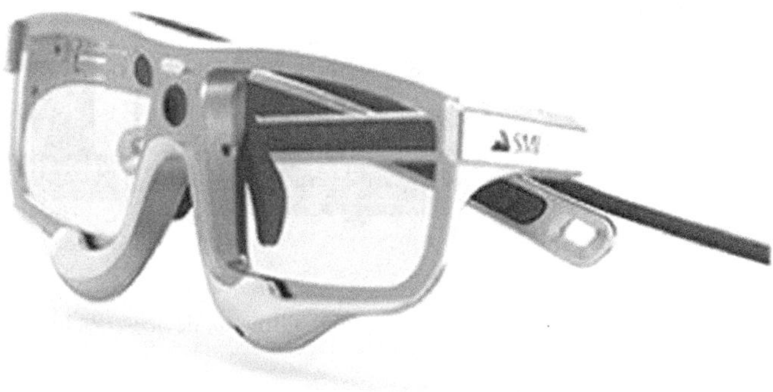

Fig. 1 Eye-tracking glasses. Picture from http://www.eyetracking-glasses.com/

The above mentioned approaches measure the position of the eyes relative to the head, while the combined use of corneal and pupillary reflections disambiguates head movement and eye rotations. This technique is based on the analysis of the light reflections (usually from infrared light sources) on the cornea and their relative position with respect to the pupil center. The eye-trackers based on this technology, usually integrate near infrared light sources that direct a beam of light on the observer's eyes. The reflection patterns produced on the cornea and the pupil, are captured by a sensor mounted on the device. The pupil is easily detected thank to the bright pupil phenomenon occurring when the eye is exposed to near infrared light [5, 6]. Advanced image-processing algorithms and a physiological 3D model of the eye are then used to estimate the position of the eye in space and the point of gaze with high accuracy. The gaze direction estimation obtained is very accurate [2].

1.2 Head-Mounted Versus Remote Eye-Tracker

Eye trackers can be divided in two major categories: head-mounted and remote eye-trackers. The advantage of the wearable solution is that little calibration is required, since the structure of the eye-tracker itself fixes the distance between the recording device and the eyes. The disadvantage is that the user is required to wear the device, making this approach intrusive and suitable only in a limited number of scenarios. Modern wearable eye-tracker integrate all components in a device very similar to normal eye glasses, see Fig. 1.

Remote eye-trackers are less intrusive, although nowadays solutions usually require to stabilize the head by chin and/or forehead resting. In this category of eye-tracker, we can further distinguish the monitor embedded solution from the standalone eye-tracker. The advantage of the monitor embedded eye-trackers, is the simple system set-up, since the screen is already in the correct position with respect to the eye-tracker. The most employed devices belonging to this category are from

Fig. 2 Remote eye-tracker used in combination with different devices. Pictures from https://theeyetribe.com/

Tobii AB.[1] As an example, the Tobii 1750 integrates all components into a 17′ LCD monitor (1280×1024 resolution) and the frame rates are about 50 Hz.

Standalone eye-trackers have the further advantage of possibly being used in combination with any kind of device, e.g. in combination with a TV, a tablet or a PC (see Fig. 2). They include the Tobii X120 and the EyeLink 1000. The latter provides two different setups: head supported (2000 Hz Monocular/1000 Hz Binocular) and remote/head free (500 Hz Monocular).

In [7], a 50 Hz infrared Dual Purkinje CRS eye-tracker (Cambridge Research Systems) has been employed. The eye-tracker illuminates the eye with an infrared beam and employs a complex combination of lenses and servo-controlled mirrors to continuously locate the positions of the first and fourth Purkinje images. These Purkinje images are formed by light reflected from surfaces in the eye.

Beyond the eye-trackers, other devices, originally developed for other purpose, have been used for tracking eye movements. A simple webcam has been employed in [8], while a device designed for PlayStation 3, namely the PlayStation Eye, with a temporal resolution of 75 Hz and average calibration accuracy of 1.1°. This has been possible thanks to specific software to process the output of "simple" video cameras for eye-tracking, such as the open-source ITU Gaze Tracker software [9]. Poole and Ball [5] provide a succinct but helpful introduction to eye-tracking in human-computer interaction and usability research. More recently, Duchowski [3] has authored an important and throughout monograph on eye-tracking methodology. Hornof and Halverson [10] discuss calibration for better accuracy and how to remove systematic errors in eye-tracking data.

1.3 Applications

Gaze analysis is employed in a wide number of fields, in particular for Human-Computer Interaction (HCI) to analyze the usability of computer interfaces. The idea is to analyze the saliency map, i.e. the map representing the visual saliency of a

[1] http://www.tobii.com/.

corresponding visual scene [11], in order to better arrange the content of a web page. The same analysis can be applied in other contexts, as for example to organize the products on the shelves in a supermarket.

In [12], eye-tracker technology is used to assess the usability of the University of Zilina website. A similar approach is adopted in [13] to study the banner positioning of online newspapers. In [14] demographics data and behavioral data, namely gender recognition and gaze analysis, are combined to investigate the possibility of providing more adaptive and interactive advertising. Cantoni and Setti [15], present a preliminary qualitative analysis, of the image boundary detection performed by human gaze, by recording eye movements during the observation of an illustration of two combined entities. Activity Recognition, in particular of the following tasks: read, browse, write, watching a video, and copy; is obtained by the use of a EOG based system in [16], with an average precision of 76.1% and recall of 70.5%. Visual processes occur both overtly and covertly, and it is just the covert modality that is strictly connected to a person's cognitive and psychological processes [17]. In [18], eye movement analysis is used for diagnostic assessment of disorders of consciousness. In [19], a novel eye-gaze driven technique for surgical assessment and workflow recovery is presented. An example of more recreational application of eye-tracking is given in [20], where eye-tracking is used to facilitate aiming in computer gaming. In [21], an eye-tracker is employed to capture the iris for real-time continuous recognition.

2 Eye Movement Analysis in Biometrics

Covert attention is strictly correlated to eye behavior and the orienting mechanisms used to make inferences about spatial layout and object identity. Attention mechanisms, which vary from person to person (due to personal background, physiological make-up, and training), can therefore be gainfully used for biometric identification [22]. Eye movement recognition systems are mainly composed by three components: the observer, an eye-tracker (or any recording device employed for the same purpose), and a stimulus used to trigger eye movements (Fig. 3).

It has been demonstrated that systematic tendencies in viewing behavior during scene exploration exist. For instance, saccades are known to follow a positively skewed, long-tailed distribution, and to be more frequently initiated in the horizontal or vertical directions. In [23], it is showed that these viewing biases are not universal, but are modulated by the semantic visual category of the stimulus.

A wide number of different stimuli have been investigated over the years. One of the most employed stimuli is the so-called "jumping point." It can be a LED or a point displayed on a screen. The point appears (or the LED lights up) in predefined positions, e.g. at the intersection points of a 9×9 grid, and the human observer is asked to follow with her gaze the jumping point. Since the fixation positions are fixed, i.e. where the jumping point appears, the main information exploited here is related to saccades. The second traditional stimulus is the text. The observer is involved

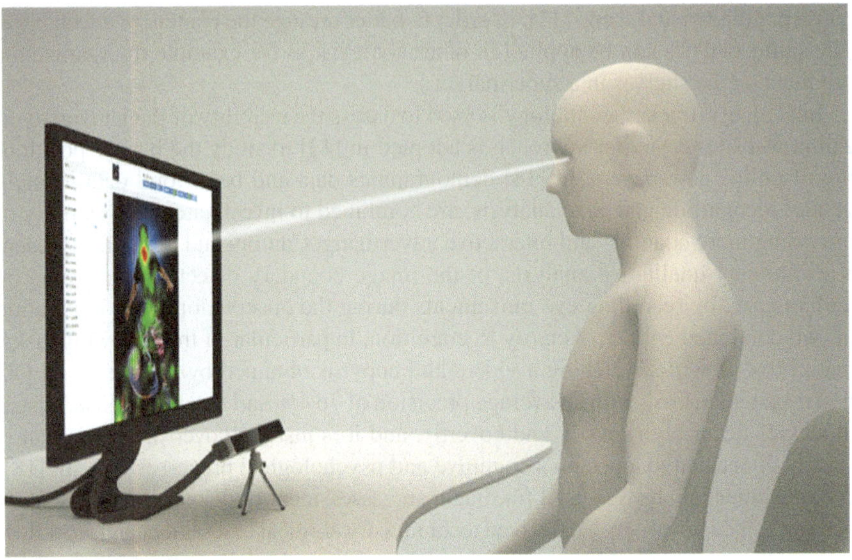

Fig. 3 Eye movement recognition system set-up. Picture from https://theeyetribe.com/

in a text reading task; the text can be displayed on a screen for better eye-tracker calibration. The reading task provides that the observer scans the text line after line. Thus, the scanning path is somehow defined. However, the horizontal scan path can vary (e.g. in terms of speed) from person to person. The unconscious eye movements (saccades) are mainly exploited for the eye movement analysis.

Even simpler than the jumping point, the static cross stimulus has been successful and achieved 90% correct classification [24]. It displays a small cross at the center of the screen and records the saccades originated from the observer's eyes.

Another way used to trigger eye movements would require following a moving object. This stimulus is usually realized by displaying the moving object (e.g. a cross, a circle, or in general a geometrical shape) on a screen. The above mentioned stimuli share the fact that the gaze path is defined by the stimulus itself. The eye movements recognition systems based on those kind of stimuli leverage on saccadic movement analysis to perform biometric recognition. An illustration of the above mentioned stimuli is given in Fig. 4.

Grey-scaled or colored images have been successfully used as stimuli in eye movement analysis. Two possible modalities can be adopted: free viewing and task based observation. In the free viewing of scenes, the observer can freely look at the displayed image for a determined time. In the task based modality the observer accomplishes a predetermined task, e.g. finding an object in the image, see the photographs and decide if she knows the face by pressing yes/no button [25].

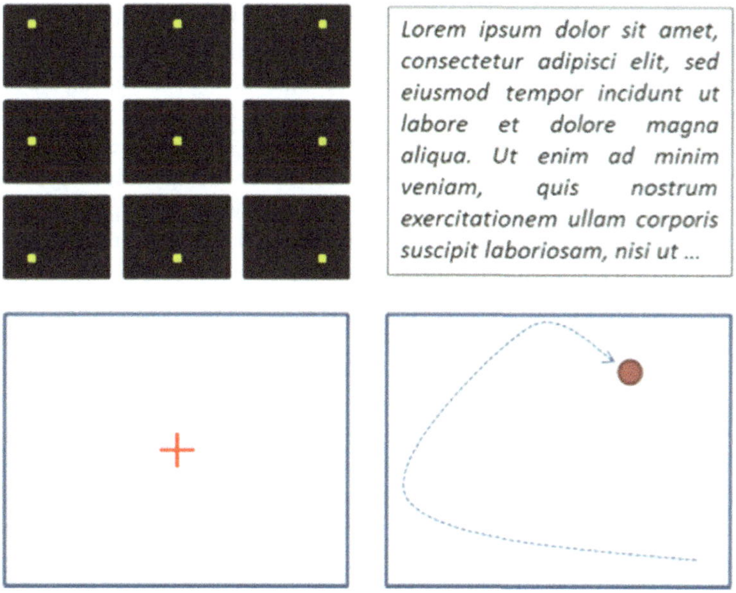

Fig. 4 Eye movement stimuli: top-left, jumping-point; top-right, text; bottom-left, static cross; bottom-right, moving object

Video provides complex information about the observer's gaze behavior because they display dynamic scenes and can trigger emotional reactions. Images and videos are mainly employed as stimuli when not only the saccadic movements have to be analyzed but also the fixations points, i.e. the gaze path. In fact, some works, e.g. [7], only rely on the fixation point coordinates.

2.1 Head-Mounted Eye-Tracker Based Approaches

The use of head-mounted eye-tracker is preferable when the saccadic movements are used for biometric recognition. Since saccades are very fast, the recording process require more controlled conditions and accurate eye-tracking. Moreover, simple stimuli are more suitable, since saccades are unconscious eye movements that continuously occur even whilst fixating a single point. In fact, the most employed stimuli are the jumping point, static cross, and text. The features that can be extracted from the saccades include amplitude, accuracy, latency and maximum velocity, acceleration, and deceleration.

Kasprowski and Ober [26], survey previous research on eye movements of subjects while following a jumping point on the screen and confirmed that behavioral biometrics characteristic of data-driven eye movements (e.g., saccades), are difficult to spoof, and can be leveraged for subject identification. Cuong et al. in [27], using the Ober2 eye-tracker and the jumping point as stimulus, comparatively evaluated the Mel-Frequency Cepstral Coefficients (MFCCs) for encoding various eye-tracking features for classification. They further showed that MFCC compare favorably to using any of the Fourier transform, cepstrum, or raw representations. Zhang and Juhola in [28] differentiated between different eye movement types such as saccade, nystagmus, smooth pursuit and vestibulo-ocular reflex eye movements, and noted that saccades are predominant. More recently, Juhola et al. [29] compared eye movements measured by EOG and movement signals given by VOG system, using the jumping point stimulus, with EOG performing the best (Identification Rate of 97%).

Another interesting application of eye-tracking is for the realization of continuous intrusion detection methods. The ultimate goal here is to identify impostors as those users showing "abnormal" behavior, which is different from behavior shown by legitimate users. Substantial discrepancies in eye activities may alert the system and possibly lock-up its further use to prevent nefarious activities. Holland and Komogortsev [30] have presented a variety of eye movement-based biometric features acquired while the user is reading, e.g., fixation count, average fixation duration, average saccade amplitudes, average saccade velocities, average saccade peak velocities, the velocity waveform, scan path length, scan path area, scan path inflections, regions of interest, the amplitude-duration relationship, the main sequence relationship, and the pairwise distance between fixations, and their capability to correctly differentiate between subjects. The same authors further comment that the "eye movements are uniquely counterfeit resistant due to the complex neurological interactions and the extraocular muscle properties involved in their generation". In their experiments, they employed the Eye Link II head-mounted eye-tracker and the text as stimulus.

2.2 Remote Eye-Tracker Based Approaches

Remote eye-trackers have the great advantage of capturing the eye movements at a distance, allowing less intrusive eye-tracking. Bednarik et al. [24] exploited various kinds of eye data, including pupil size and their dynamics, gaze scan speeds, and distances of infrared reflections on the eyes. They also investigated the use of a number of different stimuli, including text and moving object, with the static cross stimulus performing the best and achieving an identification rate of 90%. The eye-tracker employed in the experiments is a Tobii ET-1750. Deravi and Guness [8] recorded gaze trajectory data of human subjects while looking at some images for about five seconds each. Gaze durations, pupil positions, pupil sizes and gaze points were measured and then fed to feature selection algorithms. An interesting aspect of their work is the use of a simple webcam for eye movements recording.

Holland and Komogortsev [31] assessed the effects of eye-tracking specification and stimulus presentation on the biometric feasibility of complex eye movement patterns. They tested three different remote eye-tracker, namely the Tobii TX300, the Eye Link 1000, and the Playstation Eye. Through two experiments, they examined the effects of varied temporal resolution, stimulus type, and spatial accuracy. The authors found that, for biometric purposes, eye-trackers with spatial accuracy of less than $0.5°$ and a sampling frequency greater than 250 Hz are most suitable.

Eye-movement analysis can be used alone or in combination with other biometrics. A multi-modal data fusion is presented by Komogortsev et al. in [32], in order to provide better recognition rates. The approach combines and exploits three different eye features, that are the eye anatomical properties (using oculomotor plant characteristics—OPC), visual attention strategies (using Complex Eye Movement patterns), and the physical structure of the iris. Results indicate that combining three ocular traits improves accuracy, yielding a HTER (Half Total Error Rate) of 19%. The EyeLink 1000 eye-tracker was employed in combination with the jumping point stimulus.

Kinnunen et al. in [33] considered video clips as stimuli and employed the Tobii X120 eye-tracker. They proposed a task-independent rather than a task-dependent scenario as it is usually the case, in which short-term eye gaze direction is used to build feature vectors modeled by means of Gaussian mixtures. Eye movements, recorded while watching a 25 min long video, are described as a histogram of all the angles the eye spans during a certain time period. The results reported suggest that in eye movements there are person specific features that can be modeled in a task-independent way. Videos were employed also by Liang et al. [34] for a biometric identification method achieving an identification rate of 82%.

The work by Rigas et al. [7] focuses on the free observation of face images. Participants, lacking prior experience with an eye-tracking device, watch ten photos depicting human faces for four seconds each. The similarity of spatial distributions of fixation points was quantified by means of a graph theoretic measure based on the multivariate generalization of Wald-Wolfowitz runs test. The results obtained indicate the existence of characteristic patterns that can be potentially exploited to discriminate among different subjects. 15 volunteers participate in the study, 12 males and 3 females. More recently, Rigas et al. [35] investigated the efficiency of a multi-stimulus and multi-biometric fusion scheme. Diverse visual stimuli (jumping-point-of-light, text, and video), were analyzed with different and appropriated techniques and efficiently combined via a weighted fusion scheme. The best result achieved is a Rank-1 Identification Rate of 88.6% and an Equal Error Rate (EER) of 5.8%. The technique has been tested on a large database of 320 subjects.

In [36], Cantoni et al. extend their previous work [37] by developing a novel graph-based technique, namely GANT, that exploits the information relative to the time spent by the user gaze in a specific area of the face. Since the observed faces are all different, the position of a given face landmark, e.g. the pupil center, can be slightly different from a picture to another. Thus, the fixation point coordinates are normalized using the distance between the eyes and the distance between eyes' middle point and the mouth of the observed face. This assures that all the points from different images relative to the same face area (e.g., left eye) will fall around the same point (corresponding to the normalized position of the left eye) in the normalized cloud of points. The cloud of fixation points is then subdivided with a grid. A graph representing the number of fixation point in each cell, the sum of the duration of the fixation points in each cell, and how many times the user's gaze moved from a cell to another, is built. It is interesting to note that this graph-based approach can be applied on any kind of image. The graph obtained is represented by matrices, with one matrix for each feature (density, duration and arcs). To compare different observers based on densities, durations or arcs, the distance between couples of matrices of the same feature is measured through the Frobenius norm of the matrices' difference.

2.3 Eye Movement Analysis for Demographic Categorization

The aim of biometric recognition is to uniquely identify a person. However, biometrics can also be used for other purposes such as demographic categorization, e.g. to assess the age or the gender of a person as one of the steps needed for dynamic stratification leading to partitioned galleries suitable for ultimate identification. An interesting question worth additional study would test if gaze analysis supports gender identification or age estimation. Some steps in this direction have been made in [38] by Galdi et al., where gaze features extracted by the GANT technique [36], are fed to two learning methods, namely Adaboost and support vector machines (SVM). The experiments conducted show that SVM-based protocol obtained better results (about 60% of correct classification) than the Adaboost-based when applied on gender categorization and vice versa for age categorization (about 55% of correct classification). Further innovation and development is required to make gaze analysis a viable alternative for demographics categorization, on its own or together with other biometrics.

3 Conclusion

This chapter discusses about the different kinds of approaches in eye movement analysis developed over the years. It focuses mainly on the use of eye movements for biometric recognition. The most used techniques and eye-trackers are reviewed and the pros and cons on their usage are illustrated. As shown in the previous sections, many kinds of stimuli, eye-trackers and techniques have been investigated but further studies are required to identify the best setup for an eye movement recognition system with a good trade-off between accuracy and intrusiveness. The ultimate goal of a recognition system, is to perform the identification in a transparent, fast and accurate way. Biedert et al. [39], studied an intrusion detector for active authentication on learning effects, while assuming that legitimate users become progressively more accustomed with the execution of certain tasks. In some experiments, the students tested were asked to perform common tasks such as check for emails, requiring to read something on the screen. The authors conclude that "gaze behavior is most discriminative with static and repetitive stimuli." The latter can be considered a transparent application of eye movement analysis, since users are only required to perform some tasks they are used to.

References

1. R. Dodge, Visual perception during eye movement. Psychol. Rev. **7**(5), 454–465 (1900)
2. V. Cantoni, M. Porta, in *Eye Tracking as a Computer Input and Interaction Method*, ed. by B. Rachev, A. Smrikarov, in Proceedings of the 15th International Conference on Computer Systems and Technologies (CompSysTech'14) (ACM, New York, NY, USA), pp. 1–12 (2014)
3. A.T. Duchowski, *Eye Tracking Methodology—Theory and Practice*, 2nd edn. (Springer, London, 2007)
4. Y. Zhang, Biometric Verification of a Subject Based on Data Mining of Saccade Eye Movement Signals (Tampere University Press). URN: ISBN: 978-951-44-9356-0 (2014)
5. A. Poole, L.J. Ball, in *Eye Tracking in Human-Computer Interaction and Usability Research: Current Status and Future Prospects*, ed. by C. Gahoui. Encyclopedia of Human-Computer Interaction (Idea Group, 2006), pp. 211–219
6. Z. Zhu, K. Fujimura, Q. Ji, Real-time eye detection and tracking under various light conditions, in *Proceedings of the 2002 Symposium on Eye Tracking Research & Applications (ETRA'02)* (ACM, New York, NY, USA), pp. 139–144 (2002)
7. I. Rigas, G. Economou, S. Fotopoulos, Biometric identification based on the eye movements and graph matching techniques. Pattern Recogn. Lett. **33**(6), 786–792 (2012)
8. F. Deravi, S.P. Guness, Gaze trajectory as a biometric modality, in *Biosignals 2011, Paper 128* (2011)
9. J.S. Agustin, H. Skovsgaard, J.P. Hansen, D.W. Hansen, Low-cost gaze interaction: ready to deliver the promises, in *Proceedings of CHI*, pp. 4453–4458 (2009)
10. A.J. Hornof, T. Halverson, Cleaning up systematic error in eye tracking data by using required fixation locations. Behav. Res. Meth. Instrum. Comput. **34**(4), 592–604 (2002)
11. O. Le Meur, Z. Liu, Saccadic model of eye movements for free-viewing condition. Vision. Res. **116**, 152–164 (2015)
12. T. Kvasnicová, I. Kremenová, The use of eye tracking to assess the usability of university website, in: *Proceedings of the CBU International Conference on Innovation, Technology Transfer and Education, 2015, Prague, Czech Republic* (2015)
13. V. Cantoni, M. Porta, S. Ricotti, F. Zanin, in *Banner Positioning in the Masthead Area of Online Newspapers: An Eye Tracking Study*, ed. by B. Rachev, A. Smrikarov, Proceedings of the 14th International Conference on Computer Systems and Technologies (CompSysTech'13) (ACM, New York, NY, USA), pp. 145–152 (2013)
14. W. Zhang, M.L. Smith, L.N. Smith, A. Farooq, Gender and gaze gesture recognition for human-computer interaction. Comput. Vis. Image Underst. **149**, 32–50 (2016)
15. V. Cantoni, A. Setti, in *Image Segmentation Scrutiny by Eye Tracking*, Proceedings of the 2014 Tenth International Conference on Signal-Image Technology and Internet-Based Systems, pp. 554–559 (2014)
16. A. Bulling, J.A. Ward, H. Gellersen, G. Troster, Eye movement analysis for activity recognition using electrooculography. IEEE Trans. Pattern Anal. Mach. Intell. **33**(4), 741–753 (2011)
17. L. Itti, C. Koch, A saliency-based search mechanism for overt and covert shifts of visual attention. Vision. Res. **40**, 1489–1506 (2000)

18. W.K.-C. Ting, J.L. Perez Velazquez, M.D. Cusimano, Eye movement measurement in diagnostic assessment of disorders of consciousness. Front. Neurol. **5**, 137 (2014)
19. A. James, D. Vieira, B. Lo, A. Darzi, G.-Z. Yang, in *Eye-Gaze Driven Surgical Workflow Segmentation*, ed. by N. Ayache, S. Ourselin, A. Maeder. MICCAI 2007, Part II, LNCS 4792 (Springer, Berlin, Heidelberg, 2007), pp. 110–117
20. J. Leyba, J. Malcolm, *Eye Tracking as an Aiming Device in a Computer Game*. Technical report, Clemson University, Clemson (2004)
21. K. Jain, S.C. Dass, K. Nandakumar, Soft biometric traits for personal recognition systems, in *Proceedings of the International Conference on Biometric Authentication, LNCS*, vol. **3072**, pp. 731–738 (2004)
22. C. Galdi, M. Nappi, D. Riccio, H. Wechsler, Eye movement analysis for human authentication: a critical survey. Pattern Recogn. Lett. Available online 8 November 2016, ISSN 0167-8655 (2016)
23. O. Le Meur, A. Coutrot, Introducing context-dependent and spatially-variant viewing biases in saccadic models. Vision. Res. **121**, 72–84 (2016)
24. R. Bednarik, T. Kinnunen, A. Mihaila, P. Franti, in *Eye-Movements as a Biometric*, ed. by H. Kalviainen, J. Parkkinen, A. Kaarna. Image Analysis, LNCS, vol. 3540 (Springer, Berlin, Heidelberg, (2005)), pp. 780–789
25. P. Kasprowski, K. Harezlak, The second eye movements verification and identification competition, in *Proceedings of the 2014 IEEE International Joint Conference on Biometrics (IJCB)*, pp. 1–6 (2014)
26. P. Kasprowski, J. Ober, Eye movements in biometrics, in *Proceeding of the 2004 International Workshop on Biometric Authentication (BioAW 2004)*, pp. 248–258 (2004)
27. N.V. Cuong, V. Dinh, L.S.T. Ho, Mel-frequency Cepstral coefficients for eye movement identification, in *Proceedings of the IEEE International Conference on Tools with Artificial Intelligence (ICTAI 2012)*, pp. 253–260 (2012)
28. Y. Zhang, M. Juhola, On biometric verification of a user by means of eye movement data mining, in *Proceedings of IMMM 2012, The Second International Conference on Advances in Information Mining and Management*, pp. 85–90 (2012)
29. M. Juhola, Y. Zhang, J. Rasku, Biometric verification of a subject through eye movements. Comput. Biol. Med. **43**, 42–50 (2013)
30. C.D. Holland, O.V. Komogortsev, Biometric identification via eye movement scan-paths in reading, in *Proceedings of the 2011 International Joint Conference on Biometrics (IJCB 2011)*, pp. 1–8 (2011)
31. C.D. Holland, O.V. Komogortsev, Complex eye movement pattern biometrics: the effects of environment and stimulus. IEEE Trans. Inf. Forensics Secur. **8**(12), 2115–2126 (2013)
32. O.V. Komogortsev, A. Karpov, L. Price, C. Aragon, Biometric authentication via oculomotor plant characteristics, in *Proceedings of the IEEE/IARP International Conference on Biometrics (ICB)*, pp. 1–8 (2012)
33. T. Kinnunen, F. Sedlak, R. Bednarik, Towards task-independent person authentication using eye movement signals, in *Proceedings of the 2010 Symposium on Eye-Tracking Research & Applications (ETRA'10)*, pp. 187–190 (2010)
34. Z. Liang, F. Tan, Z. Chi, Video-based biometric identification using eye tracking technique, in *Proceedings of the 2012 IEEE International Conference on Signal Processing, Communication and Computing (ICSPCC)*, pp. 728–733 (2012)
35. I. Rigas, E. Abdulin, O. Komogortsev, Towards a multi-source fusion approach for eye movement-driven recognition. Inf. Fusion **32**(B), 13–25 (2016). ISSN 1566-2535
36. V. Cantoni, C. Galdi, M. Nappi, M. Porta, D. Riccio, GANT: Gaze ANalysis Technique for human identification. Pattern Recogn. **48**(4), 1027–1038 (2015)
37. V. Cantoni, M. Porta, L. De Maio, R. Distasi, M. Nappi, Towards a novel technique for identification based on eye tracking, in *Proceedings of the 2012 IEEE Workshop on Biometric Measurements and Systems for Security and Medical Applications (BIOMS)*, pp. 1–4 (2012)

38. C. Galdi, H. Wechsler, V. Cantoni, M. Porta, M. Nappi, Towards demographic categorization using gaze analysis. Pattern Recogn. Lett. **82**(Part 2), 226–231 (2016). ISSN 0167-8655
39. R. Biedert, M. Frank, I. Martinovic, D. Song, in *Stimuli for Gaze Based Intrusion Detection*, ed. by J.J. Park at al. Future Information Technology, Application, and Service. Lecture Notes in Electrical Engineering, vol. 164, 757–763 (2012)

Biometric Legged Locomotion Fundamentals

Rami Alkhatib, Mohamed Diab, Christophe Corbier and Mohamed El Badaoui

Abstract This chapter can be classified in the category of Behavioral Biometrics. The chapter is entitled, "Biometric Legged Locomotion Fundamentals". The general purpose approach is to either analyse and identify pathologies, or to identify individuals from their gait ground reaction force signals. Based on a classical biometric scheme, acquisition, analysis, pre-processing and feature extraction are discussed.

1 Introduction

Biometric Legged Locomotion makes use of ground reaction force signals of individuals for recognition purposes. Those signals are collected by an array of force transducers. Then observing their pattern forms the key towards individual classification. As human gait is still challenging due to its complex high degree of freedom, this chapter contributes to the better understanding of the vertical ground reaction forces component and their perspective measurement systems. We will be revisiting biomechanical gait concept from literature to inspire new machine learning algorithms. Thus this chapter will combine the acquisition systems, analysis, pre-processing, and feature extraction that can be best employed in classifiers for individual recognition based on their gait ground reaction force signals. Throughout each section, new horizons and challenges are introduced over and done with critical questions towards better gait biometric technologies.

Biometrics can be classified into behavioural and the physical systems. Fingerprint, iris/retina, and face recognition are examples of physical biometric systems that the individual would have a prior knowledge before using it. In addition, they require a physical contact or being at a short distance from the measuring system. Such biometrics can be tampered with. While gait identification is a behavioural

R. Alkhatib (✉) · M. Diab
Rafik Hariri University , Damour, Lebanon
e-mail: khatibrh@rhu.edu.lb

C. Corbier · M. E. Badaoui
University Jean-Monnet, Saint-Etienne, France

A. Nait-Ali (ed.), *Biometrics under Biomedical Considerations*,
Series in BioEngineering, https://doi.org/10.1007/978-981-13-1144-4_9

biometric that can be apprehended in either physical contact or capturing from an acceptable distance without a prior knowledge of the tracked individual. However, gait parameters can be affected by the wearing shoes, vision, disease, emotional status, aging and many more, and this indicate that gait can vary with time. As a result, such systems must be updated continuously with the restructured parameters for each individual. Though, the power of gait in multi-biometric system by combining gait with other biometrics like face recognition is reported to be efficient [1]. If verification is achieved then the possibility of recognition is feasible. A critical question rises if whether we can find some fixed parameters that characterizes gait as unique?

- To assess the effectiveness of the Biometric legged locomotion system, we ultimately hope at one point to achieve all the following properties.
- Universality: Individuals use their biological motion that allow them to move from one place into another. Still few number of persons (e.g. paralyzed) can't be enrolled in such biometric system and this limitation covers all biometric systems.
- Uniqueness: Individuals have their unique gait signature.
- Permanence: Finding invariant gait parameters over time that characterize each individual is still under development and research.
- Collectability: Various measuring technologies of gait parameters are introduced as shown in the section of measurement systems.
- Acceptability: VGRF will be collected with no prior knowledge of individual and will not affect sociological situation of individuals. In addition this will not raise concerns like unhygienic, harm of a biometric system, religious objections and so on.
- Circumvention: The level at which the biometric system can be attacked or fooled by means of fraudulent approaches.

This chapter will focus on the vertical component of the force exerted by ground on foot during walking (VGRF: Vertical Ground Reaction Force). For instance, [2] gave a good summary for the work done in literature by using footsteps in biometrics between 1997 and 2006 as shown in Table 1. Table 1 shows different sizes of databases being used with a relatively good identification rate. This is accompanied with different features being used like cadence, step length, and power spectral density. Moreover, different classifiers like Euclidian Distance, multilayer-perceptron neural network and Nearest Neighbour are used [2]. It is also reported that the recognition performance increase with the increase of training set of the footsteps.

To be more precise, study [3] used a combination of geometric and wavelet features of GRF profile for each step as follows:

$$GRF_T[T] = \frac{1}{S}\sum_{i-1}^{S}\left(\sum_{t=0}^{T}(S_i[t])\right)$$

where

S number of pressure sensors

Table 1 Footsteps in biometrics from literature

Group/year	Database (total steps/persons)	Technology	Features	Classifier	Results
The ORL active floor/1997	300 steps, 15 persons	Load cells	Subsampled GRF	HMM	ID rate: 91%
The smart floor (USA)/2000	1680 steps/15 persons	Load cells	Geometric feat from GRF	NN	ID rate: 93%
ETH Zurich/2002	480 steps/16 persons	Piczo force sensors	Power spectral density	Euclidean distance	Verif.EER: 9.4%
Ubifloor (Korea)/2003	500 steps/10 persons	Switch sensors	Position of several steps	Multilayer-perceptron neural network	ID rate: 92%
EMFi floor (Fin-land)/2004	440 steps/11 persons	Electro mechanical film	Geometric feat from GRF	Learning vector quantization	ID rate: 70%
Southampton University (UK)/2005	180 steps/11 persons	Resistive (switch) sensors	Stride length, stride cadence and heel-to-toe ratio	Euclidean distance	ID rate: 80%
Southampton University (UK)/2006	400 steps/11 persons	Load cells	Geometric feat, from GRF	NN	ID rate: 94%

I the number of the sensor being labelled
Si[t] GRF (in newton) measured at a given from ith sensor.

In this way the feature generated preserve in its components spatial parameters opposed to averaging the GRF profiles as follows:

$$GRF_{avg}[t] = \frac{1}{S}\sum_{i-1}^{S}(S_i[t])$$

Then the first moment statistical features like standard deviation, mean, minimum and maximum can be easily extracted from both time domain and the continuous wavelet transform (CWT) [3]. The study was able to obtain an error rate of $16.3\% \pm 0.7\%$ using Random Forest classifier by implementing the wavelet analysis.

Throughout this chapter, VGRF is our concern. A new database is introduced to be used in legged locomotion biometrics. The potential of this database is concerned in having VGRF from both normal and patient subjects. A deep analysis is then provided and related to clinical aspects. We will end up this chapter by introducing

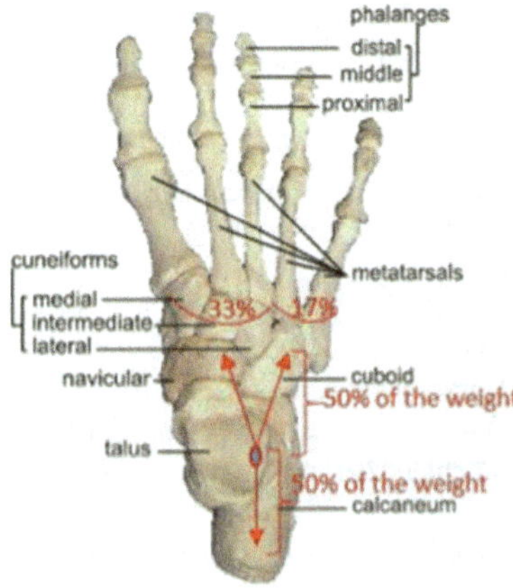

Fig. 1 Dorsal view of foot with weight distributions

some efficient features that are rarely used in literature. A conclusion is then made by highlighting the new perspectives in gait recognition based on VGRF signals.

2 Anatomy of Foot

The foot is a combination of Bones, Joints, Muscles and Skin. Each foot consist of 26 bones, two are in the rear foot, 5 in midfoot and 19 in forefoot. Figure 1 shows a dorsal view of the foot [4]. There are 33 joints in the foot. Joints add several degrees of freedom allowing multiple directions of movement of the foot. Those joints also serves as shock absorbers. There is also more than 100 tendons, ligaments and muscles. This all indicate the complex structure of foot.

Figure 1 also indicates weight distribution during a stance phase while the subject is standing. The talus of a foot receives a part of weight through the talocrural joint that connects the talus to the fibula and tibia. Then this weight is equally divided so a portion passes backward to be supported by calcaneus while the other portion passes forward to other tarsal bones [4] as in Fig. 1. It is worthy to mention, that the momentum around centre of mass is negligible which avoid rotational motions of body and this is a good indicator that the resultant force passes through centre of mass.

3 Gait Factors and Terminology

In order to achieve an efficient and balanced gait while maintaining upright posture in addition to bearing the weight of the body accompanied with brake impulse and

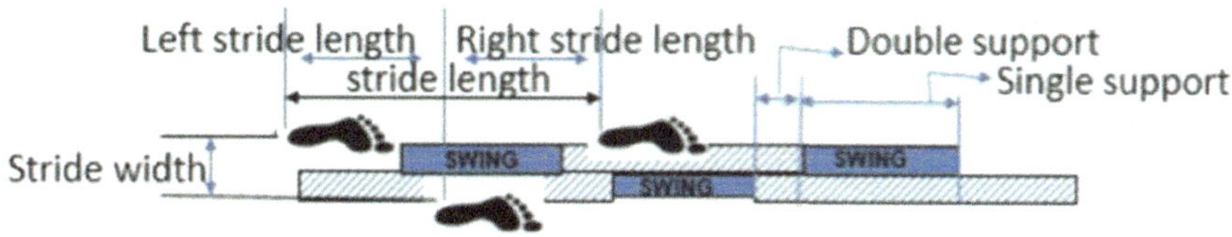

Fig. 2 Phases of walking cycle

Table 2 Temporal and spatial gait parameters are in the first two columns respectively. 3rd column shows factors that interfere in the measured parameters

Temporal parameters	Spatial parameters	Interfered parameters
Stance time	Stride length	Age, height, sex
Single support time	Stride width	Shape and size of bones
Double support time	Step length	Muscles' strength
Swing time	Foot angle placement	Joint mobility clothes, shoes…
Stride time		Psychological status
Step time		
Cadence speed		

propulsive stage of gait, one must consider the determinants and the terminology of gait.

Human gait is a manner or style of walking and a suited medical term to describe human locomotion [5]. Walking gait defined as sequences of repetitive cyclic gestures as shown in Fig. 2. It consists of both periodic movement of each foot from one position of support to a next position of support by sufficient ground reaction forces applied through the feet [6]. This bipedal locomotion is enhanced by different parts of body like bones, muscle, nervous system and others.

Thus the three main phases of gait can be summarized as stance phase, swing phase and double support phase. The time and distance parameters that can be extracted are shown in the first two columns of Table 2 [7]. The third column explores some supplementary aspects that can affect the time and distance factors.

Therefore Human locomotion is determined by displacement over time. Scientists focus on the displacement of the center of gravity (COG). Moreover, human movement done is three planes: sagittal, frontal, and transverse. As most of the movement will occur in the sagittal plan, this chapter will focus only on the VGRF component. Factors that can alter the gait in general and VGRF in particular vary from Injury [8] to cloth changing and backpacks [9], Shoe type.

4 Databases for Gait Recognition

In order to discriminate people based on the way they walk, a set of datasets are collected for this purpose. The database must consist of enough number of participants

Table 3 Datasets used in gait recognition

Database	9 sub	Descriptors	URL
CMU MoBo	25	6 views, 2 speeds, 2 slopes, baggage	http://www.hid.ri.cmu.edu
University of Maryland database	44	Four views, T-shape pathway, walking at different angles	http://degas.umiacs.umd.edu/hid
Georgia tech database	20	4 speeds, indoor and outdoor gaits, different view angles	http://www.cc.gatech.edu/CPL/PROJECT
MIT database	25	Time, several gaits for same subjects with different cloths, two backgrounds	http://www.ai.mit.edu/PROJECTS/GAIT/
USF	71	Outdoor, elliptical path, grass and concrete surfaces	http://marathon.csee.usf.edu/GaitBaseline/
University of Southampton database	11 8	Indoors and outdoors and treadmill, different view angles, footwear, clothes	http://www.gait.ecs.soton.ac.uk/database/
Arizona State University database	5	3d GRF, joint-angles trajectories	http://cubic.asu.edu/projects/DRAG.html
CASIA gait part D	20	Foot scan: pressure	http://kylezheng.org/gait-recognition/
OU-ISIR Biometric database	10 35	Treadmill, speed, inertial sensor, similar actions	http://www.am.sanken.osaka-u.ac.jp/BiometricDB/GaitLP.html

associated with some other criteria that could affect their gait. Table 3 summarizes some of those databases.

In order to serve the goal of this chapter, a database is used in which eight sensors (Ultraflex Computer Duyno Graphy, Infotronic Inc.) were placed underneath each of the subject's feet to collect VGRF in Newton as a function of time, while each subject walked at his/her free speed of gait, back and forth for two minutes at their self-selected pace level ground without any secondary task in a well-lit, obstacle free, 25-m long, 2-m wide corridor. the sensors location inside the insole as lying approximately at the following (X, Y) coordinates measured as a person is comfortably standing with both legs parallel to each other are shown in Fig. 3. The origin (0, 0) is located between the legs and the person is facing towards the positive side of the Y axis . The sampling rate is 100 Hz.

This dataset were captured from 29 patients with idiopathic PD (disease stage was 2–3 on the Hoehn and Yahr scale, mean age: 66.3 years; 63% men), and 18 healthy controls (mean age: 66.3 years; 55% men). Figure 4 displays a sample of the data captured by the array of the eight sensors in addition to their summation. It displays a sample of the data for both Normal and Parkinson gaits. This database has been drawn from physionet gait database [10]. Subjects provided written informed consent prior to performing the experiment [10].

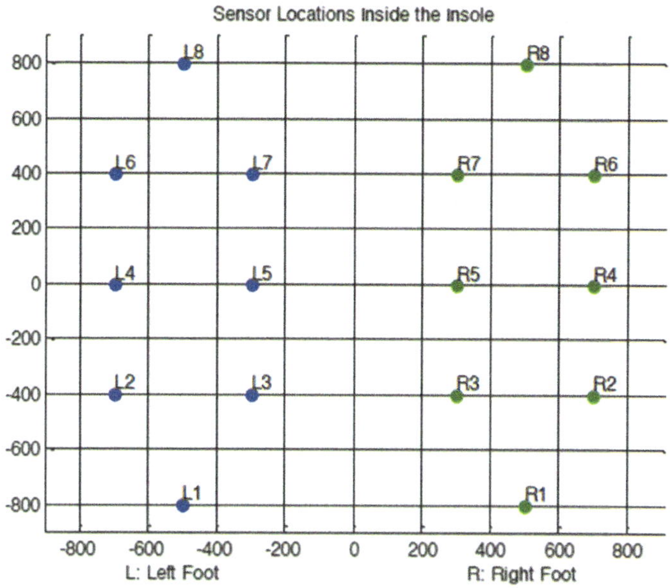

Fig. 3 Sensor's position as distributed underneath both feet

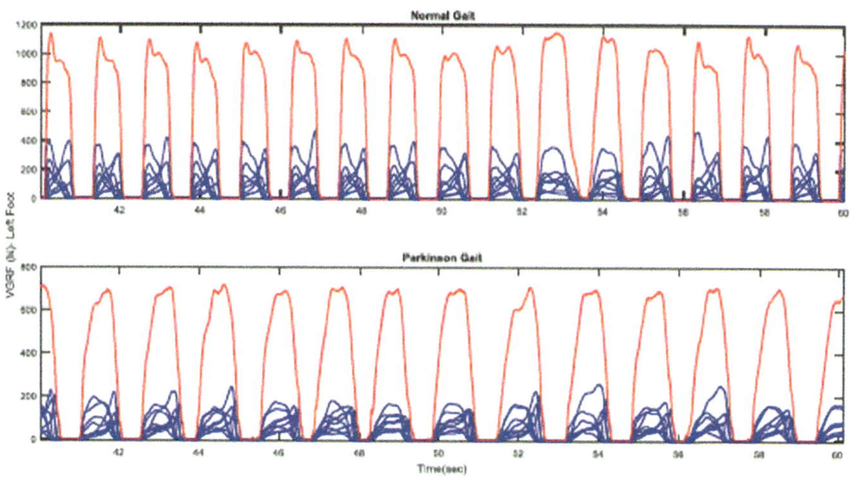

Fig. 4 Sample of VGRF data captured by the eight sensors underneath the left foot. The red curve represents the resulting signal of their summation. The first row corresponds to normal gait and second raw corresponds to Parkinson gait subject

5 Gait Measurement Systems

This section will tackle how the gait measurement systems available without dwelling too much on how they are being instrumented and how they are connected to sensors though this is very important step before handling the data. However, it is important to state that the level of processing of gait signals is highly based on the level of mastering gait signals. Deciding the level of processing of gait signals will simplify and indicate the amount of data to be gathered from a sensor.

1. Vision approach: For example this done using Multi-camera gait-acquisition system: where a camera captures video of the subject. Then processing the video by subtracting the background will make it easier to extract features by for example Silhouette Representation. However, this method is still facing some complications like the view from different angles will complicate the analysis in addition to the different environmental backgrounds. Moreover, still the end user show objection and consider it as a kind of privacy intrusion.
2. Acoustic approach by collecting the sound generated by footsteps during walking [20].
3. Biometric height sensors that can be employed in identifying few residents at home. This can be done with ultrasonic distance sensor mounted at the entrance of each room [11]. This is best used for elderly tracking.
4. Wearable approach: different methods are available and some of them are presented here:
    - Sensory loaded shoes which collect the pressure signals while walking.
    - Accelerometer: User-Authentication using phone's accelerometer [12] which requires a continuous record of person's acceleration data.
    - Gyro sensor: Measuring the rotation and tilting of a specific location on the body like arm, neck… while walking.
5. Floor based sensor [13]: for instance the force plate sensor are widely used to collect pressure signals. For instance, the hardware in [13] is made up of three components load cells, steel plate in addition to data acquisition hardware. Figure 5 illustrates the hardware.

We will focus on VGRF measured by a pressure sensor as to flow in coherence with the database between our hands. It is important to highlight their treatment. For instance, various researches rely on summation of force signals being measured from different sensors. Figure 6 shows the Hilbert spectrum of the original VGRF. The U-shaped pattern characterizes all Hilbert spectrum plots of subjects in time frequency evolution as shown in the considered example in Fig. 6 for a normal subject. If the instant just after heel contact with the ground is taken, indicated by a middle dotted black line in Fig. 6, VGRF exerted by the subject starts to increase.

The frequency as shown in Hilbert spectrum is at high and starts to decrease as the foot becomes plantar with ground. From the colour bar, the amplitude of frequency starts from minimum and starts gradually increasing as inferred by time interval (*l*1).

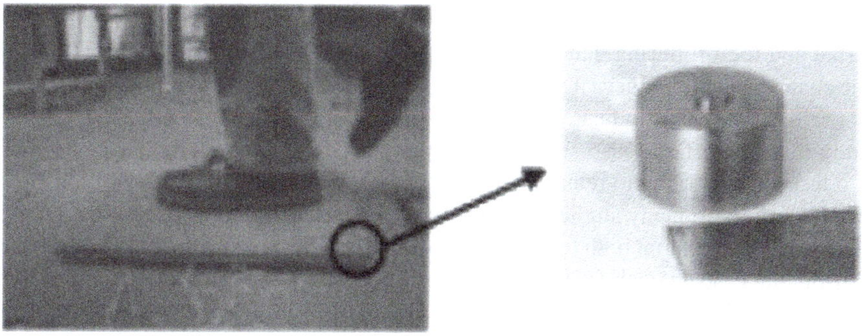

Fig. 5 Floor based sensor [13]

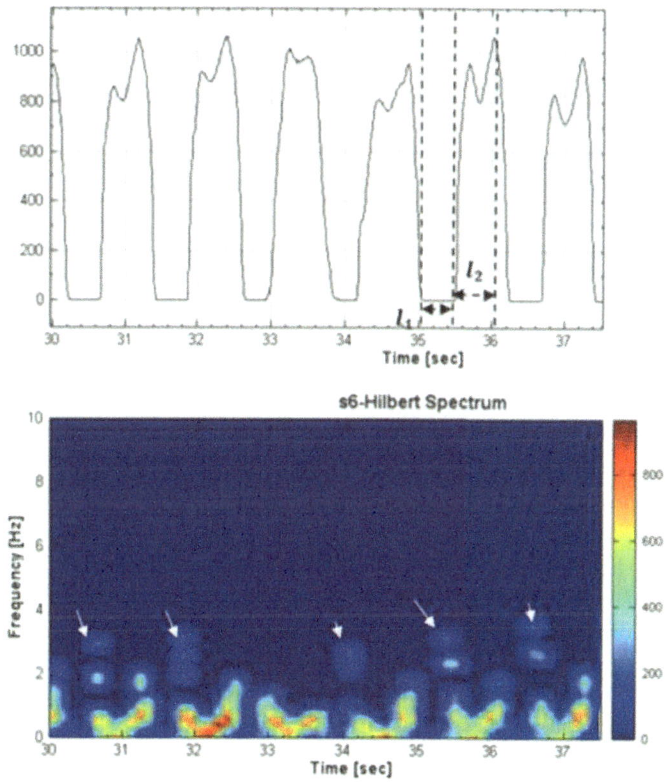

Fig. 6 A magnitude version of Hilbert spectrum of the total VGRF and the raw signal for a normal subject

When the foot is in plantar position with ground, the force is at maximum while the frequency is at minimum, however, its magnitude is at maximum suggesting the long time and this can be inferred $l2$. As the force decreases during the toe-off, the frequency starts to increase. As a result the maximum frequencies exist at the heel strike and toe-off during gait. In addition, this suggests that the frequency opposes the change which produces it as a trying to retain the body in balance and therefore of not falling [14]. This indicates that the frequency of the signal is high at heel-strike and at the toe-off and also suggests that the frequency opposes the change which produces it as a trying to retain the body in balance and not falling.

However, still many studies consider the sensor location to measure the Ground reaction force either at both toe and heel or usually by analyzing the total force from sensors underneath each foot [14]. Though, a need to investigate more exploration of choosing the correct position of a sensor or set of sensors in the insole becomes crucial. This is for the reason that some features are used for classification and better work when they are extracted from sensors data at a given location. ROC curve is then used to evaluate each feature for all sensors. This could contribute to better manufacturing of GRF acquisition systems [15]. In this part, simple features from literature were extracted then their performance was tested. In this section the 11 most relevant features were retained. The commonly wide used features [14, 16–18] are: Mean, Median, standard deviation, range, Interquartile range, 95% percentiles of the distribution of the signal, Skewness, Kurtosis, Power of the signal, Mean power frequency and Magnitude of peak frequency. As too many statistical features could be extracted and evaluated in time domain analysis and frequency domain, in this section one feature tested using the Receiver Operating Curve (ROC). However this is done for all features among all sensors for the 47 subjects. For instance the skewness is chosen since it has been used widely in various studies related to VGRF [14] and has demonstrated its capability in distinguishing normal and Parkinson's disease person. When area under the curve (AUC) equals to one, this refers to a perfect discrimination and has a ROC curve that passes through the upper left corner i.e. 100% sensitivity and 100% specificity with no overlap in the two distributions. Figure 7 shows a plot of the ROC curve for skewness feature tested over the mid-sensor numbered 5 in the right foot among normal and Parkinson. Each point on the ROC curve represents the sensitivity and specificity pair corresponding to a particular decision threshold. The diagonal line dividing the ROC space is also called line of no-discrimination in which a point on this line corresponds to a completely guess. When points are above the diagonal, this indicates a good classification results and on the other side points below this line indicate a poor predictors. Therefore the distance from the random guess line is the best indicator of how much predictive power a method has.

The evaluation of skewness shows the following results:

ACC	85.1064
MCC	0.6822
Sensitivity	0.8966
Specificity	0.7778
Area under curve	0.902

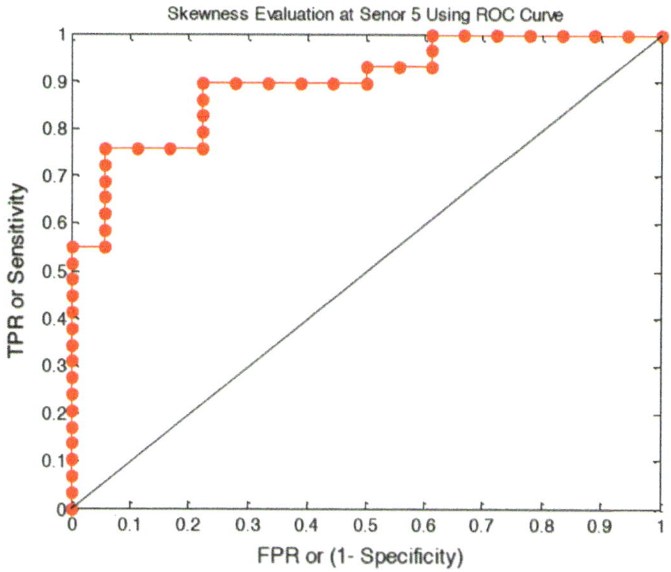

Fig. 7 Skewness ability in discrimination between normal and Parkinson gait using ROC curve

Table 4 ROC evaluation of skewness among all sensors

Sensor 9 in foot	Left foot	Right foot
1	0.645594	0.655172
2	0.609195	0.637931
3	0.609195	0.613027
4	0.701149	0.676245
5	0.904215	0.842912
6	0.787356	0.764368
7	0.850575	0.808429
8	0.62069	0.703065
VGRF corresponds to summation of the 8 sensors	0.605364	0.611111

Using the above listed data, 0.9 as area under the curve indicates an excellent performance of skewness in discriminating normal from Parkinson using VGRF from sensor 5. The results of ROC evaluation among all the 47 subjects are shown in Table 4 for each sensor.

Analysing Table 4 indicates that unlike other studies similar to [14] that consider total summation of force signals from all sensors as the most important, however, its clearly shown that sensor 5 is the most important sensor (AUC = 0.9) to consider in building acquisition system to acquire data for analyses. Figure 8 shows the ROC curve of the total ground reaction force from sensors of the right foot. The sensitivity

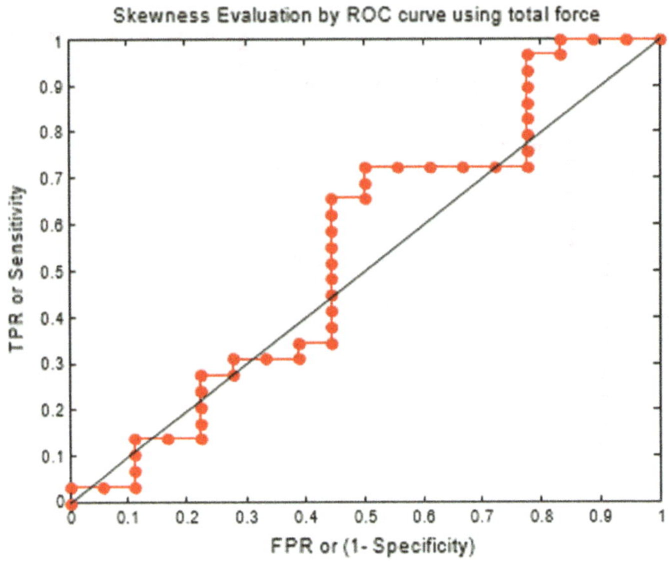

Fig. 8 Skewness is extracted from the total force from sensors located under the right foot and its performance in binary classification is evaluated using ROC curve

is recorded to be 0.6552 while the specificity is 0.5556. This yield AUC = 0.5460 which refers a fail level of accuracy in classification.

This conclusion is generalized as the same procedure is applied over the rest of the features chosen in this study. If expanding data is needed, then adding sensor 7 and 6 corresponds as main sensors also to be considered for classification as shown in Fig. 9.

Not to add, the average of the strides of the 2 classes corresponding to the 120 s of walking is also considered. As a result, each series of strides are represented by their average. Next, 3 features were extracted: the amplitude of the first peak, that is, the peak that corresponds to the heel contact (in case of total force), time to the first peak, stride time. The ROC evaluation also infers a better accuracy for sensor 5 compared to other sensors.

Then the question raised is whether it is superior to have a topographical pressure diagram rather than sensory location investigation? IoT-based biometrics technology and infrastructure is rapidly evolving. So shall we start to think about secure, scalable and interoperable gait data acquisition systems with the right communication schemes? How could we shift our acquisition systems from VGRF signal storage and then analysis to real-time analysis?

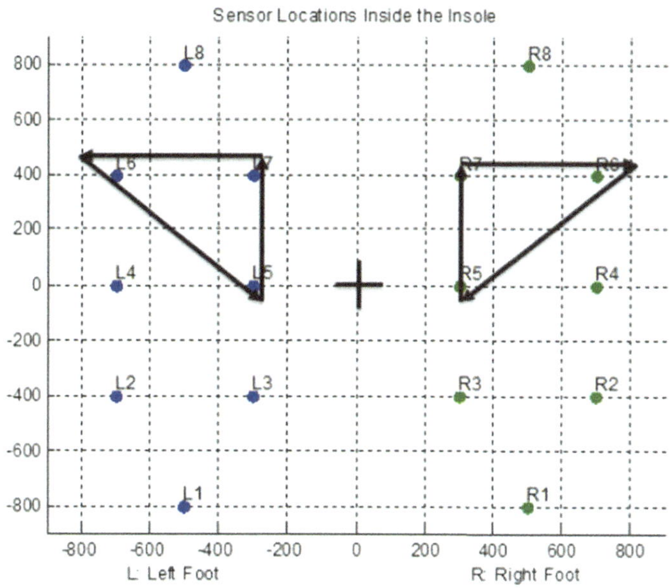

Fig. 9 The significant locations to acquire VGRF data for gait recognition

6 Processing

The analysis and manipulations of the GRF data will be carried out to give an insight about the VGRF signals themselves. This would help in extracting significant features and specific information that help in verification and even may be in identification of subjects based gait. The synchronization of consecutive gait steps in subjects in both normal and Parkinson simplify the analyses.

1. Periodizing VGRF Signals

On-off controller simply drives the acquisition of VGRF during walking from fully closed to fully open depending on the location of the foot. In other words, when the foot hits the ground entering the stance phase then VGRFs signal do exist. The acquired data will be saved into an array with preliminary fixed size ending up with a matrix of stance phases. Its size is controlled by the preliminary defined stance interval depending on the sampling rate. This would eliminate any difference in the stride interval between subjects stemmed from their difference in height, weight, gender and so on. The signal then have fixed time periods (i.e. the time from one heel strike into the successive heel strike is fixed). That's why the VGRF signals are divided as shown in Fig. 10.

The matrix of stance phases is reshaped into 1D array vector for analysis. Such fixing of sampling points interval for stance phase of the gait would definitely change

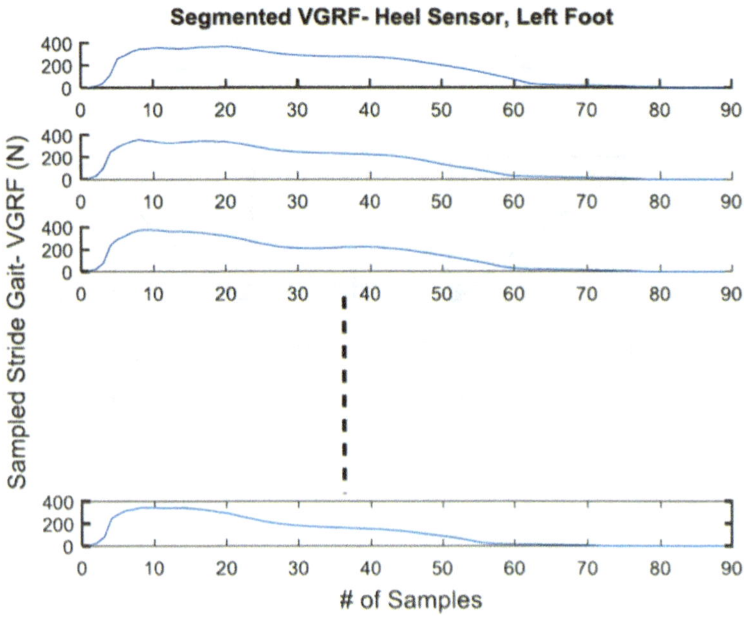

Fig. 10 Step isolation of VGRF stance phases at the heel. They are saved over 89 samples equivalent to 0.89 s

frequency content analysis. Certainly, they will exist at different harmonics of the saved interval as shown in Fig. 11.

The Autocorrelation Function (ACF) of the altered signal indicates a very slow decay over lags and thus the signal becomes more non-stationary. This is inherited from the periodicity being added to the signal. Figure 12 is an illustration of the ACF.

Differencing the new signal by two consecutive times as an intention to remove periodicities signifies the interval chosen by implementing Partial Autocorrelation Function (PACF) as shown in Fig. 13. It agrees with the predefined interval of 89 lags.

2. Filtering

While standing at rest, the VGRF is the only one existing. However at the time of heel-strike that separates the swing phase from the stance phase in addition to the toe-off, the vertical force is no longer vertical; it tilts over to produce shear force. When the foot hits the ground as termed heel strike, the VGRF associated with tangential forces slanted from GRF vector component parallel to the ground acting backwards. It is formed by an exchanging of frictional forces with ground that leads to a brake impulse and therefore the body slow down. This prevents foot sliding forward along the ground. Figure 15 shows the horizontal component affecting the vertical ground reaction force contaminated by noise. However, the dynamic characteristics of gait reaction forces are usually exploited by filtering as shown in Fig. 15. Filtering at

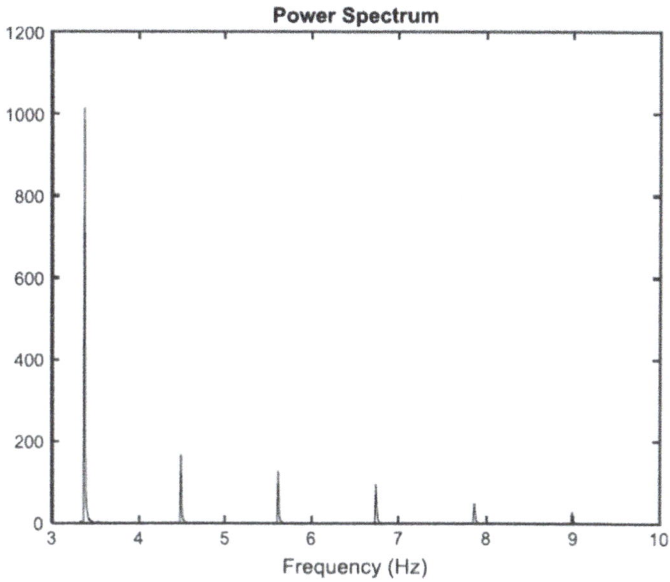

Fig. 11 Power spectrum of the new generated signal

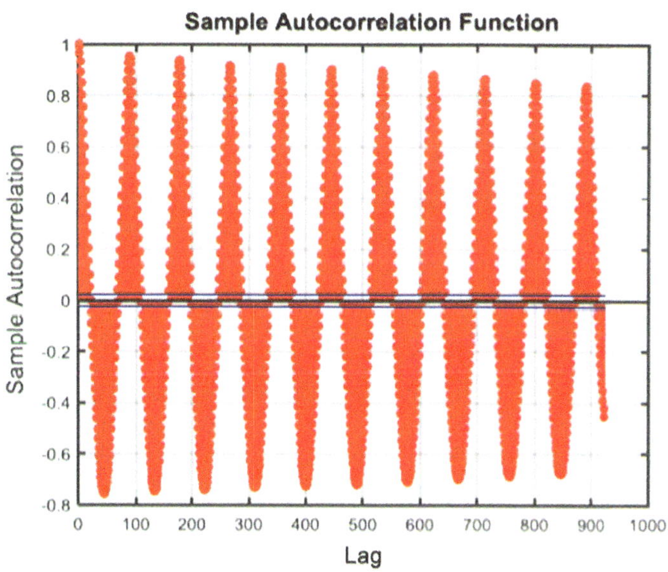

Fig. 12 Sample autocorrelation function

25 Hz is useful for certain data while it is not applicable for other subjects as in Fig. 15. Given that the foot will act also as a shock absorber as to disperse the force

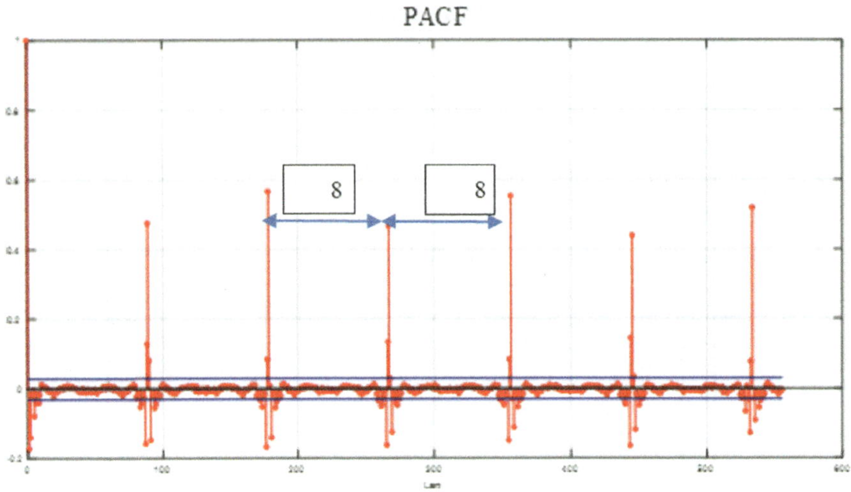

Fig. 13 Partial autocorrelation function

Fig. 14 Vertical GRF vector during gait cycle

of the body during landing. The GRF vector is illustrated by a black arrow on Fig. 14 during contact, midstance and propulsive phase.

Likewise, during toe-off there will be an appearance of propulsive impulse to stimulate motion due to tilting of the force over forwards. This helps accelerating the body.

Knowing that concavity upward horizontal and normal forces indicate brake impulse which is followed by deceleration motion while concavity downward signifies propulsive impulse followed by acceleration motion. This specifies an important coefficient to consider which named static friction. Static friction is defined as the ratio of the magnitude of the horizontal frictional force to the normal force. This coefficient could yield when slippage could occur. Moreover, part of this noise reflects the speed horizontal speed of the foot during the touchdown of the heel with the ground.

Using Fig. 16 and by Pythagoras theorem the measured ground reaction force is given by Eq. (1):

$$F_m^2 = F_v^2 + F_h^2 \tag{1}$$

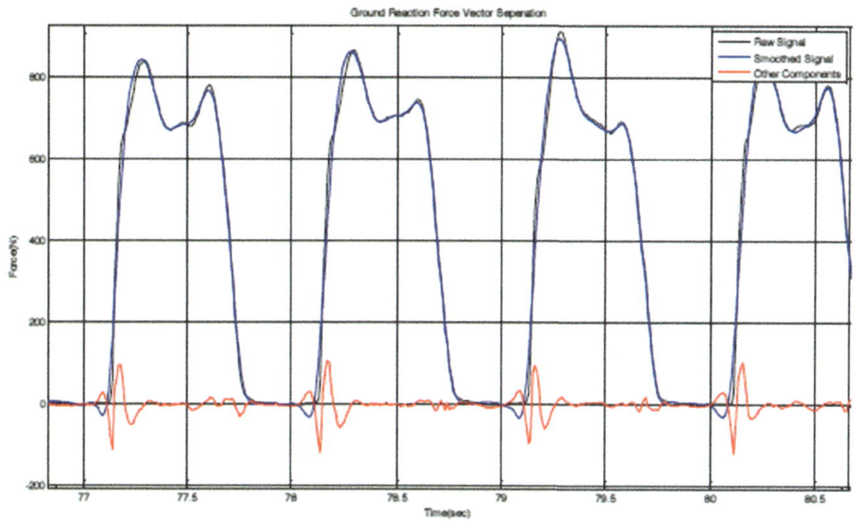

Fig. 15 Ground reaction force filtering by second order Butterworth filter of 25 Hz shows a vital part being attenuated coloured in red

Fig. 16 During heel strike, the measured VGRF consist of horizontal frictional component

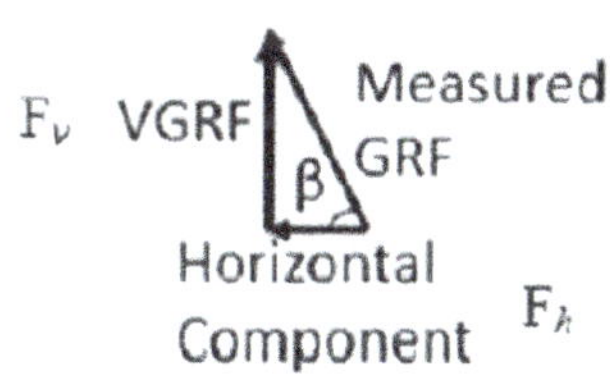

$$\tan\beta = \frac{F_v}{F_m} \tag{2}$$

where

Fm The measured ground rcaction force
Fv The vertical component of ground reaction force, load
Fh The horizontal frictional component exerted by each surface on the other indicating the direction of the measured GRF

The angle β indicating the direction of the measured GRF.

This horizontal friction is considered as non-fundamental force but is a result from intermolecular and interatomic kinetic dry friction between ground and foot which make it complicate to be calculated and considered as highly stochastic. This energy that examined as frictional forces by subject is lost as heat. An empirical law termed as Coulomb's Law of Friction can approximate this model by Eq. (3):

$$F_h \le \mu F_v \tag{3}$$

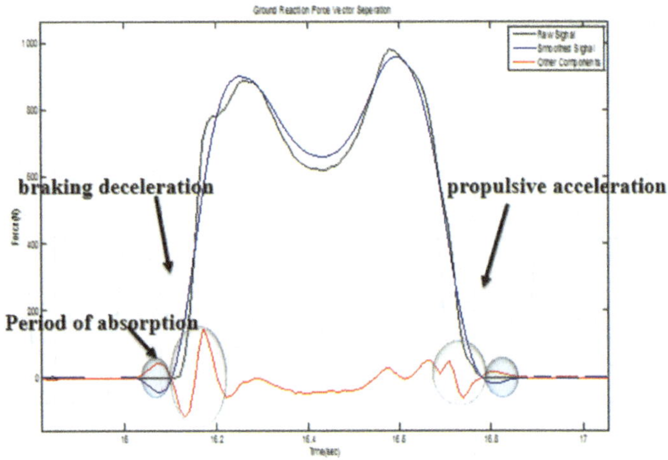

Fig. 17 Loading response and the push-off are circled

where, μ is the dimensionless coefficient of friction. It can be defined from Eq. (3) as the ratio of the force of friction between foot and ground and the pressing normal force.

To sum up, and by newton's law the force is given by: F = ma where "m" is the mass of the parts contributed to this force, and "a" is the acceleration. Since frictional backward force exists, this suggests definitely the existence of backward acceleration as a braking action on the body, slowing it down.

From Fig. 17, at instant just prior to the collision of the heel with ground i.e. as it start to touches the ground the normal force is in negative due to the fact that a forward directed force termed as "claw back" due to initial parameters [18] and in addition the skeletal system act as a shock absorber. At those moments slippage is occurring and shown by the small red peak and this indicate the existence of force with direction of motion. However, this pronation activate information record by the lower central nervous system through sensory neurons that registered in central nervous system which in turns activate the muscles contraction to prevent the forefoot from slapping down and therefore generate forces and moments at synovial joints to invoke the movement regulated by rigid links. Figure 4 can be used for illustration. The last exert ground reaction force which followed by a decreasing in magnitude of the horizontal component as shown in Fig. 17 to serve a friction in the opposite direction of motion preventing the subject from slipping and therefore falling. As a result, at the period of absorption shown in Fig. 17 indicates that the horizontal and frictional force is equal in magnitude but in opposite directions where still there is no movement. This can be verified by newton's third law that states that for every action i.e. force there is an equal and opposite reaction i.e. counter force.

As a summary, in this thesis instead of using a fixed filtering bandwidth of frequencies over all gait VGRF signals (x[n]), we have developed a new filter using the EMD technique as shown in Fig. 18.

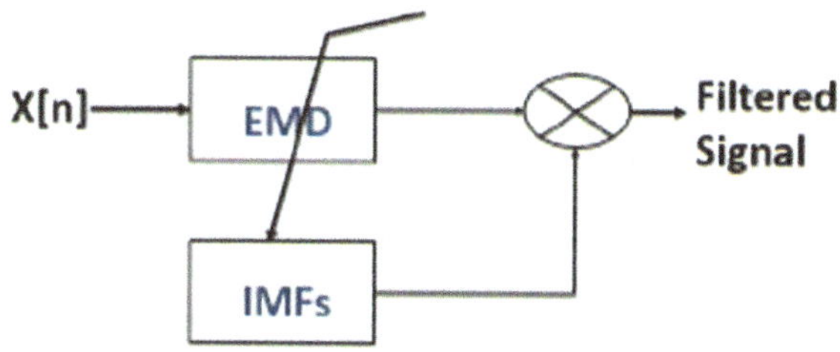

Fig. 18 Proposed filter

Table 5 Intrinsic mode function characteristics

Channel	Zero crossings	Extienia counts	Mean freq. (Hz)	Power (%)
IMF_h1	7050	4056	29.09068	0.045846
IMF_h2	2506	1676	10.34326	0.354629
IMF_h3	736	472	3.04068	9.128521
IMF_h4	494	316	2.042248	4.668593
IMF_h5	229	114	0.948923	84.29544
IMF_h6	158	86	0.655995	1.052419
IMF_h7	75	40	0.313557	0.241007
IMF_h8	35	18	0.148527	0.089646
IMF_h9	18	10	0.078389	0.041991
IMF_h10	10	6	0.045383	0.008957
IMF_h11	2	4	0.012377	0.031212
IMF_h12	3	1	0.016503	0.041738
IMF_residual	0	1	0.004126	

Then we choose to remove certain intrinsic mode functions (example is shown in Table 5) according to their weighted energy and preferred number of intrinsic mode function.

An example of EMD applied to a data from a control subject is shown in Fig. 19. The first subplot represents the raw signal. Figure 19 illustrates the idea that the first IMF captures the largest frequency components. The second IMF has a lower oscillation and so on to reach a trend with the lowest component as shown in the last row of Fig. 19. Therefore EMD acts as an adaptive filter to extract the components present in the signal. It's worthy therefore to mention that the first IMF extracts most of noise present in signal.

As a result, it is better to filter the signal at a mean frequency of 29.09 Hz instead of filtering it by 25 Hz second order Butterworth filter and losing crucial part of the signal. This filtering could be at 28 Hz on another gait if it represents the mean frequency of the highest oscillations. This is because we have assumed in this case that noise exist in the IMF1. Figure 18 therefore indicates that EMD is applied to the signal then a choice of the number of intrinsic mode functions that must be removed from signal must be made. This would remove same number of oscillations between different gait subjects but not necessary the same frequency content.

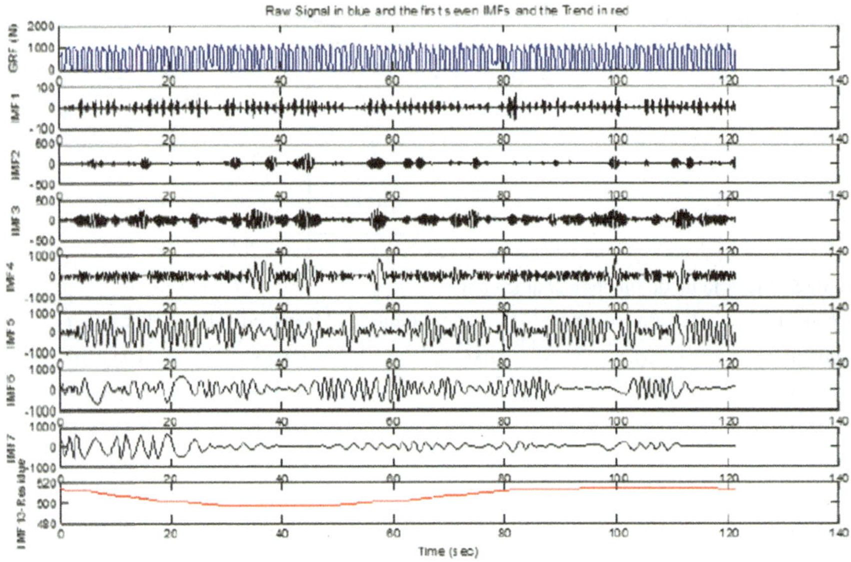

Fig. 19 The first raw is the original signal. Seven IMFs plus the trend are plotted

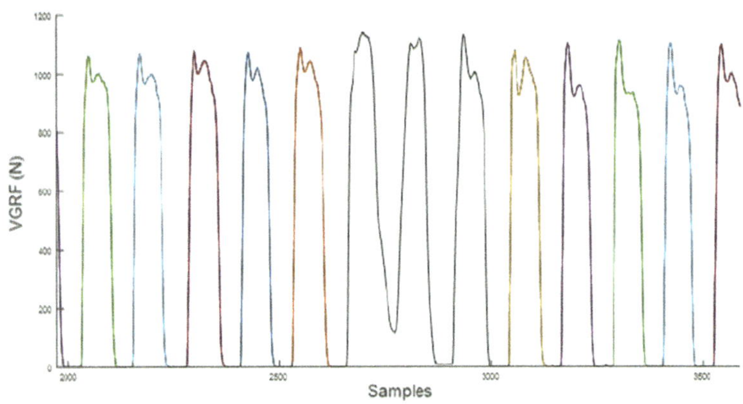

Fig. 20 Segments coloured in black are being omitted

The foot-ground contact part of the VGRF time series signal are then extracted forming steps. Some steps are being eliminated when their statistical properties form outlier in the vast of other segmented step signal, mainly this is done by computing the mean and standard deviation. For instance the black segment of VGRF shown in Fig. 20 forms an outlier. Those are so called turning points.

3. Normalization Failure

Dual task induced a significant increase sway parameters variability in both control and Parkinson subjects [17]. In this part, two sets of measurements for VGRF derived

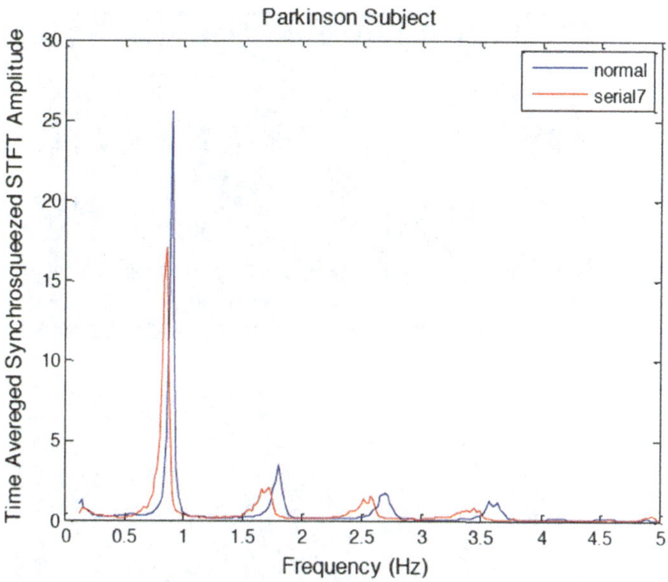

Fig. 21 Time averaged synchrosqueezed STFT

from six elderly normal control subjects and six elderly patients with Parkinson are used due to database limitations. Participants then walked and performed a second task: serial 7 subtractions. The effect of dual task gait training is pointed out by comparing gait properties before and after training. It improves balance and gait abilities of chronic stroke patients [16]. Furthermore, another study shows the rhythmic walk is not affected by cognitive tasks in healthy subjects and have an influence on pathological gaits [18]. In addition, during specific dual task the gait speed decreased ($p < .001$) and swing time variability increased ($p < .001$) [19].

To sum up, a bunch of literature directly pointed out some features that would be affected as a result of adding cognitive task during walking. However, a need for deep analysis of main properties (like instantaneous frequency) of gait changes while performing a dual task as this part will focus on. In addition, VGRF consist of close frequency components contaminated by noise. Such signals are time varying frequency and amplitude. However, major stationary component is dominating the VGRF signals that mask the detection of such non-stationary components that would give the most relevant difference between two signals.

Figure 21 shows two main important phenomena. A shift in the frequency and this is explained by change in the speed of subjects performing a certain task while walking. The second fact is a change in the amplitude in the time averaged synchrosqueezed STFT while performing dual task. This is due to divergence of the power from certain frequency component into other frequencies. Therefore, it would be beneficial to extract the instantaneous frequencies and track their variation.

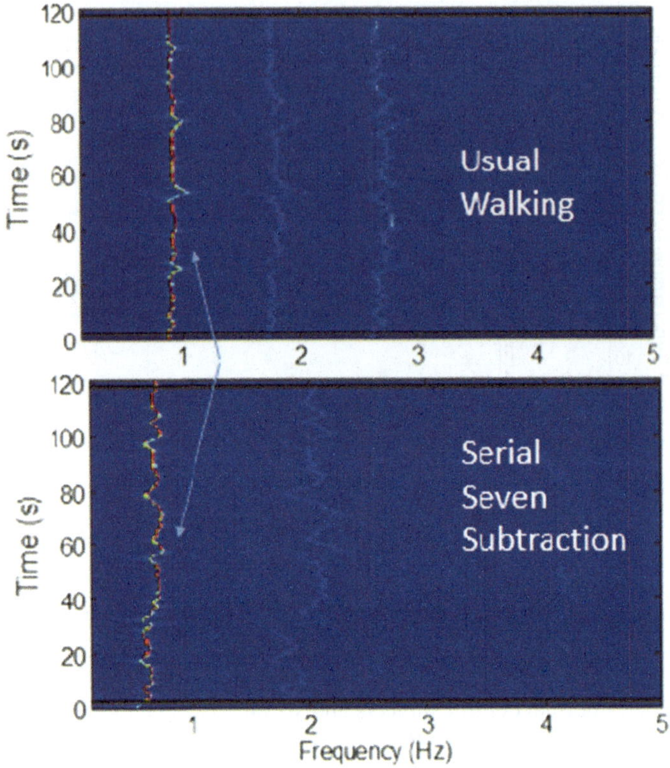

Fig. 22 Synchrosqueezed STFT for total VGR signal extracted from the gait of Parkinson subject on both walking conditions

In order to summarize the ability of synchrosqueezing in revealing valuable info about the gait, Fig. 22 is obtained for the total ground reaction force so we can generalize the following differences:

- Fluctuation in the instantaneous frequency as a function of time
- Some components disappear
- A shift in the frequency content is obtained
- The power is relatively different

As a result, one important recommendation is pointed out: dual task while walking affects the gait in a tremendous way in all subjects whether normal or Parkinson. The shift in the value of the frequencies is due of the voluntary or automatic movement during dual task. This is revealed by hypokinetic. Hypokinetic disorders usually appeared in Parkinson subjects' gait.

That is why filtering and normalization must be given specific attention. While we perform gait pathological assessment, it is hard to normalize signals from cognitive tasks or even when it comes to physiological or other situations. Therefore, significant content of the signal will be altered. That is why the adaptive filtering model best fit in such cases. Keeping in mind it is always to go over the frequency content of the signal. In addition, normalizing VGRF from parameters like cognitive tasks for comparison purposes among subjects became problematic. In order to overcome this debate, we focused mostly on this thesis on inter subject—VGRF comparison. This is done by comparing VGRF from right and left foot of the same subject in addition to comparing two consecutive steps with each other from the same foot.

7 Hints for Feature Extraction

Extracting unique features from VGRF signals will increase the rate of successful recognition among people. For instance, [20] extracted the body mass from VGRF measured without an active participation of the individual being measured in a fraction of second with an error less than 1 kg standard deviation. However, it is reported as a weak biometric feature. Methods of Gait Recognition:

- Model-based approach
- Model free or Holistic approach that is based on appearance.

It is reported in [2] that holistic feature over perform the geometric features in both K-NN and SVM classifiers. As the ultimate goal is to build a device that can reach accurate biometric legged locomotion system, then this device must be reliable and have the ability to measure time features on real time basis. For instance,

- Rise time (Fig. 23): The time required for the VGRF signal being measured to change its state from a certain low level into a given high value. It reflects how much fast the ground contact is being conducted during walking.
- Fall time (Fig. 24): The time required for the VGRF signal being measured to change its state from a certain high level into a given low level value. It reflects how much fast the ground contact is being conducted during walking.
- Duty factor: The ration of the stance time to the stride interval. This can be easily computed using the period. For instance the period can be chosen to be the time between two consecutive crossings of the halfway between low level and high level then duty cycle can be determined (Fig. 25).

Certainly there are too many features to extract that we are not going to dwell into them, however, make them as simple as possible and give them the real explanation that would enhance the performance of the chosen classifier.

8 Conclusion

This chapter provided the reader with up-to-date analysis of gait through different research orientation. A methodology of thinking is presented. However, one chapter

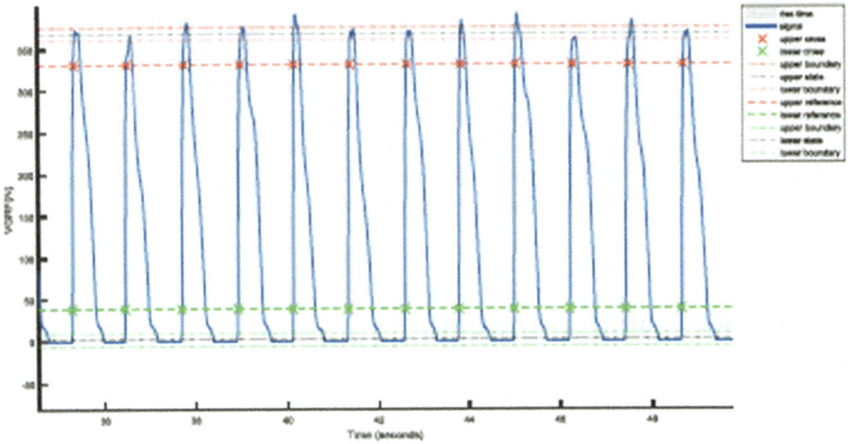

Fig. 23 Rise time (10% of the low level and 90% of the high level)

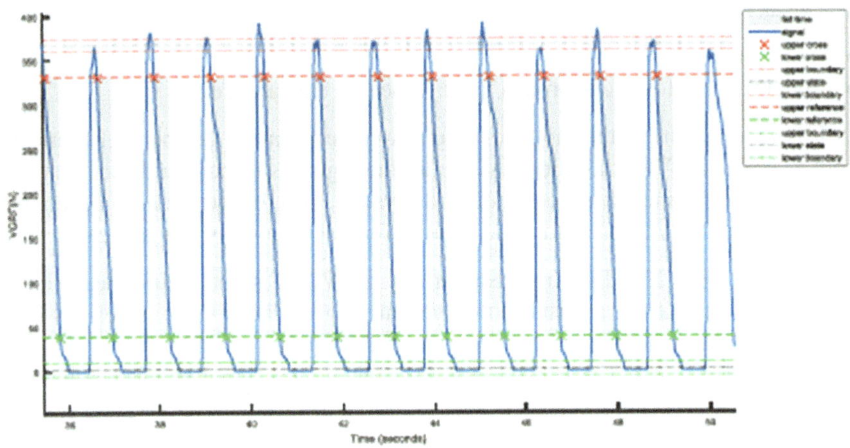

Fig. 24 Fall time

is not enough to cover all aspects of gait and their identification and recognition. However, we were able to highlight the importance of correlating the signals and their analysis to the physiology of gait and manner of walking. Second, it is recommended to pay attention to the acquisition system by focusing on the location of sensors being used to collect VGRF. Third, the analysis indicates that pre-processing without attention could hurt the raw signals. For instance, filtering and normalization were treated and recommendations are made. Finally, the feature selection must start from the signals themselves before taking the VGRF into higher level of transformations! Rise time, fall time, duty cycle and many more could be realized as important features for legged locomotion recognition. If this is not the case, at least they will provide an intuition into how we must handle the signals and therefore facilitates the proposed

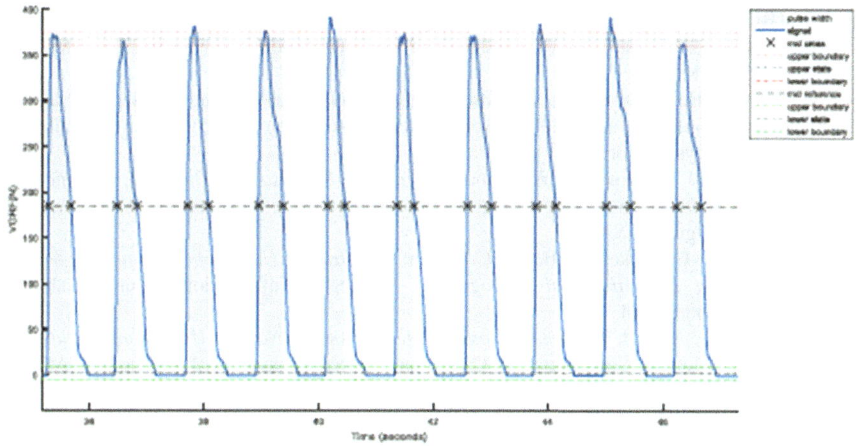

Fig. 25 Period

methodologies in treating those signals. Still the analysis is young and need further elaboration and investigation. Based on the above analysis, better gait analysis would improve learning algorithms that maximizes the performance of the biometric legged locomotion machine!

References

1. N.V. Boulgouris, D. Hatzinakos, K.N. Plataniotis, Gait recognition: a challenging signal processing technology for biometric identification. IEEE Signal Process. Mag. **05**(1053-5888), 78–90 (2005)
2. R.V. Rodriguez, N.W.D. Evans, R.P. Lewis, B. Fauve, J.S.D. Mason, An experimental study on the feasibility of footsteps as a biometric," in *15th European Signal Processing Conference (EUSIPCO)*, Poznan, Poland (2007)
3. M. Edwards, X. Xie, Footstep pressure signal analysis for human identification, in 7th *International Conference on BioMedical Engineering and Informatics* (Swansea University, Swansea, SA2 8PP, United Kingdom, 2014)
4. ARCHES OF FOOT, University of Glasgow, University of Strathclyde, Glasgow Caledonian University and the University of Paisley [Online]. Available http://www.gla.ac.uk/ibls/US/fab/tutorial/anatomy/arch1.html9T1. Accessed 26 Dec 2016
5. R.L. Williams II, Engineering biomechanics of human motion, in *NotesBook Supplement for ME 4670/BME 5670* (Ohio University, Dr. Bob Productions, 2014)
6. C.L. Vaughan, B.L. Davis, J.C. O'Connor, *Dynamics of human gait, second edition, Cape Town* (Kiboho Publishers, South Africa, 2009)
7. C.L. Vaughan, B.L. Davis, J.C.O. Connor, *Dynamics of Human Gait*, Howard Place, Western Cape 7450, 2nd edn. (Kiboho Publishers, South Africa, 1999)
8. K. Sato, M. Butcher-Mokha, Relationship between ground reaction force and stability level of the lower extremity in runners, in *Proceedings of the Fifth Annual College of Education Research Conference: Section on Allied Health Professions* (Florida International University, Miami, Apr 2006)

9. S.A. Birrell, R.H. Hooper, R.A. Haslam, *The Effect of Military Load Carriage on Ground Reaction Forces*, Department of Human Sciences, Loughborough University, Leicestershire, LE11 3TU, UK
10. M. Jeffry. F. Research, Gait in Parkinson's Disease, National Institutes of Health, National Parkinson Foundation, and the Parkinson's Disease Foundation , (2005). [Online]. Available: http://www.physionet.org/.
11. V. Srinivasan, J. Stankovic, K. Whitehouse, *Using Height Sensors for Biometric Identification in Multi-Resident Homes*, Department of Computer science, University of Virginia, Charlottesville, Virginia, USA
12. M.O. Derawi, C. Nickely, P. Bours, C. Busch, *Unobtrusive User-Authentication on Mobile Phones using Biometric Gait Recognition*, Norwegian Information Security Lab and Hochschule Darmstadt
13. R.J. Orr, G.D. Abowd, *The Smart Floor: A Mechanism for Natural User Identification and Tracking*, College of Computing and GVU Center, Georgia Institute of Technology, Atlanta, GA 30332-0280 USA
14. A.I. Bazin, M.S. Nixon, *Probabilistic Combination of Static and Dynamic Gait Features for Verification*, School of Electronic and Computer Science, University of Southampton, SO17 1BJ, UK, (2005)
15. D.S. Bloswick, D.B. Chaffin, An ergonomic analysis of the ladder climbing activity. Int J. Ind. Ergonom. **6**(1):17–27 (1990)
16. J.T. Worobets, D.J. stefanyshyn, *Normalizing Vertical Ground Reaction Force Peals to Body Weight in Heel-Toe Running*, Human performance Laboratory, Faculty of Kinesiology, University of Calgary, (1990)
17. D.A. Winter, Human balance and posture control during standing and walking, Gait Posture, 3, 193–214 (1995)
18. D.W. Powell, B. Long, C.E Milner, S. Zhang, Effects of Vertical Loading on Arch Characteristics and Intersegmental Foot Motions, J. Appl. Biomech. **28**(2012) Human Kinetics, Inc., pp. 165–173, (2012)
19. B. Kluitenberg, S.W. Bredeweg, S. Zijlstra, W. Zijlstra, I. Buist, Comparison of vertical ground reaction forces during overground and treadmill running. A validation study. BMC Musculoskel. Dis. **13**(1) (2012)
20. J. Jenkins, C. Ellis, *Using Ground Reaction Forces from Gait Analysis: Body Mass as a Weak Biometric*, Western Carolina University & Duke, University, USA

Wearable Technologies in Biomedical and Biometric Applications

Sherif Said, Samer Al Kork and Amine Nait-Ali

Abstract The purpose of this chapter is to discuss the challenging and emerging technology namely "Wearable technology". Based on bio-signal acquisition numerous applications in biomedical and biometrics are discussed. After highlighting some available systems in the market, the chapter focuses on a particular technology using an arm band. Biometrics is then considered from different points of view.

1 Introduction

The application of wearable electronics in the biomedical, biometrics research and commercial fields has been gaining great dramatic interest over the last several decades [1]. Small-sized, light weight monitoring systems with low-power consumption and off course wearable allow for the collection of physiological and behavioral data in ecological scenarios (e.g., at home, during daily activities like driving or sleep, during specific tasks, etc.) with a maximal comfort for the end user. As a result, outpatient-monitoring care can be associated with improved quality of life, especially for patients with chronic disease, possibly preventing unnecessary hospitalizations and reducing direct and indirect healthcare costs. Within this context, wearable sensors and systems have evolved to the point that they can be considered ready for clinical application [2]. This is due not only to the tremendous increase in research efforts devoted to this area in the past few years but also to the large number of companies that have recently started investing aggressively in the development of wearable products for clinical applications. Stable trends showing a growth in the use of this technology suggest that soon wearable systems will be part of routine clinical evaluations.

S. Said · S. A. Kork
American University of Middle-East (AUM), Eqaila, Kuwait

S. Said · A. Nait-Ali (✉)
LiSSi, University-Paris-Est Créteil (UPEC), Vitry-sur-Seine, France
e-mail: naitali@u-pec.fr

A. Nait-Ali (ed.), *Biometrics under Biomedical Considerations*,
Series in BioEngineering, https://doi.org/10.1007/978-981-13-1144-4_10

The interest for wearable systems originates from the need for monitoring patients over extensive periods of time. This case arises when physicians want to monitor individuals whose chronic condition includes risk of sudden acute events or individuals for whom interventions need to be assessed in the home and outdoor environment. If observations over one or two days are satisfactory, ambulatory systems can be utilized to gather physiological data. An obvious example is the use of ambulatory systems for ECG monitoring, which has been part of the routine evaluation of cardiovascular patients for almost three decades. However, ambulatory systems are not suitable when monitoring must be accomplished over periods of several weeks or months, as is desirable in a number of clinical applications.

Wearable systems are totally non-obtrusive devices that allow physicians to overcome the limitations of ambulatory technology and provide a response to the need for monitoring individuals over weeks or even months. They typically rely on wireless, miniature sensors enclosed in patches or bandages, or in items that can be worn, such as a ring or a shirt. They take advantage of hand-held units to temporarily store physiological data and then periodically upload that data to a database server via a wireless LAN or a cradle that allow Internet connection. The data sets recorded using these systems are then processed to detect events predictive of possible worsening of the patient's clinical situation or they are explored to assess the impact of clinical interventions.

In this chapter, the wearable technologies used in biomedical and biometrics application will be discussed. starting in Sect. 2 surveying the latest work done on the field of wearable technologies. Section 3 presents available systems on the market to acquire different biosignals types and differentiate between them. In Sect. 4 the applications of wearable technologies in the arm band in robotics field, medical field and biometrics field is discussed. Followed by a conclusion about the chapter.

2 A Brief Review on Wearable Technologies

These days most of researchers are focusing on the wearable technologies in which the users should wear a smart devices equipped with biosensors embedded inside them (e.g., smart-textile or contactless electrodes) and sensing platforms, as well as effective algorithms for embedded signal processing and data mining. In this section a few researches in the field of wearable systems will be presented briefly.

Furthermore, significant endeavors have been related to small-scale integration of analog and digital sensor signal conditioning and energy harvesting, especially in the case of wireless body area/sensor networks.

The high impact of wearable technologies in the frame of multidisciplinary scientific research is also confirmed by the significant number of studies published throughout the last several decades.

Carbonaro et al. [3] proposed a sensorized shoe for gait analysis. The shoe has built-in force sensors and a triaxial accelerometer and can transmit sensor data to the smartphone through a wireless connection. Experimental results confirmed a reliable detection of the gait phases.

A novel system to monitor the driver's health status is presented by Said et al. [4]. Wearable Bio-Sensors Bracelet that detects heath emergency threatening situations for drivers. The system collects live bio-signals data through a Wearable Bio-Sensors Bracelet that is embedded with multiple Bio-sensors such as Electromyography (EMG), Electrocardiography (ECG), Electrodermal Activity (EDA), Accelerometer (ACC) and Temperature. A robotics rover car model has been used as a testing vehicle scenario. The developed system detects an emergency and takes control over the rover maneuver when the driver lost his capacity, and automatically avoids any obstacle by sensing the surrounding area.

The wearable technologies aims comfortable feeling for the user is improving. A sensorized T-shirt able to acquire multi-channel, long term (ECG) using active electrodes embedded inside it has been proposed by Boehm et al. [5], therefore avoiding the use of adhesive gel electrodes. Experimental results validated the proposed wearable monitoring system as compared with a commercial Holter ECG in healthy volunteers during movement phases of lying down, sitting, and walking. ECG can be captured by the electrodes embedded on the shirt and as presented before it can be captured from the human hand. The design of the wearable device is optimized by the application itself. Greco et al. [6] proposed a wearable system for monitoring the (EDA) signals during emotional elicitation. EDA was studied at different frequency sources through data gathered from healthy subjects undergoing visual affective elicitations. Based on the users nervous the sweating level is changing and can be detected by EDA sensors to monitor the nervous level for humans. Caldara et al. [7] developed a potentially implantable blood pressure telemetry system, based on an active Radio-Frequency IDentification (RFID) tag, aiming to continuously measure the average systolic and diastolic blood pressure of small/medium animals. RFID energy harvesting has also been investigated. The authors present an experimental laboratory characterization and in vivo tests. Saponara et al. [8] present a scalable remote model for telemedicine scenarios using wireless biomedical sensors, an embedded local unit (gateway) for sensor data acquisition/processing/communication, and a remote e-Health service center. The use of a mix of commercially available sensors and new custom-designed ones was also presented.

An integrated sensing wearable glove combining a low number of knitted piezo resistive fabrics to reconstruct both hand posture and tactile pressure sensing. Bianchi et al. [9] proposed the glove which can be used in different applications. To this end, a priori information of synergistic coordination patterns in grasping tasks was employed. In the frame of a wireless body area network. Researches in wearable technologies different fields require sensors for monitoring a certain physiological signals. In the next section different platforms available on the market will be differentiated showing the advantages and disadvantages for each.

3 Wearable Systems on Market

There are a lot of systems available on the market that offer to the user the ability to acquire bio-signals. Several requirements and criteria differentiate between these systems. The cost of the system is one of the most important criteria. The connectivity with the processor either wireless or with wires. The accuracy of the bio-sensors in the system. The consumption rate of the battery which gives the system the ability to stay powered for long time. The Software Development Kit (SDK) available with the system which allows the researchers to develop their systems easily and can have access to the raw signals for further development.

BITalino development bio-medical kit is one of the most powerful kits available on the market. It gives the user the ability to work on physiological data.

This kit comes with EMG sensor which can monitor the muscle activation using 3 wet bipolar surface electrodes (plus a ground lead), Measuring the electrical activity in muscles and nerves. Surface electromyography (sEMG) is a technique which is used in many clinical and biomedical applications, in areas like HCI, neurology, rehabilitation, orthopedics, ergonomics, sports, etc. It is widely used as biofeedback tool to assess muscle fatigue; disorders of motor control and low-back pain is also possible with the EMG sensor. Sensing isometric muscular activity, where no movement is produced, enables a definition of classes of subtle motionless gestures to control interfaces without being noticed and without disrupting the surrounding environment. These signals can be used to control prosthetic devices such as prosthetic hands, arms, and lower limbs or as a control signal for an electronic device such as a mobile phone or PDA.

Conduction of action potentials through the heart generates electrical currents that can be picked up by electrodes placed on the skin. A recording of the electrical changes that accompany the heartbeat is called an electrocardiogram (ECG). Variations in the size and duration of the waves of an ECG are useful in diagnosing abnormal cardiac rhythms and conduction patterns. The ECG works mostly by detecting and amplifying the tiny electrical changes on the skin that are caused during the heart muscle cycle during each heartbeat. The ECG sensor provided by BITalino use only 2 electrodes to acquire the signal. The most important applications of ECG sensor focus on wellness of the patient and include heart rate and stress monitoring, biometric verification and life monitoring.

Electrodermal Activity (EDA) can be defined as a transient change in certain electrical properties of the skin, associated with the sweat gland activity and elicited by any stimulus that evokes an arousal or orienting response. The EDA sensor can measure the skin activity with high sensitivity measurement power in a miniaturized form factor. With low noise signal conditioning and amplification circuitry, EDA sensor provides accurate sensing capability and detect even the feeblest electrodermal skin response events by using 2 electrodes. Some of the applications of this sensor include detection of changes in the attentive, cognitive and emotional states. EDA sensors were also used for relaxation biofeedback, sympathetic nervous system reaction detection, among many others.

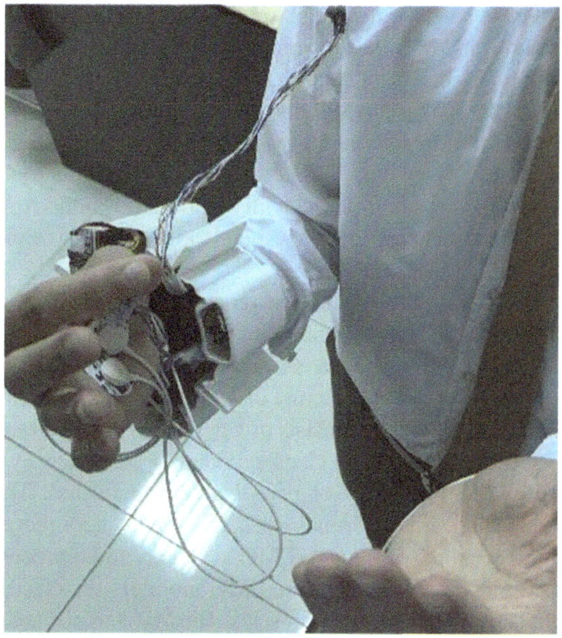

Fig. 1 A person wearing BITalino freestyle kit

The triaxial accelerometer is based on MEMS (Micro Electro-Mechanical Systems) technology and has been developed for biomedical applications where kinematic and motion measurements are required. This sensor can measure accelerations relative to free fall and the model available can detect magnitude and direction of this same acceleration, as a vector quantity. This resulting vector can then be used to sense position, vibration, shock, fall, etc. Attaching the accelerometer to a limb for example, an acceleration can be measured within the dynamic range of the sensor.

The revolutionary kit comes with a Microcontroller Unit (MCU) in a very small size that can connect all the sensors to it. The MCU has 6 analog input ports 4 (10-bit), 2 (6-bit) and it has 8 digital ports 4 ports digital input (1-bit), 4 ports digital output (1-bit). All acquired signals can be transmitted wirelessly using Bluetooth V2.0. The price of the freestyle BITalino kit is 196 $ [10].

The BITalino kit has some advantages in terms of the number of sensors embedded inside it and the wireless communication with the main processor as shown in Fig. 1. Also, the size of the sensors is very small to be embedded inside any wearable device. The shortcoming of this system is that the electrodes are wet since it needs a gel and are connected via wires to the human skin which makes uncomfortable feeling to the user.

MySignals is a development platform for medical devices and eHealth applications. It is used to develop eHealth web or even to build new medical wearable devices. MySignals HW Complete Kit includes 17 sensors which allows the user to measure 20 different biometric parameters such as pulse, breath rate, oxygen in blood, electrocardiogram signals, blood pressure, muscle electromyography signals, glucose levels, galvanic skin response, lung capacity, snore waves, patient position, airflow and body scale parameters (weight, bone mass, body fat, muscle mass, body water, visceral fat, Basal Metabolic Rate and Body Mass Index). These broad sensing portfolio makes MySignals the most complete eHealth platform in the market.

All the data gathered by MySignals is encrypted and sent to the user's private account at Libelium Cloud through WiFi or Bluetooth. The data can be visualized in a tablet or smart phone with Android or iPhone Apps. The price of MySignals kit is 1757 $. The price of Mysignals kit is expensive in comparison with most of the available kits in the market.

Myo Armband is the basic concept of (HCI) in which humans interact with computers and design technologies that let humans interact with computers in an interactive way. Hence Myo is a new way that is used to control the real-life applications by the human. The structure of Myo consists of the EMG (electromyography) sensor and an IMU which consists of a gyroscope, accelerometer and a magnetometer [11].

EMG signal is the electric potential generated by muscle cells [12] when these cells are electrically or neurologically activated. The signals can be analyzed to detect medical abnormalities, activation level, or recruitment order, or to analyze the biomechanics of human. EMG signal are to be processed for multiple hand gesture and movement recognition. EMG monitors the electrical signals under human skin that are produced by the muscles. The Myo has no cameras, no keyboard and mouse, it just involves the hands. The Myo can detect any tiny electrical activation (contraction) produced by the muscle. Myo is a wearable bracelet that can be used as a commander for any connected device. It is like a thought being Implemented into a command for any device [13] (Fig. 2).

Myo Armband was developed by Thalamic Labs in summer 2014 [14]. Myo is an armband that can be worn on the forearm below the elbow which is controlled by human gestures and movements. The Myo involves five basic gestures wave left,

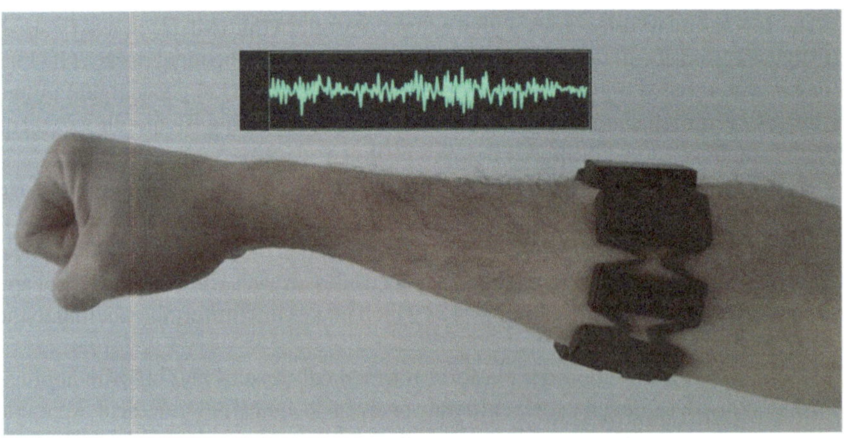

Fig. 2 A person wearing Myo armband in his forearm

wave right, Double tap, fist and fingers open. With the help of Myo a lot of tasks can be done easily like controlling of lights, robots, drones and change slides of presentation by just waving hand in lectures. With the help of Myo armband people can control movies, music from anywhere in their homes. Myo can be used to connect to YouTube [15], operate iTunes and VLC media player just by their gestures and hands movement. Myo plays an important role in the medical field, it can be used by doctors to examine the EMG reports and control their electronic devices.

Myo Armband detects the electrical activity in forearm muscles just below the elbow. The human forearm has different types of muscles, each of which has a different arrangement, and these muscles control the movement of the wrist, such as moving fingers, making a fist, turning left or right [16].

Myo armband designed on a wearable way it can fit in the human forearm easily. Sizing clips are available which allow for a more constrained grip, better suited for smaller arms. The sizing clips give it the ability to expand between 7.5 and 13 in. (19–34 cm) [15] forearm circumference. Unlike other EMG sensors, the Myo armband does not require the wearer to shave the area around which the armband electrodes will be worn, this allows for easier setup procedures in real world environment. The Myo armband weighs 93 g which gave it the ability to be wearable for a long time without uncomfortable feeling to the wearer. The Myo armband design is thin with a thickness of 0.45 in. (1.14 cm) which allows it to be worn under the shirts.

The Myo armband gives the user the ability to perform five gestures to interact with the environment. These five gestures include fist, double tap, finger spread, wave left, and wave right as shown in Fig. 3.

Myo developed applications focus on using five gestures only for different functions; this limitation might be considered positive because a lot of gestures can generate memorization difficulty for the user and difficulty to understand for what gesture should be used for a given situation [17].

Myo users can map each gesture into an input event to interact and communicate with the paired devices. According to the Myo SDK [18], the Myo provides gestural and spatial data to the application. The data is provided to the application by three types of events spatial, gestural and auxiliary event. The work of spatial data is to provide information to the application about the orientation and arm movement of the user, where the orientation points to the way in which the band is pointing. Thus,

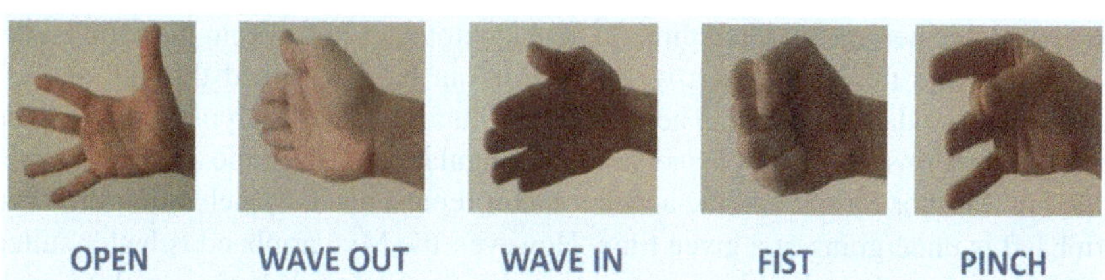

Fig. 3 Myo hand gestures

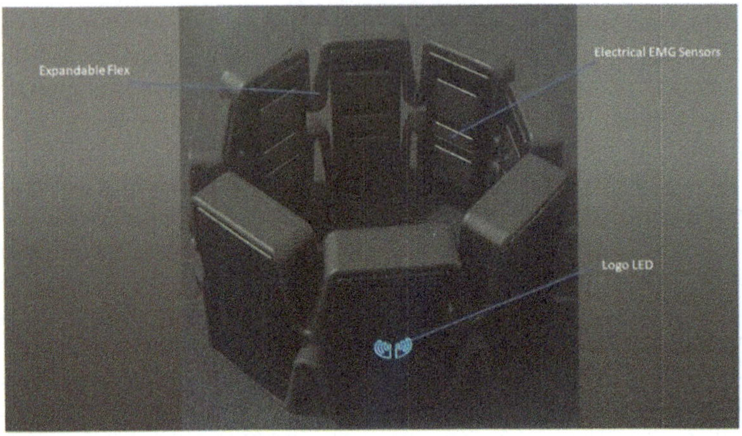

Fig. 4 Myo armband structure

the Myo SDK provides several kinds of spatial data: orientation in term of pitch, yaw and roll, acceleration vector data which represents the acceleration of the armband, and angular velocity data provided by the gyroscope.

The gestural event involves the data of what the user is doing with his hands. The SDK provides the gestural data in the form of preset gestures whereas the auxiliary events occur infrequently and corresponds to situations like Myo being connected or disconnected by the device.

The Myo armband structure shown in Fig. 4 has eight medical dry grade stainless steel EMG sensors like other surface electrodes (sEMG), the EMG signals returned by the sensors represent the electric potential of the muscles because of muscle activation [19]. These electrodes don't need any gel to be added to human skin to acquire the signal. However, since the electric potential of muscle is small in the range of sub millivolts, signals are sensitive to other sources of electric noise such as electric noise induced by wall-electricity. The range of potentials provided by the Myo armband is between −128 and 128 in units of activation. These units of activation are integer values of the amplification of the potentials measured by the EMG sensors. The Myo armband can pull sEMG data at a sample rate of 200 Hz.

The Myo armband is equipped with ARM Cortex M4 Processor. It includes a nine-axis inertial measurement unit (IMU) which contains a three-axis gyroscope, three-axis accelerometer and a three-axis magnetometer [14]. From these units, the orientation and movement of a wearer's arm can be determined through analyzing the spatial data provided. The orientation data indicates the positioning of the armband in terms of roll, pitch and yaw. The angular velocity of the armband is provided in a vector format and the accelerometer represents the acceleration the Myo armband is undergoing at a given time. However, the Myo armband is better suited for determining the relative positioning of the arm rather than the absolute position, a consideration to be aware of when applying pattern recognition algorithms. Currently, the Myo armband can pull IMU data at a sample rate of 50 Hz.

The Myo Armband has a Bluetooth 4.0 module embedded inside it which ease the communication between the ARM M4 processor of it and the communicated device. This wireless communication adds an advantage to Myo in comparison of other EMG sensors that needs wires to acquire the bio-signals.

A built-in rechargeable lithium ion battery which gives the Myo Armband the power to work for a whole day with single charge. A micro USB charging port allows to charge the internal battery of the Myo Armband by using a USB charging adaptor.

Among the previously discussed systems, Myo Armband has advantages since the user feel very comfortable while wearing it, the price is low in comparison with its features. The Myo Armband has 8 EMG signals with 8 channels. This chapter will focus on the applications of Myo Armband in biomedical and biometrics applications.

4 Wearable Armband Research Fields

The Myo Armband used in applications involving things like playing video games, controlling devices like drones or motorized vehicles, starting and stopping a music player, etc. In recent years, the word Myo has created a revolution in the technological world by its astonishing features and its utility in various fields. Numerous studies on Myo armband-based gesture recognition have been found in different fields such as: medical applications and robotics applications.

4.1 Wearable Armband Used in Robotics Field

Robotics has invaded almost every field but its tie-up with Myo is more promising. Myo Robotics which witnesses the merge of Cognitive Systems and Robotics has come into existence to make musculoskeletal robots readily available to researchers working in robotics and other domains (e.g. cognition, neuroscience), educators and the industry.

The structurally simple but electronically complicated armband can avoid our dependence on huge control systems. Gone are those days when we had to maintain and manage separate controllers for the robots. This simple wearable band enables us to control and command our robots with our gestures. With the birth of Myo dimensions of research on robotics has broadened.

In real world robotic systems, communication exist between humans and computer-based systems. Hand gesture is one of the imperative methods of communication for people to control the systems called gestural control. There is an active research being carried out in hand gesture recognition systems because of its applications for human computer interaction and virtual environments.

Sathiyanarayanan et al. [20], initially developed a prototype command-controlled robot, automatic controlled robot and gesture-controlled robot using gloves. In such situations in which Unmanned Ground Vehicle (UGV) are not allowed to be operated with base station assistance (manual and auto control), an alternative to tackle such problem is to provide gesture control mode for soldiers to maneuver UGV.

The gestural control mode has two sides on side is from Myo armband side in which the user wraps the Myo armband around his arms then by moving his hand this provides horizontal and vertical values based on the inclination along x and y axis of the ground. The values are serially monitored and transmitted by Arduino and Zigbee respectively while on the other side (Robot Side), UGV monitors serial input for the received characters and makes the subsequent decisions.

A research is done on the application of Myo Armband in the development of a controlling interface for PeopleBot robot [21]. This kind of application could be useful in many applications like reaching bio-infected or radioactive places or any place that would be harmful to human beings. Another application is to control a robot for helping elderly or disabled people on tasks involving drag and drop objects, PeopleBot robot is made for human-robot interaction problems. It has wheels to move and a grip to get objects on the top of a table.

To integrate Myo and PeopleBot, ROS (Robot Operating System) is used to allow the connection of the commands interpreted by Myo and sent to the robot controller. To allow the controlling of the robot, is necessary for Myo to be synchronized. This is done with the standard gesture from the Myo manufacturer. The movement of PeopleBot robot depends on the state of the motors, all the controlling of the robot depends on the angular position of the user's arm, defined by: Above 0.5: robot's gripper control, below 0.5: robot's movement control as shown in Fig. 5.

To analyze the precision in detection of movements by the software, some tests were executed on five volunteers. Users made 20 times each gesture in a random sequence. The mean classification rate to distinguish among movements was 93.6%.

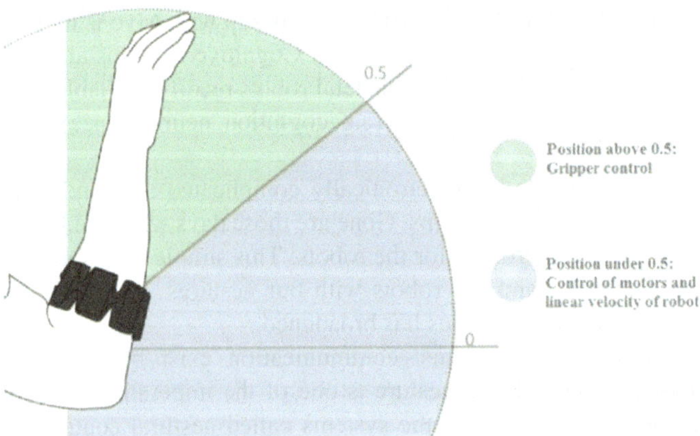

Fig. 5 Illustration of the control area [21]

Some difficulties while using Myo interface with the PeopleBot robot. The first one is the few number of gestures available. Thus, it can be concluded that the gestural options offered by Myo are limited for controlling more complex devices. Another point noted was that the continuous use of the equipment can cause some muscle discomfort due to fatigue generated by the repetition of gestures, which also causes problems in the recognition of the data. Apparently, muscle fatigue causes different myoelectric signals to the gesture performed or just turns it more difficult to be recognized by the algorithm. Physical activity before the use of Myo seems to cause difficulty on determining the gestures as well.

Swarms of robots will revolutionize many industrial applications, from targeted material delivery to precision farming. Controlling the motion and behavior of these swarms presents unique challenges for human operators, who cannot yet effectively convey their high-level intentions to a group of robots in application. Ferrer [22] proposes a new human-swarm interface based on novel wearable gesture-control. This work seeks to combine a wearable gesture recognition device that can detect high-level intentions, a portable device that can detect Cartesian information and finger movements, and a wearable advanced haptic device that can provide real-time feedback.

The wearable armband has a great impact on the research in robotics field specially programming by demonstration applications. In which the user has to perform some movements to control a robot. In the next section the Myo armband which revolutionized the research in medical field will be discussed.

4.2 *Wearable Armband Used in Medical Field*

Through studying the relationship between surface electromyography and hand kinematics, hand amputees will help recovering a significant part of their lost functionality using noninvasive methods. By investigating the accuracy of the Myo gesture control armband, more people may be able to afford devices which help them recover their hands functionality. Experiments were conducted using a data acquisition interface to gather surface electromyography to classify movements, the accuracy of the Myo armband can be compared to other devices that were used in similar studies. By analyzing the results obtained, the Myo armband may be a viable replacement for other more expensive devices that can analyze surface electromyography [19].

Muscular dystrophy is a group of genetic diseases that cause the loss of muscles and hence weakening the muscle strength. According to an investigation by the Centers for Disease Control and Prevention, 1 out of 7250 males aged 5–24 suffer from muscle dystrophy. A typical treatment for muscular dystrophy patients is routinely performing weight exercises to slow down the loss in muscles. Ho et al. [23] proposed a system and they called it MyoBuddy to help both physical therapists and patients to keep track of the weights in workout activities based on electromyography (EMG) sensors embedded in Myo armband.

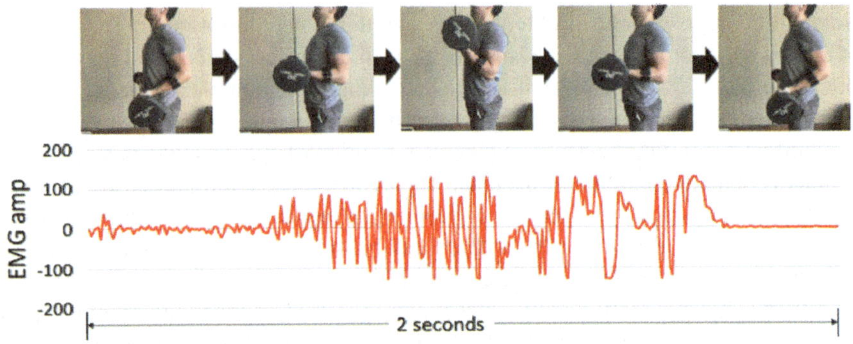

Fig. 6 The motion of one repetition of the barbell curl exercise [23]

The system measures EMG signals from users. A mobile phone application serves as a proxy to receive EMG data from Myo via bluetooth low energy and forwards the data to the server over Transmission Control Protocol (TCP). All the processing is done on the server.

Barbell bicep curl as the exercise in collecting experimental data the procedure of a standard barbell curl is shown in Fig. 6. Two persons tried their best to complete each barbell curl repetition in 2 s: 1 s for lifting the barbell and 1 s for returning to the start position. 102 sessions of data over a week have been collected. All features are computed separately for the 8 channels of Myo device. Support Vector Machine (SVM) with RBF kernel and Random Forest (RF) as learning classifiers have been used in this work.

The results showed that MyoBuddy can distinguish weights with a 10-pound increment with 73.4% repetition-level and 77.1% session-level accuracy. The system has potential to enable different interesting applications, such as a tool for bodybuilders to track their weight limits and adjust their training plan accordingly.

Myo Armband has been used in a very efficient way by researchers at Johns Hopkins University [24] to control a prosthetic limb using electric impulses transmitted from an amputee's mind to his limb. The armband works by reading the electromyographically (EMG) impulses triggered by a thought from a person's brain sending a signal to a limb, which causes a movement.

Medical researchers at the University's Applied Physics Lab (APL) have attached a couple of Myo armbands to an artificial prosthetic limb of Johnny Matheny, a man who lost his arm to cancer, just above his prosthetic limb. When he thinks about making a gesture, as he would have done with his original limb, the prosthetic makes the same gesture.

A research project aims to expand on the Non-Invasive Adaptive Hand Prosthetics (NinaPro) project started in January 2011 with the aim of controlling prostheses through sEMG channels and to provide a public sEMG database suitable for analysis of sEMG and its application to hand prosthetics [25].

In an attempt to further encourage research in this field that supports the motives of the NinaPro database. Abduo et al. [19] provide an open source solution which various devices can be integrated into. This solution would take care of all experimental procedures required such as which gestures need to be recorded and how they would be stored. Ultimately, researchers and developers would need to integrate their respective drivers into this solution given an easy to use API and they would be ready to go.

The interface screen of the developed program displays all the data that can be obtained from the Myo armband. In the upper half, the interface displays a video and image of the exercise that a participant should perform. The software contains three predefined sets of exercises. Recording functionality is also provided so that sEMG and IMU data can be recorded against the current exercise and repetition performed. Once enough repetitions of an exercise have been recorded, the next exercise in the selected exercise set is automatically loaded and the recording procedure is repeated. Data recorded is stored in a database file for each subject and exercise set performed. Various tables are used to record sEMG, accelerometer, orientation and gyroscope data.

Myo Armband has been used as well by Arnatanon et al. [26] to develop a system that transform the sign language into alphabet spelling based on EMG signals produced by the muscle. Deaf people use gestures to communicate every day. Therefore, it would be interesting to map such gestures to textual input in applications.

4.3 *Wearable Armband in Biometrics*

The growing popularity of wearable devices is leading to new ways to interact with the environment [27], with other smart devices, and with other people. Wearables equipped with an array of sensors can capture the owner's physiological and behavioral traits, thus are well suited for biometric authentication to control other devices or access digital services. However, wearable biometrics have substantial differences from traditional biometrics for computer systems, such as fingerprints, eye features, or voice.

Biometric recognition can be viewed as a pattern recognition problem in which a user who wants to be authenticated provides a set of physiological and/or behavioral characteristics to match a previously registered signature (or reference). Biometrics takes advantage of the fact that humans have natural diversity and certain traits are unique for everyone. Biometric systems, whether traditional or not, are usually composed of the three main functional components: (i) a sensor or set of sensors that capture raw biometric signals (r); (ii) a signal-processing unit that preprocesses and extracts feature vectors from the signals; and (iii) a recognition system, which usually includes a signature (or template) database and implements a pattern recognition function.

The matching phase depends on the mode of operation, either verification or identification. Biometric verification systems are configured by a sole user to verify the user's identity at a later stage. In biometric identification, the system is presented with a biometric signal and must decide who is the owner of that signal from a pool of registered users.

A biometric system should fulfill the following requirements:

- Performance: The system should respond promptly to queries with satisfactory accuracy.
- Acceptance: The system must be accepted by its intended users to be practical. If a sensor or device is not comfortable enough, it will not be used.
- Circumvention: The system should not be easy to circumvent. This implies that the system should be protected against unauthorized access to any of its components.

Wearable biometric system in which its primary user is in control of all the system components, including the signature database. A wearable biometric system requires owners to constantly wear the sensor that captures their bio signals. The signal processing and recognition units can also be embedded in the same wearable device or can be in a different smart device (e.g., a smartphone, pc). The resources unlocked when the user is successfully recognized by the wearable might include the rest of the services provided by the wearable or a cryptographic key that can be used to prove the identity of the user to other systems [28]. In any case, the process triggered after authentication is out of the scope of this survey.

In this configuration, wearable sensors are capable of reading signals from the subject at any time. This enables the biometric system to continuously authenticate the wearer.

Wearable biometric systems are generally used for identity verification processes. In this case, the biometric traits of the subject never leave the control of the user; they are stored in the wearable or a smart device in the user's possession. This avoids other entities accessing the biometric traits of the user provided that the devices are properly configured and protected against external attackers. An example of a commercial product implementing this philosophy is Nymi [29]. Nymi is a biometric verification wristband that includes one electrode in direct contact with the wrist and a second electrode that the user must touch with a finger from the opposite hand. When the user identity is verified, it has access to previously stored security tokens that can be used to authenticate against other devices, such as a car or a lock (Fig. 7).

Myo Armband presented in this chapter can be very useful in biometrics authentication system in which a user can wear an armband in his forearm to save a signature of this person. This signature will consist of some movements by the hand and muscles contraction. With the help of EMG signals and IMU embedded inside the armband. The signature will be a mixture of EMG, gyroscope, accelerometer and orientation signals as shown in Fig. 8. This signature can be a passkey that is used to access a PC, home and even cars. Also, it can be used to categorize a certain action taken by different users by creating a database from a lot of users performing certain movements (Gesture recognition). These movements can be classified to perform a predefined action.

5 Conclusion

Whatever the application, the "wearables technologies" is a hot topic. There is still more work to be done. Many devices are still in their testing phase and performance, benefits of use and long-term reliability need to be confirmed. The concept and new developments in this field are what will continue in the future and with the variety of devices possible will come the vast array of requirements which will define the need for suitable protection medium. Wearable technology is designed with the intention to make everyday tasks easier and more accessible. Wearable devices in biomedical and biometrics fields are growing rapidly. This will encourage the researchers developing different methods of interaction and communication, make it popular for use for all the people soon.

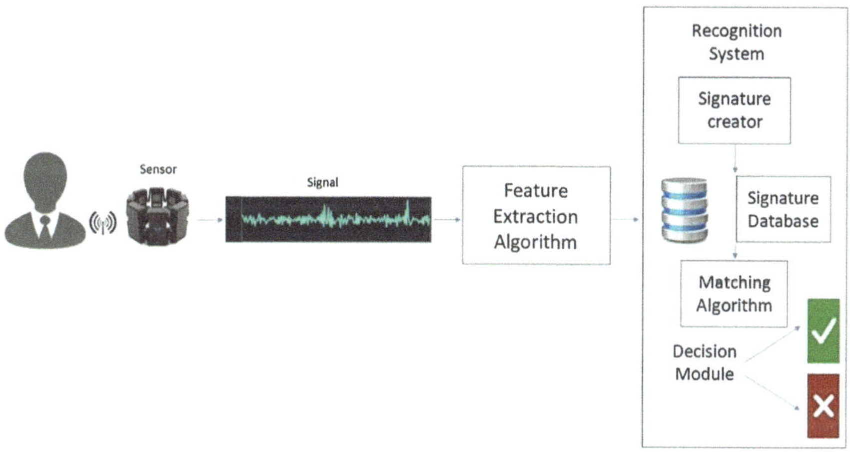

Fig. 7 Wearable armband for biometrics authentication

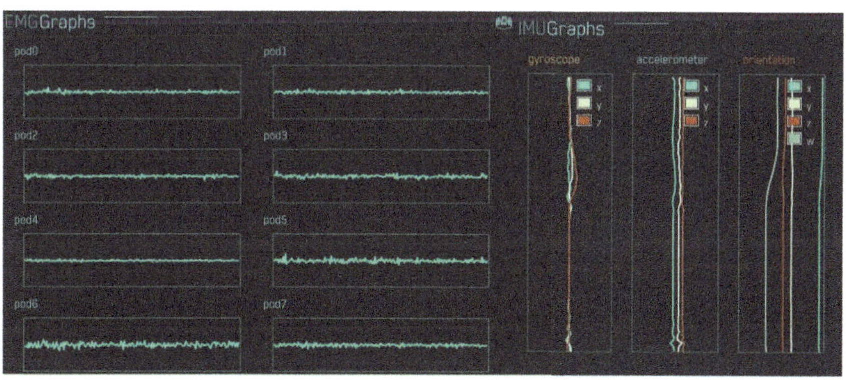

Fig. 8 Signals created using Myo

References

1. E.P. Scilingo, V. Gaetano, Recent advances on wearable electronics and embedded computing systems for biomedical applications. Electronics MDPI (2017)
2. P. Bonato, Wearable sensors/systems and their impact on biomedical applications. IEEE Eng. Med. Biol. Mag. (2003)
3. N. Carbonaro, F. Lorussi, A. Tognetti, Assessment of a smart sensing shoe for gait phase detection in level walking. Electronics MDPI (2016)
4. S. Said, S. Alkork, T. Beyrouthy, M. Fayek, Wearable bio-sensors bracelet for driver's health emergency detection, in *Biosmart*, Paris, France (2017)
5. A. Boehm, X. Yu, W. Neu, S. Leonhardt, D. Teichmann, Novel 12-lead ECG T-shirt with active electrodes, Electronics MDPI (2016)
6. A. Greco, A. Lanata, L. Citi, N. Vanello, G. Valenza, Skin admittance measurement for emotion recognition: a study over frequency sweep. Electronics MDPI (2016)
7. M. Caldara, B. Nodari, V. Re, B. Bonandrini, Miniaturized blood pressure telemetry system with RFID interface. Electronics MDPI (2016)
8. S. Saponara, M. Donati, L. Fanucci, A. Celli, An embedded sensing and communication platform and a healthcare model for remote monitoring of chronic diseases. Electronics MDPI (2016)
9. M. Bianchi, R. Haschke, G. Büscher, S. Ciotti, N. Carbonaro, A. Tognetti, A multi-modal sensing glove for human manual-interaction studies. Electronics MDPI
10. "http://www.biosignalsplux.com/downloads/bitalino_manual/manual.html Last checked 25 Feb 2018
11. S. Rawat, S. Vats, P. Kumar, Evaluating and exploring the MYO armband, in *International Conference on System Modeling & Advancement in Research Trends*, Moradabad, India (2016)
12. G. Robertson, G. Caldwell, J. Hamill, G. Kamen, S. Whittlese, *Research Methods in Biomechanics*, 2nd edn. (Human Kinetics, 2014)
13. https://store.myo.com/. 3 Feb 2018
14. http://blog.thalmic.com/. 3 Feb 2018
15. https://market.myo.com/. 2 Mar 2018
16. A. Ganiev, H.-S. Shin, K.-H. Lee, Study on virtual control of a robotic arm via a Myo armband for the self manipulation of a hand amputee. Int. J. Appl. Eng. Res. **11**, 775–782 (2016)
17. A. Pereira, J.P. Wachs, K. Park, D. Rempel, A user-developed 3-D hand gesture set for human-computer interaction, Hum. Factors J. (Human Factors and Ergonomics Society) (2014)
18. https://developer.thalmic.com/docs/api_reference/platform/the-sdk.html
19. M. Galster, M. Abduo, *Myo Gesture Control Armband for Medical Applications*, Department of Computer Science and Software Engineering, University of Canterbury (2015)
20. M. Sathiyanarayanan, T. Mulling, B. Nazir, Controlling a robot using a wearable device (MYO). IJEDR (2015)
21. G. Morais, L. Neves, A. Masiero, M. Castro, Application of Myo armband system to control a robot interface, in *Proceedings of the 9th International Joint Conference on Biomedical Engineering Systems and Technologies* (BIOSTEC 2016), vol. 4. BIOSIGNALS (2016)
22. E.C. Ferrer, *A Wearable General-Purpose Solution for Human-Swarm Interaction*. arXiv:1704.08393v2 (2017)
23. B.-J. Ho, R. Liu, H.-Y. Tseng, M. Srivastava, MyoBuddy: detecting barbell weight using electromyogram sensors, in *Proceedings of the 1st Workshop on Digital Biomarkers*, New York, USA (2017)
24. https://www.cantechletter.com/2016/01/thalmic-labs-myo-armband-allows-amputee-to-control-prosthetic-limb/
25. M. Atzori, A. Gijsberts, S. Heynen, A.-G.M. Hager, O. Deriaz, P. van der Smagt, Building the Ninapro database: a resource for the biorobotics community, in *Proceedings of IEEE International Conference on Biomedical Robotics and Biomechatronics* (2012)
26. V. Amatanon, S. Chanhang, P. Naiyanetr, S. Thongpang, *On Sign Language Thai Alphabet Conversion Based on Electromyogram* (2014)

27. J. Blasco, T.M. Chen, J. Tapiador, P. Peris-Lopez, *A Survey of Wearable Biometric Recognition Systems*, vol. 49 (ACM, 2016)
28. C. Rathgeb, A. Uhl, A survey on biometric cryptosystems and cancelable biometrics. EURASIP J. Inf. Secur. (2011)
29. https://nymi.com/
30. "https://www.pehub.com/canada (2017)
31. A. De Marcellis, E. Palange, L. Nubile, M. Faccio, G. Di Patrizio Stanchieri, T. Constandinou, A pulsed coding technique based on optical UWB modulation for high data rate low power wireless implantable biotelemetry. Electronics MDPI (2016)
32. A. Guidi, A. Lanata, P. Baragli, G. Valenza, E. Scilingo, *A Wearable System for the Evaluation of the Human-Horse Interaction: A Preliminary Study*

Biometrics from Cellular Imaging

Alice Ahlem Othmani, Sreetama Basu, Amulya Nidhi Shrivastava, Sinem Aslan, Francesco De Carli, Amesefe Delase Afua, A. S. M. Shihavuddin and Amine Nait-Ali

Abstract In this chapter, cellular imaging is considered from Medical Biometrics of cells. In particular, the chapter brings together different aspects of the cellular imaging from microscopy to cell biology, and from image processing to genomics.

1 Introduction

Cells are the fundamental building units of life. Since the establishment of this theory in early 19th century, scientists have been striving to study and understand the mechanisms underlying fundamental cellular and molecular functions. To obtain a better understanding of biological function of cell, cell imaging was pioneered in the first decade of the 20th century [1–4]. Since then, several microscopy methods have been developed that allow researchers to study cells with more precision and in an automated way. Consequently, heterogeneous, complex and large number of

A. A. Othmani (✉) · A. Nait-Ali
Université Paris-Est, LISSI, UPEC, 94400 Vitry-sur-Seine, France
e-mail: alice.othmani@u-pec.fr

S. Basu
Complex Cellular Phenotype Analysis Group Bioinformatics Institute (BII), Singapore, Singapore

A. N. Shrivastava
Paris-Saclay Institute of Neuroscience, CNRS, Paris, France

S. Aslan
Department of Informatics, Systems and Communication, Milano, Italy

F. De Carli · A. D. Afua
Institut de Biologie de L'Ecole Normale Supérieure (IBENS), Paris, France

A. A. Othmani · S. Basu · F. De Carli · A. D. Afua · A. S. M. Shihavuddin
Ecole Normale Superieure, CNRS, INSERM, PSL Research University, 75005 Paris, France

A. S. M. Shihavuddin
DTU Compute, Technical University of Denmark, Lyngby, Denmark

A. Nait-Ali (ed.), *Biometrics under Biomedical Considerations*,
Series in BioEngineering, https://doi.org/10.1007/978-981-13-1144-4_11

images are generated in modern cell imaging using the advanced microscope systems which require computerized techniques for quantitative and automatic analysis. These computational methods allow to achieve biometrics through measurements or characteristics extraction of the cells, quantify cell migration and interaction (morphodynamics) and cells organization (intracellular structures) [5, 6].

In this book chapter, we present an overview of cellular imaging, by highlighting the Medical Biometrics (in terms of measurement) of cells. In particular, the chapter brings together different aspects of the cellular imaging from microscopy to cell biology and from image processing to genomics. It is novel to combine all those aspects and discuss them from biometrics point of view. Our goal here is to gather and summarize some aspects of cellular imaging from the microscopy to the analysis of generated data. The rest of this chapter is structured as follows: the techniques of cellular imaging are presented in Sect. 2, the generated data types and issues are presented in Sect. 3, some specific applications of cellular imaging are presented in Sect. 4, the biometric approaches of cellular images analysis are presented in Sect. 5 and the unique biometry of the cell, the DNA in Sect. 6. In the last Sect. 7, we talk about the future of cell imaging and how past and actual technologies and developments can be expected to extrapolate into the future.

2 Cellular Imaging Techniques

The optical microscopes, invention which dates back to 16th century, can be accepted as the oldest design for microscopes [7]. Two main components of the device are an objective lens which is used to amplify the object to be observed and eyepieces which are used to further magnify the projected image on the intermediate image plane. In this pipeline, the optical lenses determine the spatial resolution which is limited to 0.2 μm using the visible light with the wavelength in the range of 390–760 nm. Optical microscopes can be grouped under different categories based on a variety of criteria. According to the lighting method they can be grouped under two categories, i.e. transmission and reflection microscope. In a transmission microscope, the light is transmitted through the specimen to the viewer from bottom to top, while in reflection microscope the specimen is illuminated from the top and is reflected from the specimen to the viewer. Optical microscopes can also be grouped according to the observation methods, i.e. bright field microscopes, dark field microscopes, phase difference microscopes, polarized light microscopes, interference microscopes, and fluorescent microscopes [8–11] where each can use transmission of reflection approach.

An important invention in 20th century was development of electron microscopy [12] where the magnification power of microscopic imaging was improved, i.e. spatial resolution reached up to 0.3 nm, by using electron beam and magnetic lenses instead of the initial optical lenses. The magnification power is obtained by accelerating the electrons to very high speeds which yields to reduction in the optical wavelength. A disadvantage of the electron microscope is only non-living specimens can be observed since it works under the vacuum environment. Another invention in 20th was Scanning Tunnel Microscope (STM) [13]. The working scheme of this microscopic imaging was totally different, i.e. no lenses are employed but a probe is used. Voltage is added between the probe and the observed object and when the distance between the probe and the surface of the observed objects is sufficiently small, tunnel effect occurs, i.e. electrons pass through the object surface from the probe and weak electronic current occurs. Strength of the current changes according to the distance between the probe and the observed surface, thus measuring the current change yields to observe the surface morphology of the object. Spatial resolutions reach less than 0.001 nm at STM which provides to locate single atoms.

Fluorescence microscopy is a branch of optical microscopy that has been a very common approach for studying dynamic cellular events of live cell [14–16]. While a conventional optical microscope uses visible light to illuminate and magnify the observed sample, a fluorescence microscope uses a much higher intensity light source for the same purpose. Once this high intensity light excites a fluorescent species in the observed sample, it emits a light of longer wavelength and magnified image of the observed sample is produced. Fluorescence microscopy provides better resolution than the conventional optical microscopy by increasing contrast between neighbouring structures in the cells and allows to label different structures with different colours. The working scheme of the fluorescence microscopy starts with illuminating the sample which are labelled by fluorophore with high energy light source (such as Mercury arc lamp) through the lens. Fluorophores that are attached to the sample absorbs the illumination light and emits lower energy light with longer wavelength and the emitted fluorescent light is separated by a specialized filter that can be visualized.

Three types of fluorescence microscope systems exist: widefield, scanning confocal and spinning disk confocal systems, schematics of which are given in Fig. 1, No single microscope system is suited to every experiment and compromises must be made depending on the experiment type. Therefore, three parameters are to kept in consideration: the sensitivity of detection, the speed of acquisition and the viability of the specimen [15]. The whole sample is permanently illuminated and emitted fluorescence is detected simultaneously by eye or camera in widefield microscopy, while confocal systems illuminate and record only one focal spot at once. Therefore, the acquisition is much fast at widefield microscopy. On the other hand, since the confocal systems remove the out of focus light, the obtained resolution is higher. Scanning confocal microscopes use a pinhole to eliminate the out-of-focus light whereas the spinning disk confocal microscopes exploits a rotating array of micro lenses to focus illumination. Acquisition rate is higher at spinning disk confocal microscopes than scanning confocal microscopes since specimen is scanned by thousands of points of light in parallel.

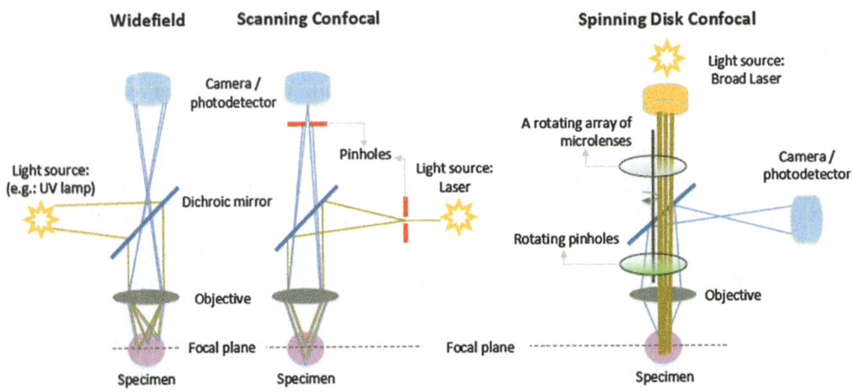

Fig. 1 Types of fluorescence microscope systems. Schematics demonstrate three types of fluorescence microscope systems. On the left, widefield microscopy is depicted where the whole sample is permanently illuminated and emitted fluorescence is detected simultaneously. Scanning confocal microscopy that is depicted in the middle panel uses a pinhole to eliminate the out-of-focus light. Right panel depicts the spinning disk confocal microscope which exploits a rotating array of micro lenses to focus illumination

With the best optics, the resolution of fluorescence microscopy is limited to 200 nm and in the following years many efforts are attained for super-resolution fluorescence microscopy [17], examples include but not limited to saturated structured-illumination microscopy (SSIM), stimulated emission depletion microscopy (STED), photoactivation localization microscopy (FPALM), photoactivated localization microscopy (PALM) and stochastic optical reconstruction microscopy(STORM). A very good review on basic principles and their applications in biology of these schemes can be found in [17]. As mentioned in [17], each technique has its own strengths and weaknesses and experimental designs are limited by the available fluorescence probes. For example, while STED employs traditional dyes and fluorescent proteins, PALM and STORM require special photoactivatable fluorophores. Moreover, while STED is more complicated and providing lower resolution (50–100 nm), PALM and STORM are technically easier and providing higher resolution (20–40 nm).

3 Data Types and Issues

In many cellular imaging applications, the signal level is weak especially in fluorescence microscopy where the fluorophore bleaches after exposure to light. Consequently, the number of detected photons is limited and the quality of image is degraded. To estimate the quality of the image, the signal-to-noise ratio (SNR) is considered: the more noisy the image it is, the lower the ratio of signal-to-noise [18]. The noise is not the only problem in cell-based imaging but also its complexity. The bio-logical diversity increase the intraclass heterogeneity of cellular morphology. When studying the living cells using time-lapse microscopy, the number of cells could not be the same in all the time-series images, it could for example be multiplied after the cellular division. Designing algorithms which deal with the noise of the microscope and the intraclass variability of the cells is not trivial task and need a lot of expertise. Image acquisition techniques presented in Sect. 2 generate large image data sets especially time-series of large volume data (time-lapse) imposing high requirements on storage capacity [19]. Furthermore, it is not only difficult to archive such huge data sets but also to visualize them [20]. It is also challenging to process and extract information from those big data sets to provide much deeper insight in detailed cellular phenomena. Thanks to multi-core processors, programs are properly designed to take advantage of parallelism that they could be executed faster than their sequential counterparts. This multiprocessing architecture increase the overall speed for programs and accelerate the process of big data. Biomedical image processing is one of the fields that benefit from this technological progress. However, much remains to be done especially that the data volume growth is extraordinary.

The low ratio of signal-to-noise, biological diversity, abundance, big dimensionality, heterogeneity and complexity of the data generated in modern cell imaging experiments make the handling, storage, visualization and computing more difficult, more demanding, more costly and needs more scientific requirements and efforts.

4 Cell Analysis Applications

4.1 Imaging of Biomolecules to Study Cell Behaviour

Cells exhibit various shapes (round, tubular, branched, elongated, etc.) and have the ability to morph (constrict, dilate, retract, elongate, etc.) under different biological conditions. This dynamic morphing of cells is controlled by the biomolecules present within a cell as well as external factors. Therefore, it is necessary to understand the underlying molecular processes that regulate the bulk behaviour of a given cell.

Several biomolecules such as lipids, proteins, carbohydrates, and/or nucleic acids as present within cells. Within a cell, these biomolecules form and are localized within different subcellular compartments; plasma membrane, endoplasmic reticulum, golgi bodies, cytosolic fluid, nucleus, etc. There are several tools and reagents to label either of these biomolecules and organelles. For example certain lipid types labeling using recombinant cholera-toxin B conjugated with a dye [21], mitochondria labeling using Mitotracker dyes [22], telomerase/proteins labeling using halotag technique [23, 24] has been achieved. Labeling of cellular proteins by antibodies, transfection and/or by transgenic expression is most commonly used to study the behavior of labeled proteins and the cells expressing them. Antibodies based labeling are primarily detected using organic fluorophores such as Alexa dyes. Transfec-

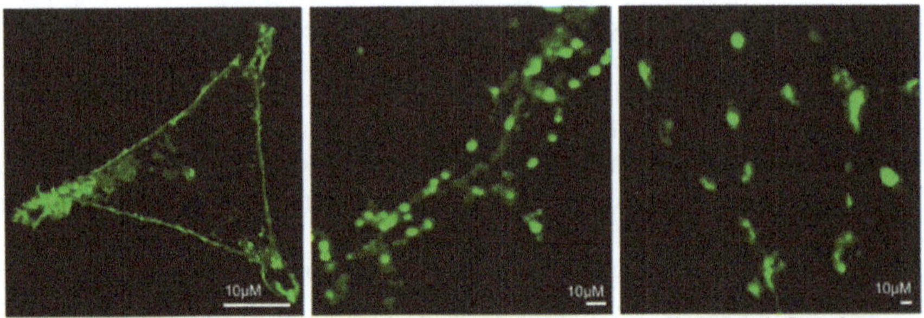

Fig. 2 Molecular labelling in cells. Different examples showing how molecular labelling could be achieved. On left, plasma membrane is identified using a pHluorin-tagged plasmid in HEK cells; middle panel identifies endocytosis pits using clathrin-mRFP plasmid in cultured hippocampal neurons; on right are mitochondria labelled using Mitotracker-green dye in live cultured hippocampal neurons

tion/transgenic expression involves conjugation and expression of protein of interest with a fluorescent protein such as GFP (Green Fluorescent Proteins). These fluorescent images like Fig. 2 are typically imaged using epifluorescence or confocal microscopes as described above.

Independent of the tool used to label cells (organelles or biomolecules), it can be argued that cell imaging is in fact (bio)-molecular imaging. The data extracted following image analysis can be used to not only obtain information about the biomolecules but also that of the cells [21, 25–27]. When imaging is performed at cellular resolution (1 μm, e.g. using low magnification objectives), it is not possible to define the precise localization and/or composition of biomolecules. However, in this case, it is possible to separate between two closely spaced cells by segmentation (see Sect. 5.2). It is noteworthy that segmentation of individual cells is only possible when the staining is bright and uniform with high SNR. However, this is not often the case as biomolecules exist in multiple forms: free unbound (diffused labeling-segmentation possible) and localized within certain sub-compartment (segregation does not separate individual cells but organelle). Further complexity may arise due to homomeric and/or heteromeric super-structure formation. Therefore, in order to study the localization of biomolecules, often, high-resolution (100 μm, e.g. high magnification confocal) imaging approaches are applied. 3D imaging allows nearly accurate localization of molecules within the cellular sub-compartments. When biomolecules exist as clusters (several molecules bound together), their localization within a given cellular compartment can be determined. For biomolecules that are free (single molecules) in the cytosol/membrane; a diffuse labeling like background is seen, making it difficult to quantify their precise localization and number as explained in next paragraph.

Even at high magnifications, confocal microscopes cannot separate between two individual biomolecules due to the diffraction limitation of light first noted by Ernst Abbe. This means that a microscope could not resolve two objects located closer than $\lambda/2NA$, where λ is the wavelength of light and NA is the numerical aperture of the imaging lens [28]. Visible light has a diffraction limit of 200–300 nm and it is not possible to separate molecules that are within this distance. This limitation has been overcome by the development of super-resolution imaging techniques such as PALM/STORM/SIM as described in Sect. 2 [29]. Following the advent of super-resolution imaging approaches, it has now become possible to understand precise molecular organization/dynamics within a cell that allows the modeling of the behavior of entire cell. Now it is possible to study cell shape dynamics [30], cytoskeletal dynamics [31], protein dynamics [23, 26], cell-to-cell-contact [32] and many other biological events that define cell behavior. Super-resolution techniques also allow us to understand the physico-pathology of neurodegeneration, which implies studying the interaction of pathological proteins such as amyloid β and α-synuclein, with the plasma-membrane [33].

4.2 Cell and Molecular Imaging of Neuronal Cells and Role of Image Processing

Studying the morphology of cells expressed in the central nervous system (brain) at high resolution has always remained a challenge for neuronal cell biologist and biophysicists. This is because of the highly complex architecture and dynamic morphology of neuronal cells [34–36] as shown in Fig. 3. Different cell-types within the brain are identified using cell-specific antibodies (neurons, astrocytes or microglia) or by transgenic expression of cell-specific fluorescent proteins. For simple cell counting, 3D, z-stacks can be obtained using low-resolution objectives (e.g. 10X). If the SNR is good, segmentation based algorithms can be used as cells appear round at low magnifications allowing cell counting. For neuronal morphology determination, imaging is performed at higher magnifications as discussed below.

It is increasingly becoming clear that neuronal cells and not round and within the brain they make hundreds and thousands of connections with other cells, and therefore scientists now are trying to identify these connections in order to understand the functioning of the brain. BRAIN Initiative [37], Allen Brain atlas [38], and Human Brain Project [39] are some of the sources providing information on brain cell connections. Extracting any information from these databases is impossible for man and requires complex image processing algorithms. Using image processing, it is possible to model certain circuits and correlate with the function of cells involved. Albeit of these existing databases, the information about neuronal cells architecture and function is far from complete and requires the development of new tools and software to analyze huge data [36].

One focus has been to develop new methodologies that would allow the imaging of single cell at molecular resolution within an entire brain. This is because currently it is only possible to obtain z-stacks/3D images at high cellular resolution in thick tissues using confocal microscopes (depth 100–200 μm) or two-photon microscopes (depth 100–200 μm). These distances are much shorter than the length covered by single neuronal processes. The depth limit in image acquisition is due to the fact that light is scattered by thick biological samples. This depth-limitation could be overcome by the use of tissue clearing approaches such as CLARITY [40] and iDISCO [41] that reduces the scattering of light through biological specimens, thereby allowing the imaging at higher resolution at greater depth (>1 cm) using light-sheet fluorescence microscopy. The potential of this approach was recently revealed in a study that used tissue clearing and antibody labeling protocol to analyze human embryo [42]. The volume of cellular and molecular information obtained by these studies is so high that they require the development of new computational tools to rapidly process such data.

4.3 Cell Based HCS Applications

4.3.1 Introduction to High-Content Screening

A trade-off on resolution for scale is offered by high-content screening (HCS) cellular imaging technology popularly called phenotypic HCS. Cellular imaging with HCS is a powerful tool for gaining systematic insights into biological processes through hundreds of image features (cell biometrics) at single-cell resolution on thousands of cells from various experimental conditions such as large chemical or natural substance libraries and their varying doses, or RNA interference studies. Typically, hundreds of low-level features such as texture, shape, intensity, intensity ratios of signals between subcellular compartments (e.g. cytoplasm, nucleus), spatial colocalization of signals etc. are measured in the hope that together they represent a complex phenotype which is representative of the morphology induced by drug

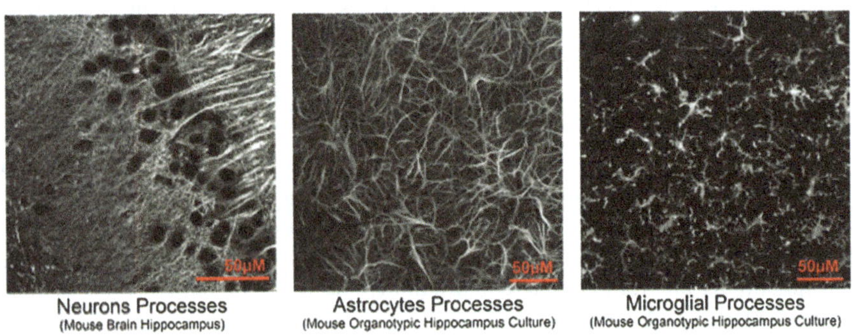

Fig. 3 Morphology of complex neuronal cells. Representative examples showing the complex morphology of neuronal cells (single z-stack) in tissue, neurons are tightly packed within the brain of adult mice hippocampus (left). Middle and right panel depict the complex morphology of astrocytes and microglial cells. Both these latter cells are highly dynamic and respond to external stimulus such as neuroinflammation

or genetic perturbation. Measuring biological markers in different cellular compartments indicate phenomenon like protein activation, translocation or colocalization. Phenotypic HCS is increasingly being adopted at all stages of drug discovery pipeline from compound screening and toxicity testing, to target identification and validation [43]. In academic research as well, imaging millions of cells with HCS is becoming routine for large scale perturbation screens of genome-wide cDNA expression systems and gene knock-down technology (siRNA interference) for understanding molecular mechanisms and pathways. Often, HCS set-ups is used to prioritise hypothesis testing which is followed up with experiments carried out by conventional higher resolution microscopy modalities for experimental validation.

HCS is at the cross-roads of multiple disciples, combining molecular cell biology, robotic sample preparation, automated high-resolution microscopy and image and data analysis. Cell-based HCS enable screening large libraries of compounds or genetic material and investigating cellular events in relevant in vitro biological context in living complete cells. This generates more accurate insights about complex, interdependent biological processes that was impossible with traditional cell extracts based screens. HCS systems come with various imaging modalities such as epifluorescence, confocal and phase contrast from different vendors.

Multichannel fluorescence microscopy, capturing differently labelled subcellular components and various proteins, can provide evidence to decipher mechanism of drug action or to understand gene functions. Phenotypic HCS rich and heterogeneous big data by imaging of various biological phenomena in millions of cells under different conditions generates. There is immense scope to exploit the rich high dimensional multi-parametric feature spaces modelling complex subcellular morphology to reveal novel chemical and biological knowledge about cellular pathways. The interested reader may refer to [44] for a nice introduction and for a recent review about the state of the art to [45].

4.3.2 HCS Applications: Drug Discovery, Biometrics and Personalized Medicine

Biometrics from cellular imaging has wide ranging applications from basic molecular biology to drug discovery, toxicity and personalized medicine. Cellular image based HCS have boosted the traditional target based drug discovery research in various ways and is noted as a more successful tool than traditional non-cellular-image-based approaches for discovery of small-molecule, first-in-class medicines [46]. Industry forecast expects cell-image based phenotypic tools to change greatly the pipeline of drug discovery research [47]. In toxicology studies in vitro phenotypic HCS assays of relevant cell models are used to predict in vivo toxicity of xenobiotics to humans [48], which is attractive for preclinical trial safety of drugs. In fact, the overall trend is a push in the community to replace existing disease and toxicity models with more physiologically relevant assays such as 3D HCS cell cultures of differentiated stem cell or patient derived primary cell instead of immortalized cell lines [49]. Hence,

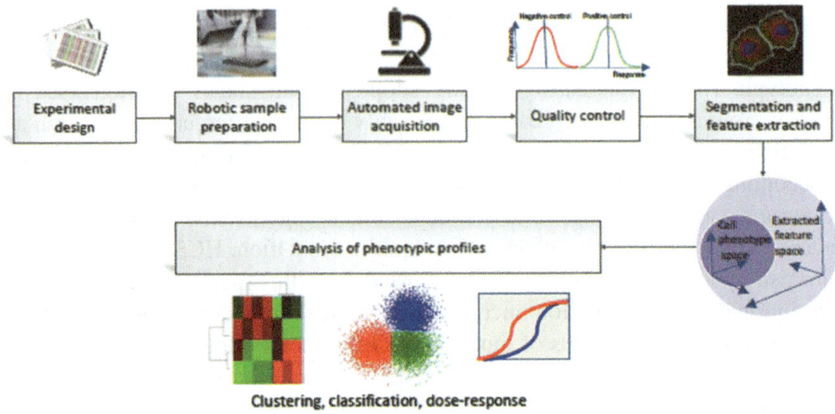

Fig. 4 High content screening workflow. It starts with experimental design and distribution of controls access the many plates for normalization and quality control. Usually with large scale screens the cell and liquid handling is performed by well calibrated robotics. Images that satisfy stringent quality control criteria are then pre-processed, segmented for single cell identification and feature extraction. The resulting phenotypic feature matrices are finally used for classification, clustering and other analysis

fast and efficient 3D cell segmentation and feature extraction algorithms are required to support analysis of 3D HCS data (see Fig. 4).

Phenotypic HCS enables aggregating of single cell biometrics across millions of cells. Therefore, it is very attractive for studying population heterogeneity such as tumor cell populations where cellular micro-environment has been observed to influence variable response to drug therapy and development of resistant subpopulations. Such heterogeneity of subpopulation can only be identified by exploiting multi-parametric cellular morphology features based on multichannel fluorescence images which is a major advantage over traditional cell count based methods and can be used to optimize treatment planning in cancer therapeutics [50].

The subpopulation studies with phenotypic HCS can even be scaled up to human population context to study the variability in cellular morphology and dynamics arising out of genetic differences. The Human Genome Project and the advancements in pharmacogenomics, have brought to light the impact of genetic composition as an significant factor of drugs metabolism and efficacy, which can sometimes manifest as severe adverse drug reactions [51]. Cell-image based HCS aids personalized medicine approaches to customize and optimize treatment planning by establish the biometric identity of patients. Given the high-throughput nature of HCS, in vitro screens on patient cells with large drug libraries and various dosage can simulate to some extent in vivo responses and help to identify drugs affecting desired responses and to minimize dangers of idiosyncratic responses.

Phenotypic HCS has the potential to completely revolutionize disease characterization and treatment planning. Alongside the benefits of phenotypic HCS come the technical hurdles arising from the complexity of highly multiplexed big data.

The image and data processing for accurate inference drawing from high content, high dimensional data places huge challenges for morphological feature representation and machine learning and requires development of methods well adapted to phenotypic HCS data analysis.

5 Cell Analysis Approaches and Softwares

It is now possible to quantify several biometrics like cell types or cell phases, cell interaction and morphodynamics and cell organization [5]. For that, we need computational approaches and algorithms to process the cell images in a stand-alone, similar and automatic way without tuning the different parameters of the algorithms. The development of smart automatic systems of cell processing were introduced in 1960s by using two-dimensional images for classifying white blood cells (leukocytes) in the base of color and morphology features [2, 6]. Since the mid-1950s, a very large technological progress allows the development of new technologies like microscopy and the development of multiple computer circuits to parallelize the tasks. Consequently, computers became powerful enough to handle big data and to propose robust, efficient and smart computational methods to answer biological issues. These approaches concern the detection and the quantification, the tracking

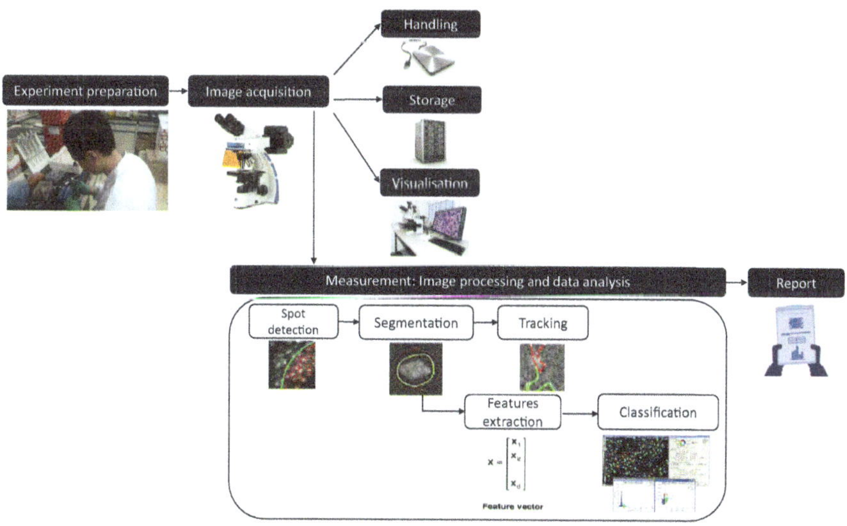

Fig. 5 Process flow for cellular images analysis starting with the biological samples preparation. Then, cellular images are acquired using one of the techniques described in Sect. 2. The large volume of data impose high requirements on storage, handling and visualization (Sect. 3). Several biometrics are extracted from the cell images and for that several cell analysis approaches and softwares are developed, more details are given in Sect. 5

and the generation of images of cells. In the following sections, we present in a non-exhaustive way the approaches most invested in cell bioimaging field (see Fig. 5).

5.1 Spot Detection

When the SNR is extremely low, especially in live cell imaging, cells may look like blobs or sub-resolution spots at low magnifications [52]. The spots are relatively blurred, small and compact, have no clear borders, sometimes touching or overlapping. Consequently, separating cells and delimiting them is very hard task and this is why researchers prefer to speak of detection rather than segmentation. Spot detection is a quite challenging task which has been well-studied in the literature. Several detectors are used for spot detection in fluorescence microscopy imaging [53] and Phase contrast microscopy imaging [54, 55]. Those detectors can be divided into unsupervised and supervised methods. The unsupervised family of approaches describe and model the cell structures from unlabeled data, like the linear or morphological filtering [56], clustering [57], model fitting [58], the spatial-frequency detectors like wavelet [59]. The supervised family of approaches learn the cell structures from training data and then can predict using the generated model the new cell structured from test data, hence, it is a classification task. Different classifiers have been used for spot detection like AdaBoost or Fisher Discriminant Analysis [52]. In [52], a quantitative comparison of spot detection methods in fluorescence microscopy was established by comparing seven unsupervised and two supervised methods. Synthetic images of three different types, as well as real images annotated manually by experts are used to perform experiments. The results of experiments suggest that for very low SNRs ($\approx$2), the supervised approaches perform better than the unsupervised ones. At high SNRs (>5), the difference in performance of all studied spot detectors becomes negligible.

5.2 Cell Segmentation and Quantification

The cell segmentation offer a better delimitation to the cell than the spot detection, on the other hand, it requires a better resolution and a lower SNR. The precise delimitation of the cells permits its quantification by the extraction of set of features. There had been lots of work done in cell segmentation with most of them being ap-plication specific [6]. For accurate cell segmentation, fundamental assumptions can vary based on which imaging method had been used, types of staining, cellular organization, and finally the relevant biological question at hand. The performance of segmentation methods depends on how high is the SNR, and how concise are the cell size and shape. Most of the segmentation method fails to perform consistently when there is large cell size and shape variation within the image and when there is significant overlap between cells especially in case of 2D images. Moreover, the task of accurate segmentation becomes even harder when cells are stained together with cytoplasm. More than that, it is usually hard to incorporate the prior knowledge of the biologists on the cell types. Therefore, semi-automated or semi supervised methods have been found to be more useful for biological studies.

Recently, deep neural networks have become very successful at various classification tasks in diverse domains including phenotypic HCS data. Empirical studies demonstrate deep learning based Convolutional Neural Networks (CNN) outperforms traditional machine learning methods such as random forests, SVMs [60] in accurate classification of single cell images. Deep learning, in contrast to traditional machine vision algorithms, presents the scope for extraction of complex, high-level abstractions of cell morphology by means of unsupervised feature learning. This is due to a hierarchical learning process in neural networks, where complex abstractions are formulated based on relatively simpler abstractions learnt in preceding levels of nodal layers. For example, in the context of face detection, the first layer may detect edges (abrupt intensity/ color change in image); the next may pool edges together to detect eyes or lips and subsequent layers may build features resembling complete faces. The main reason why deep learning outperforms traditional classifiers is due to superior feature representation learnt in the hierarchical model. The typical CNN alternates convolutional and pooling layers. The pooling layers merge semantically similar features with small variances of shift, intensity etc. in the next higher layer creating an invariant feature map [61]. However, the training of deep CNN is a demanding task given the very large number of hyper-parameters to be learned, the intraclass heterogeneity of cellular morphology and lack of single cell labels. Chemical or genetic perturbations of cells can induce very subtle phenotypic changes (compared to identifying cats versus bicycles), challenging even to domain experts. But variants of CNN have given promising results in the literature of histopathology images of cancer tissue in highlighting relevant differences in healthy and diseased cells [62, 63]. An-other rationale in favor of deep learning is the ability of HCS to generate thousands of cellular images. These large volumes of images can assist to scale up training data and has the potential to create good deep learning models.

5.3 *Tracking*

The tracking of cells concern the study of the dynamic processes of living cells at unprecedented spatial and temporal resolution and for that, we need to detect and follow single or multiple particles in a time series of images. The image processing pipeline of the tracking can be generally divided into two steps: the first step cover the spatial domain and concern the detection of the spots and the estimation of their coordinates in every frame of the image sequence, while the second step cover the temporal domain and concern the linking between the detected spots in the different frames using a set of criteria to form tracks. The first step of the pipeline was presented in earlier subsection on spot detection. Approaches of particles or spot linking ranged from simple nearest-neighbor to multiframe association, some include multiple tracking hypothesis with or without explicit use of motion models [64]. While previous studies are limited to either detection or linking and they are less objective in the sense that the evaluation is done by their original inventors and with their own data sets, in a recent study [64] a straightforward comparison of fourteen traditional and sophisticated methods of particle tracking based on an open competition was organized during ISBI 2012 conference. This comparison is unbiased since the methods are evaluated on a commonly defined data set and using commonly defined evaluation criteria. Chenouard et al. [64] believe that there study is representative of the state of the art and in our knowledge it is the more recent one and the more objective. None of the evaluated approaches performed best across all scenarios and clear differences are revealed between the various approaches.

Lately, modern microscopes like PALM/STORM offer single particle tracking (e.g. SPT-PALM) options [26, 65–67]. These methodologies allow simultaneous generation of thousands of short of trajectories. Due to high density and short length of trajectories, classical tracking methods are not ideal and new methods are being developed. For example one method is based on variational Bayesian treatment of hidden Markov models combining information from thousands of short single-molecule trajectories to extract diffusive states [68]. Another approach, multiple-target tracing, can be used to generate dynamic maps at high densities [69]. More recently, high-density SPT-PALM has been combined with statistical inference to map the diffusion and energy landscapes of membrane proteins across the cell surface for neuronal receptors [65, 70].

5.4 Generation of Images with Cell Populations

It is often difficult to decide whether obtained segmentation and tracking results are correct or not. In effect, manual generation of ground truth data is a long and tedious task, it is imprecise and highly variable among different experts. To answer that, versatile tools capable of generating image sequences of simulated living cell populations are previously introduced. Besides, dynamic cellular processes are simulated like cell motion, cell division, and cell clustering up to tissue-level density. Recently, a review is introduced on simulation methods ranging from particles up to tissue synthesis and then they have discussed the validity checking of the presented simulators [71]. In fact, the most elaborated and complete simulators of time-lapse image sequences of synthetic fluorescence-stained cell populations are: for 2D images TRAgen [72] and for 3D images MitoGen [73]. Those two simulators and their implementations are freely available [74, 75]. Both simulators simulate cell motility, shape and texture changes as well as cell divisions. To obtain realistic images, photobleaching, blur with real point spread function (PSF) and several types of noise are considered during the simulation.

5.5 *Projection*

In the microscopic image acquisition methods, it is common to take several images of the same object at different depth to capture entire part of object in focus. From those sets of images, it is important to make the optimum projection converting into one final 2D image comprising the most in focus pixels, there are some popular and simple methods such as maximum intensity projection, maximum variance projection, average intensity projection, etc. Those methods are implemented in public software like FIJI, Icy or Cellprofiler. None of these methods maintain the physical properties of the object. This issue was addressed for the first time in [76, 77] where they used wavelet analysis to compose a 2D image that captures the extended depth of field. Recently, in [78] they developed a projection technique that is locally consistent and follows the manifold of the object. This method classifies the pixels into foreground and background class and projects pixels from the manifold that goes through the estimated foreground.

5.6 *Bioimage Analysis Softwares*

At present, many software tools are available for bioimage processing. The most famous and elaborated open source ones in the field are Icy [79], Fiji [80] and Cellpro-filer [81]. For the paid softwares, Imaris [82] remains one of the most used softwares for microscopy image analysis and 3D visualization.

For data analysis and visualization of high content screening images, the HCS microscope vendors provide proprietary softwares, but alternatively there are open source platforms for image analysis and feature extraction—Cellprofiler [83], CellX-plorer [84]. In fact, Cellprofiler offers far more variety of sophisticated image analysis algorithms compared to CellXplorer, but the latter provides plate layout visualization interface, quick quality control check options and ability to select and view images in multi-and-single channel by wells and by frames. For data pre-processing, adapted quality control and analysis there exists good packages in most scripting languages such as Bioconducter [43], HCS-analyzer [85] (Fig. 5).

6 Unique Biometrics of the Cell: Its DNA

6.1 *DNA Decryption and Future Perspectives in Biometrics*

On a molecular level, DNA can be compared to a database within each cell where vital information is stored and sequentially retrieved. From this storage system originate exhaustive instructions essential for various aspects of cell life. Most of our visible traits or phenotype are coded in the series of nucleotide base composing our DNA. The shape, the size, the growth and function of every cell that make up a living organism is compiled in its genes. These coding regions of DNA are transcribed into RNA molecules and later translated to build proteins specialized in a variety of biological processes. Proteins involved in pigments synthesis for example will determine characteristics such as eye or hair color, others involved in hormonal pathways affect subtler traits such as stress resistance.

While physical and behavioral measurements provide enough information to accurately distinguish one individual from another, the analysis of genetic material opens access to a wider range of information going from genealogy to disease susceptibility. In humans, this genetic material or genome is made of 3.2 billion bases of DNA. Even though all humans have the same set of genes, small differences in DNA sequence, mostly single nucleotide polymorphisms (SNPs) account for the variety of physical features we observe making each individual unique. An increasing number of studies are using genome wide association approaches (GWAS) to correlate DNA polymorphisms with biometric features [86, 87]. Influential genomic regions associated with facial structure, weight, height and common diseases have already been highlighted using such methods [88, 89]. Whether an individual exhibits a trait or not might also depend on the abundance and panel of genes expressed in a subset of cells. Differential expression analyses are used to unveil alterations and abnormalities occurring in gene expression under given conditions. A variety of factors come into play when deciding what genes are switched on or off them genealogy to disease susceptibility. Most of our visible traits or phenotype are coded in our DNA. The shape, the size, the growth and function of every cell that make up a living organism is compiled in its genes. These codius affecting cells function and by extension that of the whole organism. The study of methylation through epigenetic approaches provides valuable insight into genes regulatory mechanisms. Understanding how genes are regulated and what stimuli affect critical genes expression is key to precisely harness the dynamics driving living organisms. With recent advances in sequencing technologies, we now have access to unprecedented volume of data to perform extensive studies on genomes (DNA), transcriptomes (RNA) and proteomes (proteins). Therefore, increasingly complex challenges occur requiring adapted computational and statistical tools to get the most out of produced biological data. On the long run, such advances will refine the way biomedical research operates, providing faster identification, more accurate diagnosing and personalized treatments of complex diseases.

6.2 DNA Imaging and Biometrics

DNA has a rich history of imaging. Although it is not known when DNA was first seen, the pioneering microscopist Antony van Leeuwenhoek was likely the first person to see it when he visualized the nucleus at the beginning of the 18th century [90]. Chromosomes were first observed in the second half of the 19th century, thanks to the introduction of fixation and staining procedures in microscopy preparations [91]. Originally fascinated by chromosome movements, scientists soon started to count the chromosome number [92]. In 1959, when numerical abnormalities of the karyotype were linked to the Down [93], Turner [94] and Klinefelter [95] syndromes, the interest in quantifying the genetic material became even more relevant. In the same decade, a huge milestone was achieved when the double-helix structure of DNA was revealed using X-ray fibre diffraction [96], bringing unprecedented resolution to DNA imaging. Visualizing the DNA backbone itself allowed us to generate detailed measurements of DNA geometry. Since then, many technical advancements have been made. From electron and atomic-force microscopy to ultra-high-resolution methods, molecular imaging techniques have provided a massive amount of information about the DNA structure, its physical and chemical properties and DNA/ligand interactions. Thus, DNA has always been an intriguing object to image and quantify because it carries the genetic instructions of all living organisms. One of the breakthroughs offered by DNA analysis was human DNA biometrics, which is the identification of an individual thanks to their DNA characteristics.

Imaging DNA in the form of electrophoretic bands allows us to measure the length of a specific polymorphic locus that varies in size among the population. The combination of multiple polymorphic loci yields a unique allelic pattern for a given person, and sets up the basis for forensic DNA fingerprinting [97]. Technological advancements such as next generation sequencing (NGS) have allowed us to reach the ultimate step in DNA analysis giving us access to the entire human genome. However,

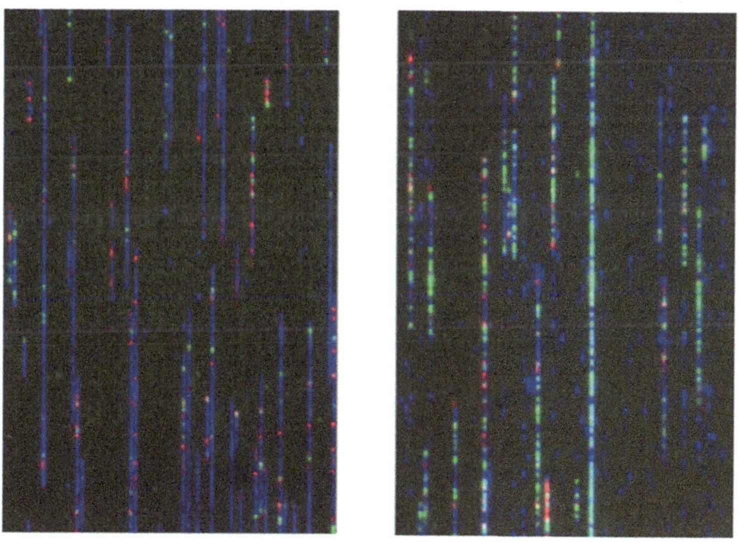

(a) Blood DNA **(b) Cell line DNA**

Fig. 6 Genomic DNA from both primary human blood cells (**a**) or a human lymphocyte cell line (**b**) was dual labeled with genetic labels (Nt.BspQl, red labels) and methylation sensitive labels (M.Taql, green labels), and stretched and imaged on a nanochannel array chip. Representative images from both samples are presented here. DNA backbone, stained with YOYO-1 is displayed in blue (figure extracted from the paper of Grunwald et al [100] and used under permission) because of the short length of NGS reads (few hundred bases), NGS cannot detect large (>1 kb) structural variants [98]. DNA imaging presents unique advantages to address this problem. For example, fluorescent labelling of long (i.e. megabase-size) DNA molecules at specific sequence motifs permits us to generate unique barcode patterns that can be revealed by high-throughput, single-molecule imaging in nanochannels arrays [99]. This generates optical maps that present several advantages over NGS: first, they can detect all structural variants beyond the kilobase scale (these are especially meaningful in human cancers characterisation); second, they increase the sensitivity in haplotype resolution (finding allelic variants at any locus); third, they allow the study of difficult genomic regions (e.g. repetitive regions, GC-rich regions). Finally, by labelling different genomic features with different colours, we can extract several types of information from the same DNA molecule. As an example, methylation analysis was recently achieved with a hybrid genetic/epigenetic barcode and high-throughput imaging in nanochannels [100], see Fig. 6. Ultimately, DNA imaging has the unique power to highlight molecular features in their native genomic context thus revealing the unique biometric signature of a human being.

7 Cell Imaging, Which Perspectives?

Cellular imaging and analysis, as seen in this chapter, is an interdisciplinary effort of researchers from multiple fields and disciplines: biology, robotics, physics, microscopy, computer vision and image processing, machine learning and statistics. The coordination and integration of expertise from multiple disciples is fundamental in addressing the major challenges facing the study of biometry of the cell. In our experience, it is often difficult to put it into practise with challenges ranging from academic to administrative. We should provide good conditions, resources and change mentalities and structures to facilitate interdisciplinary research to achieve common interests and shared goals. This effort is critical to accelerate progress of cellular imaging and biometrics for knowledge discovery.

The naked-eye visualization is never sufficient to make scientific conclusions. Recent advances of robotic sample preparations and microscopy technologies enable capturing cellular images in unprecedented resolution and volume. This has created novel demands from computational and quantitative analysis of cellular imaging data. For example, the impetus for the re-emergence and success of the field of deep learning for microscopy images is partially driven by the advances in automation of data acquisition, generating huge amounts of training data.

The plethora of modern microscopy imaging modalities and diverse markers and resolution of image has opened up new challenges for interpretation of cell images. With huge volumes of data being generated, protocols for efficient storage and retrieval of this data is now a priority. There is also a necessity for such bioimages to be made open-source to allow computer vision scientists to design new methods to provide novel insights into biological phenomena. In spite of decades of efforts, it remain challenging to design generic methods to be easily trainable for different cell imaging applications without requiring parameters tuning or algorithm adaptation. Image analysis methods seem to be locked in a stalemate where new techniques are proposed by the hundreds every year and yet are only applicable to the specific data they were developed on. To accelerate development of more robust and generic methods, we should first of all enforce the availability and the testability of the open-source model (both data and softwares) even before publication and frequently organize open challenges and hackathons to integrate new methods into popular software platforms and compare their performance on standardized test data and criteria [101]. Some very laudable efforts to integrate existing methods into one platform for ease of access for biologists and reproducibility of results are Fiji/ImageJ [80], ICY [79], and CellProfiler [81].

Aside from pushing the boundary of the state of the art a paradigm shift in expected in the field of cellular biometrics. Progress of bioimage analysis methodologies will be driven by trends in biology, for example, the current push for studying disease models in 3D even with high-throughput imaging systems due to the more physiologically relevant context it represents [49, 102]. There are also arguments for studying the dynamic cellular environment rather than static images which lead to the development organ on a chip models although at the beginning these technology is still low-throughput. Sophisticated image analysis algorithms are required to address the demands raised by such new developments. Finally, we are looking at bioimage research moving gradually into in silico experimentation supported by augmented reality and artificial intelligence in the form visualization with holographic projections where researchers can have a direct view into the 3D dynamic environments of cells, unconstrained by the complexities of markers and microscope lenses.

Acknowledgements A. Othmani would like to thank the Agence Nationale de la Recherche for funding her two years of postdoc (ANR-10-LABX-54 MEMO LIFE) at Institut de biologie de l'Ecole Normale Suprieure (Sep 2015-Sep 2017).

A. N. Shrivastava would like to thank Ronald Melki, CNRS (Paris-Saclay Institute of Neuroscience) and Antoine Triller, INSERM (Institut de biologie de l'Ecole Normale Suprieure) for their continuous support in his research and acknowledge the Agence Nationale de la Recherche (ANR-14-JPCD-0002-01) funding attributed to Ronald Melki. A. D. Afua sincerely thanks Dr. Lionel Navarro for the opportunity to join his lab as a young graduate, for making himself available whenever necessary and for the constructive research environment he provided. His precious support is a key motivator in her research work.

F. D. Carli has received support under the program Investissements dAvenir launched by the French Government and implemented by ANR with the references ANR10LABX54 MEMOLIFE and ANR10IDEX000102 PSL* Research University. F. D. Carli would like to thank Dr. Olivier Hyrien for scientific guidance and countless scientific discussions, and Benot Le Tallec and Vinko Besic for valuable comments on the manuscript.

References

1. M. Baker, Cellular imaging: taking a long, hard look. Nature **466**(7310), 1137–1140 (2010)
2. J. Prewitt, M.L. Mendelsohn, The analysis of cell images. Ann. N. Y. Acad. Sci. **128**(1), 1035–1053 (1966)
3. X. Qian, S.K. Goderie, Q. Shen, J.H. Stern, S. Temple, Intrinsic programs of patterned cell lineages in isolated vertebrate CNS ventricular zone cells. Development **125**(16), 3143–3152 (1998)
4. K. Preston, Computer processing of biomedical images. Computer **9**(5), 54–68 (1976)
5. J. Rittscher, Characterization of biological processes through automated image analysis. Annu. Rev. Biomed. Eng. **12**, 315–344 (2010)
6. E. Meijering, Cell segmentation: 50 years down the road [life sciences]. IEEE Signal Process. Mag. **29**(5), 140–145 (2012)
7. X. Chen, B. Zheng, H. Liu, Optical and digital microscopic imaging techniques and applications in pathology. Anal. Cell. Pathol. **34**(1, 2), 5–18 (2011)
8. M. Abramowitz, *Microscope Basics and Beyond* (Olympus Corporation Publishing, New York, 1987)
9. M. Abramowitz, *Contrast Method in Microscopy: Transmitted Light* (Olympus Corporation Publishing, New York, 1987)
10. M. Abramowitz, *Fluorescence Microscopy: The Essentials* (Olympus Corporation Publishing, New York, 1987)
11. M.W. Davidson, M. Abramowitz, *Optical Microscopy* (Olympus America, New York, 1999)
12. A. Bogner, P.H. Jouneau, G. Thollet, D. Basset, C. Gauthier, A history of scanning electron microscopy developments: towards wet-STEM imaging. Micron **38**, 390–401 (2007)
13. G. Binnig, H. Rohrer, Scanning tunneling microscopy. Helv. Phys. Acta **55**(6), 726–735 (1982)
14. S. Bradbury, P. Evennett, *Fluorescence Microscopy, Contrast Techniques in Light Microscopy* (BIOS Scientific Publishers Ltd., Oxford, 1996)
15. D.J. Stephens, V.J. Allan, Light microscopy techniques for live cell imaging. Science **300**(5616), 82–86 (2003)
16. E.C. Jensen, Overview of live cell imaging: requirements and methods used. Anat. Rec. **296**(1), 1–8 (2013)
17. B.O. Leung, K.C. Chou, Review of super-resolution fluorescence microscopy for biology. Appl. Spectrosc. **65**(9), 967–980 (2011)
18. C.J.R. Sheppard, M. Gu, M. Roy, Signal to noise ratio in confocal microscope systems. J. Microsc. **168**(3), 209–218 (1992)
19. C.T. Wallace, C.M. St. Croix, S.C. Watkins, Data management and archiving in a large microscopy and imaging, multi user facility: problems and solutions. Mol. Reprod. Dev. **82**(9), 630–634 (2015)
20. T. Pietzsch, S. Saalfeld, S. Preibisch, P. Tomancak, Big DataViewer: visualization and processing for large image data sets. Nat. Methods **12**(6), 481–483 (2015)
21. M. Renner, P.N. Lacor, P.T. Velasco, J. Xu, A. Contractor, W.L. Klein, A. Triller, Deleterious effects of amyloid beta oligomers acting as an extracellular scaffold for mGluR5. Neuron **66**(5), 739–754 (2010). https://doi.org/10.1016/j.neuron.2010.04.029
22. B. Chazotte, Labeling mitochondria with MitoTracker dyes. Cold Spring Harb Protoc. **2011**(8), 990–992 (2011). https://doi.org/10.1101/pdb.prot5648
23. J.B. Grimm, B.P. English, J. Chen, J.P. Slaughter, Z. Zhang, A. Revyakin, R. Patel, J.J. Macklin, D. Nor-manno, R.H. Singer, T. Lionnet, L.D. Lavis, A general method to improve fluorophores for live-cell and single-molecule microscopy. Nat. Methods **12**(3), 244–250 (2015). https://doi.org/10.1038/nmeth.3256
24. J.C. Schmidt, A.J. Zaug, T.R. Cech, Live cell imaging reveals the dynamics of telomerase recruitment to telomeres. Cell **166**(5), 1188–1197.e9 (2016). https://doi.org/10.1016/j.cell.2016.07.033

25. A.N. Shrivastava, J.M. Kowalewski, M. Renner, L. Bousset, A. Koulakoff, R. Melki, C. Giaume, A. Triller, β-amyloid and ATP-induced diffusional trapping of astrocyte and neuronal metabotropic glutamate type-5 receptors. Glia **61**(10), 1673–1686 (2013). https://doi.org/10.1002/glia.22548
26. A.N. Shrivastava, P.C. Rodriguez, A. Triller, M. Renner, Dynamic micro-organization of P2X7 receptors revealed by PALM based single particle tracking. Front Cell Neurosci. **26**(7), 232 (2013). https://doi.org/10.3389/fncel.2013.00232
27. A.N. Shrivastava, V. Redeker, N. Fritz, L. Pieri, L.G. Almeida, M. Spolidoro, T. Liebmann, L. Bousset, M. Renner, C. Lna, A. Aperia, R. Melki, A. Triller, α-synuclein assemblies sequester neuronal α3-Na^+/K^+-ATPase and impair Na^+ gradient. EMBO J. e201591397 (2015)
28. E. Abbe, Contributions to the theory of the microscope and microscopic detection (translated from German). Arch. Mikroskop. Anat. **9**, 413–468 (1873)
29. M.A. Thompson, M.D. Lew, W.E. Moerner, Extending microscopic resolution with single-molecule imaging and active control. Annu. Rev. Biophys. **41**, 321–342 (2012). https://doi.org/10.1146/annurev-biophys-050511-102250
30. J.M. Monteiro, P.B. Fernandes, F. Vaz, A.R. Pereira, A.C. Tavares, M.T. Ferreira, P.M. Pereira, H. Veiga, E. Kuru, M.S. VanNieuwenhze, Y.V. Brun, S.R. Filipe, M.G. Pinho, Cell shape dynamics during the staphylococcal cell cycle. Nat. Commun. **17**(6), 8055 (2015). https://doi.org/10.1038/ncomms9055
31. D. Li, L. Shao, B.C. Chen, X. Zhang, M. Zhang, B. Moses, D.E. Milkie, J.R. Beach, J.A. Hammer 3rd, M. Pasham, T. Kirchhausen, M.A. Baird, M.W. Davidson, P. Xu, E. Betzig ADVANCED IMAGING, Extended-resolution structured illumination imaging of endocytic and cytoskeletal dynamics. Science **349**(6251):aab3500 (2015). https://doi.org/10.1126/science.aab3500
32. W. Muranyi, S. Malkusch, B. Mller, M. Heilemann, H.G. Krusslich, Super-resolution microscopy reveals specific recruitment of HIV-1 envelope proteins to viral assembly sites dependent on the envelope C-terminal tail. PLoS Pathog. **9**(2), e1003198 (2013). https://doi.org/10.1371/journal.ppat.1003198
33. A.N. Shrivastava, A. Aperia, R. Melki, A. Triller, Physico-pathologic mechanisms involved in neurodegeneration: misfolded protein-plasma membrane interactions. Neuron **95**(1), 33–50 (2017)
34. H. Chai, B. Diaz-Castro, E. Shigetomi, E. Monte, J.C. Octeau, X. Yu, W. Cohn, P.S. Rajendran, T. M. Von-driska, J.P. Whitelegge, G. Coppola, B. S. Khakh, Neural circuit-specialized astrocytes: transcriptomic, proteomic, morphological, and functional evidence. Neuron **95**(3), 531–549.e9 (2017). https://doi.org/10.1016/j.neuron.2017.06.029
35. S. Gyoneva, D. Davalos, D. Biswas, S.A. Swanger, E. Garnier-Amblard, F. Loth, K. Akassoglou, S.F. Traynelis, Systemic inflammation regulates microglial responses to tissue damage in vivo. Glia. **62**(8), 1345–1360 (2014). https://doi.org/10.1002/glia.22686
36. N. Kasthuri, K.J. Hayworth, D.R. Berger, R.L. Schalek, J.A. Conchello, S. Knowles-Barley, D. Lee, A. Vzquez-Reina, V. Kaynig, T.R. Jones, M. Roberts, J.L. Morgan, J.C. Tapia, H.S. Seung, W.G. Roncal, J.T. Vogelstein, R. Burns, D.L. Sussman, C.E. Priebe, H. Pfister, J.W. Lichtman, Saturated reconstruction of a volume of neocortex. Cell **162**(3), 648–661 (2015). https://doi.org/10.1016/j.cell.2015.06.054
37. BRAIN Initiative, https://www.braininitiative.nih.gov/
38. Allen Brain atlas, http://www.brain-map.org/
39. Human Brain Project, https://www.humanbrainproject.eu/en/
40. K. Chung, J. Wallace, S.Y. Kim, S. Kalyanasundaram, A.S. Andalman, T.J. Davidson, J.J. Mirzabekov, K.A. Zalocusky, J. Mattis, A.K. Denisin, S. Pak, H. Bernstein, C. Ramakrishnan, L. Grosenick, V. Gradinaru, K. Deisseroth, Structural and molecular interrogation of intact biological systems. Nature **497**(7449), 332–337 (2013). https://doi.org/10.1038/nature12107
41. N. Renier, Z. Wu, D.J. Simon, J. Yang, P. Ariel, M. Tessier-Lavigne, iDISCO: a simple, rapid method to immunolabel large tissue samples for volume imaging. Cell **159**(4), 896–910 (2014). https://doi.org/10.1016/j.cell.2014.10.010

42. M. Belle, D. Godefroy, G. Couly, S.A. Malone, F. Collier, P. Giacobini, A. Chdotal. Tri dimensional visualization and analysis of early human development. Cell **169**(1), 161–173.e12 (2017). https://doi.org/10.1016/j.cell.2017.03.008. PubMed PMID: 28340341
43. Ivan Fraietta, Fabio Gasparri, The development of high-content screening (HCS) technology and its importance to drug discovery. Expert Opin. Drug Discov. **11**(5), 501–514 (2016)
44. Marc Bickle, The beautiful cell: high-content screening in drug discovery. Anal. Bioanal. Chem. **398**(1), 219–226 (2010)
45. Michael Boutros, Florian Heigwer, Christina Laufer, Microscopy-based high-content screening. Cell **163**(6), 1314–1325 (2015)
46. D.C. Swinney, Phenotypic vs. target-based drug discovery for first-in-class medicines. Clin. Pharmacol. Ther. **93**(4), 299 (2013)
47. J.G. Moffat et al., Opportunities and challenges in phenotypic drug discovery: an industry perspective. Nat. Rev. Drug Discov. **16**(8), 531 (2017)
48. R. Su et al., High-throughput imaging-based nephrotoxicity prediction for xenobiotics with diverse chemical structures. Arch. Toxicol. **90**(11), 2793–2808 (2016)
49. P. Horvath, N. Aulner, M. Bickle, A.M. Davies, E. Del Nery, D. Ebner, S.L. Shorte, Screening out irrelevant cell-based models of disease. Nat. Rev. Drug Discov. **15**(11), 751–769 (2016)
50. C.M. Garvey et al., A high-content image-based method for quantitatively studying context-dependent cell population dynamics. Sci. Rep. **6**, 29752 (2016)
51. W. Zhang et al., Pharmacogenetics of drugs withdrawn from the market. Pharmacogenomics **13**(2), 223–231 (2012)
52. I. Smal, M. Loog, W. Niessen, E. Meijering, Quantitative comparison of spot detection methods in fluorescence microscopy. IEEE Trans. Med. Imaging **29**(2), 282–301 (2010)
53. M. Hausmann, B. Schneider, J. Bradl, C.G. Cremer, High-precision distance microscopy of 3D nanostructures by a spatially modulated excitation fluorescence micro-scope, in *Proceedings of SPIE 3197, Optical Biopsies and Microscopic Techniques II*, 29 Dec 1997. https://doi.org/10.1117/12.297969
54. E. Cuche, F. Bevilacqua, C. Depeursinge, Digital holography for quantitative phase-contrast imaging. Opt. Lett. **24**(5), 291293 (1999)
55. P. Marquet, B. Rappaz, P.J. Magistretti, E. Cuche, Y. Emery, T. Colomb, C. Depeursinge, Digital holographic microscopy: a noninvasive contrast imaging technique al-lowing quantitative visualization of living cells with subwavelength axial accuracy. Opt. Lett. **30**(5), 468–470 (2005)
56. I. Smal, W. Niessen, E. Meijering, A new detection scheme for multiple object tracking in fluorescence microscopy by joint probabilistic data association filtering, in *5th IEEE International Symposium on Biomedical Imaging: From Nano to Macro, 2008, ISBI 2008* (IEEE, 2008), pp. 264–267
57. I.F. Sbalzarini, P. Koumoutsakos, Feature point tracking and trajectory analysis for video imaging in cell biology. J. Struct. Biol. **151**(2), 182–195 (2005)
58. L. Liang, H. Shen, P. De Camilli, J.S. Duncan, Tracking clathrin coated pits with a multiple hypothesis based method, in *International Conference on Medical Image Computing and Computer-Assisted Intervention* (Springer, Berlin, 2010), pp. 315–322
59. B. Zhang, M.J. Fadili, J.L. Starck, J.C. Olivo-Marin, Multiscale variance-stabilizing transform for mixed-Poisson-Gaussian processes and its applications in bioimaging, in *IEEE International Conference on Image Processing, 2007, ICIP 2007*, vol. 6 (IEEE, 2007), pp. VI-233
60. O. Durr, B. Sick, Single-cell phenotype classification using deep convolutional neural networks. J. Biomol. Screen. **21**(9), 998–1003 (2016)
61. Y. LeCun, Y. Bengio, G. Hinton, Deep learning. Nature **521**(7553), 436–444 (2015)
62. K. Sirinukunwattana, S.E.A. Raza, Y.W. Tsang, D.R. Snead, I.A. Cree, N.M. Rajpoot, Locality sensitive deep learning for detection and classification of nuclei in routine colon cancer histology images. IEEE Trans. Med. Imaging **35**(5), 1196–1206 (2016)
63. G. Litjens, C.I. Snchez, N. Timofeeva, M. Hermsen, I. Nagtegaal, I. Kovacs, C. Hulsbergen-Van De Kaa, P. Bult, B. Van Ginneken, J. Van Der Laak, Deep learning as a tool for increased accuracy and efficiency of histopathological diagnosis. Sci. Rep. **6**, 26286 (2016)

64. N. Chenouard, I. Smal, F. De Chaumont, M. Maka, I.F. Sbalzarini, Y. Gong, J. Cardinale, C. Carthel, S. Coraluppi, M. Winter, A.R. Cohen, Objective comparison of particle tracking methods. Nat. Methods **11**(3), 281–289 (2014)
65. D. Nair, E. Hosy, J.D. Petersen, A. Constals, G. Giannone, D. Choquet, J.B. Sibarita, Super-resolution imaging reveals that AMPA receptors inside synapses are dynamically organized in nanodomains regulated by PSD95. J. Neurosci. **33**(32), 13204–13224 (2013)
66. L.D. Lavis, J.B. Grimm, B.P. English, A.K. Muthusamy, B.P. Mehl, P. Dong, T.A. Brown, J. Lippincott-Schwartz, Z. Liu, T. Lionnet, L.D. Lavis, Bright photoactivatable fluorophores for single-molecule imaging. BioRxiv 066779 (2016)
67. S. Manley, J.M. Gillette, G.H. Patterson, H. Shroff, H.F. Hess, E. Betzig, J. Lippincott-Schwartz, High-density mapping of single-molecule trajectories with photoactivated localization microscopy. Nat. Methods **5**(2), 155–157 (2008)
68. F. Persson, M. Lindn, C. Unoson, J. Elf, Extracting intracellular diffusive states and transition rates from single-molecule tracking data. Nat. Methods **10**(3), 265 (2013)
69. A. Serg, N. Bertaux, H. Rigneault, D. Marguet, Dynamic multiple-target tracing to probe spatiotemporal cartography of cell membranes. Nat. Methods **5**(8), 687–694 (2008)
70. J.B. Masson, P. Dionne, C. Salvatico, M. Renner, C.G. Specht, A. Triller, M. Dahan, Mapping the energy and diffusion landscapes of membrane proteins at the cell surface using high-density single-molecule imaging and Bayesian inference: application to the multiscale dynamics of glycine receptors in the neuronal membrane. Biophys. J. **106**(1), 74–83 (2014)
71. V. Ulman, D. Svoboda, M. Nykter, M. Kozubek, P. Ruusuvuori, Virtual cell imaging: a review on simulation methods employed in image cytometry. Cytom. Part A **89**(12), 1057–1072 (2016)
72. V. Ulman, Z. Ormu, D. Svoboda, TRAgen: a tool for generation of synthetic time-lapse image sequences of living cells, in *International Conference on Image Analysis and Processing* (Springer, Cham, 2015), pp. 623–634
73. D. Svoboda, V. Ulman, Mitogen: a framework for generating 3D synthetic time-lapse sequences of cell populations in fluorescence microscopy. IEEE Trans. Med. Imaging **36**(1), 310–321 (2017)
74. TRAgen implementation, http://cbia.fi.muni.cz/projects/tragen.html
75. Mitogen implementation, http://cbia.fi.muni.cz/projects/mitogen.html
76. F. Aguet, D. Van De Ville, M. Unser, Model-based 2.5-D deconvolution for extended depth-of-field in brightfield microscopy. IEEE Trans. Image Process. **17**(7), 1144–1153 (2008)
77. B. Forster, D. Van De Ville, J. Berent, D. Sage, M. Unser, Complex wavelets for extended depth-of-field: a new method for the fusion of multichannel microscopy images. Microsc. Res. Tech. **65**(1–2), 33–42 (2004)
78. A. Shihavuddin, S. Basu, E. Rexhepaj, F. Delestro, N. Menezes, S.M. Sigoillot, E. Del Nery, F. Selimi, N. Spassky, A. Genovesio, Smooth 2D manifold extraction from 3D image stack. Nat. Commun. **8**, 15554 (2017)
79. F. De Chaumont, S. Dallongeville, N. Chenouard, N. Herv, S. Pop, T. Provoost, V. Meas-Yedid, P. Pankajakshan, T. Lecomte, Y. Le Montagner, T. Lagache, Icy: an open bioimage informatics platform for extended reproducible research. Nat. Methods **9**(7), 690–696 (2012)
80. J. Schindelin, I. Arganda-Carreras, E. Frise, V. Kaynig, M. Longair, T. Pietzsch, S. Preibisch, C. Rueden, S. Saalfeld, B. Schmid, J.Y. Tinevez, Fiji: an open-source platform for biological-image analysis. Nat. Methods **9**(7), 676–682 (2012)
81. A.E. Carpenter, T.R. Jones, M.R. Lamprecht, C. Clarke, I.H. Kang, O. Friman, D.A. Guertin, J.H. Chang, R.A. Lindquist, J. Moffat, P. Golland, Cell profiler: image analysis software for identifying and quantifying cell phenotypes. Genome Biol. **7**(10), R100 (2006)
82. Imaris, http://www.bitplane.com/
83. L. Kamentsky, T.R. Jones, A. Fraser, M. Bray, D. Logan, K. Madden, V. Ljosa, C. Rueden, G.B. Harris, K. Eliceiri, A.E. Carpenter, Improved structure, function, and compatibility for CellProfiler: modular high throughput image analysis software. Bioinformatics (2011). PMID: 21349861. PMCID: PMC3072555

84. D. Lak-sameethanasan, R.Z. Tan, G.W.-L. Toh, L.-H. Loo, cellXpress: a fast and user-friendly software platform for profiling cellular phenotypes. BMC Bioinform. **14**(Suppl 16), S4 (2013). https://doi.org/10.1186/1471-2105-14-s16-s4
85. A. Ogier, T. Dorval, HCS-Analyzer: open source software for high-content screening data correction and analysis. Bioinformatics (2012)
86. F. Liu, F. Van Der Lijn, C. Schurmann, G. Zhu, M.M. Chakravarty, P.G. Hysi, A. Wollstein, O. Lao, M. De Bruijne, M.A. Ikram, A. Van Der Lugt, A genome-wide association study identifies five loci influencing facial morphology in Europeans. PLoS Genet. **8**(9), e1002932 (2012)
87. K. Adhikari, M. Fuentes-Guajardo, M. Quinto-Snchez, J. Mendoza-Revilla, J.C. Chacn-Duque, V. Acua-Alonzo, C. Jaramillo, W. Arias, R.B. Lozano, G.M. Pérez, J. Gmez-Valds, A genome-wide association scan implicates DCHS2, RUNX2, GLI3, PAX1 and EDAR in human facial variation. Nat. Commun. **7**, 11616 (2016)
88. Wellcome Trust Case Control Consortium, Genome-wide association study of 14,000 cases of seven common diseases and 3,000 shared controls. Nature **447**(7145), 661 (2007)
89. C.J. Willer, E.K. Speliotes, R.J. Loos, S. Li, C.M. Lindgren, I.M. Heid, S.I. Berndt, A.L. Elliott, A.U. Jackson, C. Lamina, G. Lettre, Six new loci associated with body mass index highlight a neuronal influence on body weight regulation. Nat. Genet. **41**(1), 25–34 (2009)
90. A. Van Leeuwenhoek, Opera omnia, seu arcana naturae ope microscopio rum detecta, vol. 1. Langerak. Cited before Dieter Gerlach, Geschichte der Mikroskopie. Verlag Harry Deutsch, Frankfurt am Main, Germany, 2009 (1722). ISBN 978-3-8171-1781-9
91. E.B. Wilson, *The Cell in Development and Inheritance Macmillan* (New York, 1896)
92. K. Oguma, S. Makino, A revised check-list of the chromosome number in Vertebrata. J. Genet. **26**, 239–254 (1932)
93. J. Lejeune, Etude des chromosomes somatiques de neuf enfants mongoliens. L'Académie des Science Paris **248**, 1713–1727 (1959)
94. C.E. Ford, K.W. Jones, P.E. Polani, J.C. De Almeida, J.H. Briggs, A sex-chromosome anomaly in a case of gonadal dysgenesis (Turner's syndrome). Lancet **273**(7075), 711–713 (1959)
95. P.A. Jacobs, A case of human intersexuality having a possible XXY sexdetermining mechanism. Nature **183**, 302–303 (1959)
96. J.D. Watson, F.H. Crick, Molecular structure of nucleic acids. Nature **171**(4356), 737–738 (1953)
97. A.J. Jeffreys, V. Wilson, S.L. Thein, Hypervariable minisatellite regions in human DNA. Nature **314**(6006), 67–73 (1985)
98. S. Goodwin, J.D. McPherson, W.R. McCombie, Coming of age: ten years of next-generation sequencing technologies. Nat. Rev. Genet. **17**(6), 333–351 (2016)
99. E.T. Lam, A. Hastie, C. Lin, D. Ehrlich, S.K. Das, M.D. Austin, P.Y. Kwok, Genome mapping on nanochannel arrays for structural variation analysis and sequence assembly. Nat. Biotechnol. **30**(8), 771–776 (2012)
100. A. Grunwald, H. Sharim, T. Gabrieli, Y. Michaeli, D. Torchinsky, M. Juhasz, K.R. Wagner, J. Pevsner, J. Reifenberger, A.R. Hastie, H. Cao, Reduced representation optical methylation mapping (R2OM2). bioRxiv 113522 (2017)
101. E. Meijering, A.E. Carpenter, H. Peng, F.A. Hamprecht, J.C. Olivo-Marin, Imagining the future of bioimage analysis. Nat. Biotechnol. **34**(12), 1250–1255 (2016)
102. C.C. Bilgin, G. Fontenay, Q. Cheng, H. Chang, J. Han, B. Parvin, BioSig3D: high content screening of three-dimensional cell culture models. PLoS ONE **11**(3), e0148379 (2016)

Medical Devices Design: Vital Sign Monitoring Case Study

Sandy Rihana

Abstract This chapter is entitled "Medical Devices Design: Vital Sign Monitoring Case Study". It presents the proof of the concept of vital signs monitoring devices designed and conceived in laboratories. These types of designs can be used or adjusted in many applications related to Biometrics such as in medical Biometrics, clinical Biometrics and for biosignals authentication purpose.

1 Introduction

Within the field of Medical Biometrics, wearable technologies are getting more and more deployed in clinical environments. The previous chapter highlighted numerous applications of wearable technologies, requiring biosignal acquisition and processing. Therefore, in this chapter, we will basically focus in a specific study case, of wearable monitoring devices design, allowing: acquisition, monitoring and analyzing vital signs such as heart rate, blood oxygen level, respiration and blood glucose level. These devices provide non invasive acquisition, wireless monitoring and user feedback and communications. The design of the wearable device follows the five essential steps, as mentioned by the Device Development Process proposed by the FDA (Food and Drug Administration) [1]. As illustrated in Fig. 1, these steps are described as follows:

Step 1 is the Device Discovery and Concept, in which the researchers create a concept or a system of a new device, a new hardware, new software or a new formula. In general, the new device should meet an unmet medical need. Researchers build a "proof of concept", they present a concept and they analyze it making it move to later stages of development.
Step 2 concerns the preclinical research, the prototyping phase. At this stage, the device is not for human use. The prototype should be tested in controlled settings reducing thus any risk of harm in people.

S. Rihana (✉)
Holy Spirit University of Kaslik, USEK, Jounieh, Lebanon
e-mail: sandyrihana@yahoo.com; sandyrihana@usek.edu.lb

A. Nait-Ali (ed.), *Biometrics under Biomedical Considerations*,
Series in BioEngineering, https://doi.org/10.1007/978-981-13-1144-4_12

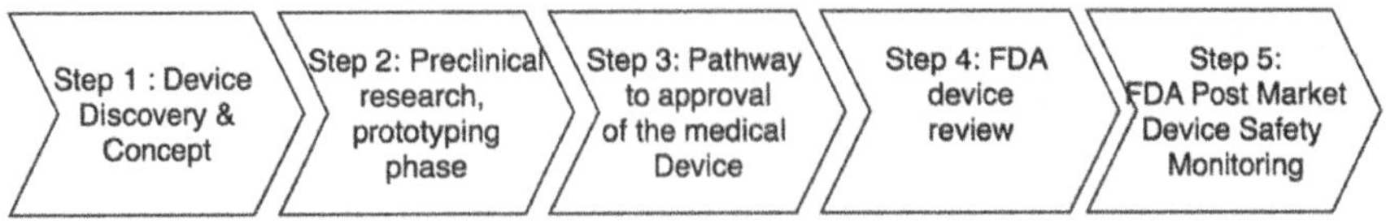

Fig. 1 FDA essential steps

Step 3 Pathway to approval of the medical device based on its class. In fact, the Federal Food, Drug and Cosmetic Act, section 513 classify the medical devices into three groups: Class I, "low risk of illness or injury", class II or "moderate risk" and class III, "unreasonable risk of illness or injury". Regulations increase from Class I to Class III, where class I devices subject to the least regulatory control and Class III devices subject to the most regulatory control.
Step 4 is the FDA device review and Step 5 is the FDA Post market device safety monitoring.

Our focus in this chapter is to present the proof of concept of vital signs monitoring devices designed and conceived in laboratories. The proof of concept includes development of prototypes of critical features. It is conceived to finalize the design requirements.

This chapter is organized as follows: a general design system architecture is presented in the next section. Then, a "divide and conquer" approach is used for simplifying the design. In the second part, the sensors and measurements are cited and described, followed by the hardware and firmware used to pilot these sensors and to analyze the measurements. The system design is described in the system block diagram that shows the interaction between the sensors, the hardware and firmware and the communications modules. Therefore, the next section is dedicated to the communication modules. Finally, the last section is dedicated to data processing and analysis, and examples of implemented algorithms are presented. By the end of this chapter, readers should be able to understand the basics of the design process and to create a design from requirements.

2 General Design System Architecture

Figure 2 illustrates the general architecture of any portable medical device. In this model, the patient serves as source of input data, biopotentials and the healthcare operator is the person analyzing and managing the collected data on a mobile device.

The sensors are the main components to acquire the physical measurement from the patient. The measurements can be and not limited to: blood pressure, heart rate, respiratory rate, pulse oximeter, voice, video recording and many others. Choosing the appropriate sensor for the specific application is very crucial. Accuracy, sensitivity, repeatability, size, price, noise level and many other characteristics should be studied in detail.

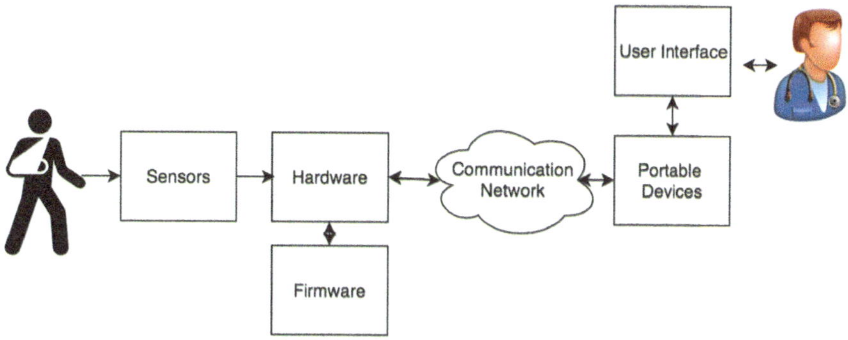

Fig. 2 General architecture of portable medical device

The hardware part is the main core of the analog front end, it includes all the modules required to acquire the analog signals. The performance of the hardware is very crucial in terms of number of analog and digital input and output, in terms of sampling rate, resolution, processing speed, bandwidth, communication interfaces, power management, etc.

The hardware is piloted by the firmware, which is the program code devoted to hardware management and to signal and data processing.

The data will be transferred through the communication network (WIFI, Bluetooth, Zigbee, or others, etc.) to the mobile device (cell phone, tablet, etc.) where a software application implements the interface with the users of the medical instrument.

In this chapter, examples of vital sign monitoring will be covered and the architectural design of a portable medical device will be described.

3 Sensors and Measurements

3.1 Pulse and Oxygen Saturation in Blood

SpO_2 is an indication of the saturation of the oxygen in the peripheral capillary. In other words, it indicates the "Concentration of Oxygen in Blood" which could be referred as the percentage of hemoglobin binding sites in the bloodstream occupied by oxygen. In fact, the oxygenated hemoglobin and hemoglobin have different absorption of red and near infrared light. Thus, the level of oxygenated blood cells can be estimated by placing two LEDs, one for red light and another for infrared light, and detectors in the opposite side of the oximeter and calculating the level of absorption of both lights.

Knowing the incident light and the transmitted light, the absorption can be deduced using the Eq. (1):

$$I_{transmitted} = I_{emitted} - Absorption \tag{1}$$

Then, according to Beer-Lambert law, the absorption differs from one solute to another and it is given by the Eq. (2) below:

$$A = \varepsilon.D.C \tag{2}$$

where A is the absorption, C is the concentration, D is the distance traveled by the light and ε is the absorptivity coefficient, which is wavelength-dependent [2].

In order to measure the saturation of oxygen, we should know the concentration of reduced and oxygenated hemoglobin. In addition, the absorption of red and infrared light should differ from one hemoglobin type to another. Fortunately, this is true for hemoglobin since oxygenated hemoglobin has peak absorption at 940 nm, and reduced hemoglobin has peak absorption at 660 nm.

Measuring the ratio Red/Infra-red or what is known as can determine the saturation of a suspension of pure hemoglobin:

$$\frac{Absorption\ at\ 660\ \text{nm}}{Absorption\ at\ 940\ \text{nm}} \tag{3}$$

In addition, the SpO_2 can be calculated using the Eq. (4):

$$\%SpO_2 = \frac{HbO_2}{HbO_2 + Hb} \tag{4}$$

As for the heart rate, it can be estimated by simple peak detection from the PPG signal.

Moreover, SpO_2 is an essential vital sign used in clinical observations especially when it comes to respiratory abnormalities like "Sleep Apnea". SpO_2 is considered as a crucial vital sign where physicians observe the variation of SpO_2 for clinical diagnosis. SpO_2 values are percentage readings and the ranges of SpO_2 are divided into three main ranges:

- $90\% \leq SpO_2 \leq 100\%$: "Normal Range". This range of percentage indicates that the patient has normal concentration of oxygen in the blood.
- $80\% \leq SpO_2 < 90\%$: "Abnormal Range". This range of percentage indicates that it is a low concentration of oxygen in the blood.
- $SpO_2 \leq 80\%$: "Dangerous Range". This range of percentage indicates that the patient is in the danger zone. It could lead to malfunctioning of organs such as heart, brain and pulmonary failure. If the concentration of oxygen continues to be that low, the patient will suffer from cardiac or pulmonary arrest [3].

A clinical application of the SpO_2 usage is the apnea detection. A person who suffers from "Sleep Apnea" faces "Apnea Events" that occur occasionally, suddenly or continuously. When an "Apnea Event" occurs the patient will suffer from irregular breathing or cessation of airflow and as a result, insufficient air will flow into the lungs. This insufficient quantity will affect the SpO_2 levels of the patient. So when "Apnea Events" occur, the concentration of oxygen in blood will drop under its normal ranges and this drop should last for 10 s to consider it as "Apnea Event". When SpO_2 value is read it is analyzed in the first place in order to detect and classify "Apnea Events":

If:

- $SpO_2 \geq 90\%$: "Normal Case" When the SpO_2 is detected for 10 s with its value higher than 90%, it is considered as a normal case so there is no "Apnea Event" detected.
- $86\% \leq SpO_2 < 90\%$: "Apnea Level 1": When the SpO_2 is detected for 10 s with its value between 86 and 90%, it is considered as an "Apnea Event" level 1 so the event is detected and registered.
- $82\% \leq SpO_2 < 86\%$: "Apnea Level 2": When the SpO_2 is detected for 10 s with its value between 82 and 86%, it is considered as an "Apnea Event" level 2 so the event is detected and registered.
- $78\% \leq SpO_2 < 82\%$: "Apnea Level 3": When the SpO_2 is detected for 10 s with its value between 78 and 82%, it is considered as an "Apnea Event" level 3 so the event is detected and registered.
- $SpO_2 < 78\%$: "Apnea Level 4": When the SpO_2 is detected for 10 s with its value between 78 and 82%, it is considered as an "Apnea Event" level 4 so the event is detected and registered. This level of SpO_2 is considered a dangerous level where the patient is required to contact his physician immediately for any medical assistance.

3.2 AirFlow Sensor

AirFlow is considered also is one of the crucial vital signs used in patient monitoring especially in ICU's "Intensive Care Units". Physicians and specialists refer to this vital sign in their clinical diagnosis considering that from AirFlow they could retrieve the "Ventilation Frequency" which is the "Respiratory Rate Interval" that is a primary parameter in respiratory abnormalities detection especially in "Sleep Apnea".

In general, "AirFlow" sensor is considered as an "AirFlow" meter. The purpose of using the "AirFlow" meter is to have an estimation of the quantity of Air that is passing along an object. As for an example, when air flows along a tube, the users use an "AirFlow" meter so that they could have an approximation of the amount of air flowing through the tube. Specifically, "AirFlow" sensor is a measurement instrument that provides its users with the mass of air flowing in an object but it is not the volume of air flowing in a tube.

Medically, "AirFlow" sensor is related to respiration. This sensor is primarily used in patient monitoring in order to monitor, analyze and evaluate the respiratory system of a patient. Any respiratory abnormalities or variations in RR "Respiratory Rate" are detected while using the "AirFlow" sensor. These variations and abnormalities are very important to detect since it leads to major physiological instabilities like "Sleep Apnea". In addition, RR "Respiratory Rate" variations are the primary indicators of these instabilities so this physiological parameter is important in the assessment of the patient health status.

RR "Respiratory Rate" is also known by many significant terms like the following: Ventilation frequency (Vf) or Respiration frequency (Rf). Specifically RR is the measurement that provides physicians and specialists with the number of breaths in 1 min which is represented by the two cycles of respiration: Inhalation and Exhalation.

RR is divided into normal ranges and abnormal ranges [4–7].

- Normal RR Ranges: Normal RR is also known by the term "Eupnea". The variation of the ranges of RR is based on the age of the person and it is divided as follow:
  - Infant at the age of six weeks: $30 \leq RR \leq 60$.
  - Infant at the age of six months: $25 \leq RR \leq 40$.
  - Child at the age of three years: $20 \leq RR \leq 30$.
  - Child at the age of six years: $18 \leq RR \leq 25$.
  - Child at the age of ten years: $12 \leq RR \leq 15$.
  - Adults: $16 \leq RR \leq 20$.
  - Elderly more than 65 years: $12 \leq RR \leq 28$.
  - Elderly more than 80 years: $10 \leq RR \leq 30$.
- Abnormal RR Ranges:

  Abnormal respiratory rate is divided into two main sub-ranges:

- An increased respiratory rate (higher than normal rates): Tachypnea–Apnea.
- A decreased respiratory rate (lower than normal rates): Bradypnea–Apnea.

In order to have an accurate detection of apnea and to classify the severity of the detected events, and additional parameter is implemented. "AHI" which is: "Apnea-Hypopnea Index". This index represents the number of detected "Apnea Events" during 60 min, and the severity of these detected events is classified as follow: [8] (Table 1).

Table 1 AHI range

AHI	Classification
5–15	Mild Apnea
15–30	Moderate Apnea
>30	Severe Apnea

3.3 Temperature Sensor

Temperature sensors could be classified into contact and contactless sensors. Based on NTC, Negative Temperature Coefficient, is a thermistor sensor commonly used in vital sign monitoring. It covers a range between 0 and 50 °C.

Non contact sensors such as infrared sensors are commonly used. It allow to acquire the temperature of the body, without having any influence on it.

As it is known, each body that has a temperature above zero degree Kelvin will emit electromagnetic radiation at its surface, proportional to its internal temperature. The infrared radiation has a range between 0.78 μm and 1000 μm. However, only the wavelengths between 0.7 and 14 μm are important for temperature measurement.

3.4 Microphone

The input (sound signal) is detected by the microphone. The microphone will transform the detected acoustic signal into an analog signal. The designed system must detect desirable signals (such as desirable sound signals or others) and suppress all the other signals. In order to do so, digital signal processing techniques such as digital filtering, adaptive filtering, time-frequential methods and many others can be used to only keep signals having frequencies within the frequency range of wanted signals. Once the signal is detected and filtered, a decision criterion based on different features extracted from the signal would be used to provoke a notification message sent via Bluetooth to the mobile application. The mentioned features could be time domain, frequency domain or time-frequency domain.

3.5 Live Streaming Using a Camera

For safety reasons, it would very important to be able to visually monitor a person (child, elderly or patient) from a distance. This feature can help caretakers ensure that the patient is doing well.

The live monitoring feature is done using a sensor. As a result, the activation of the camera should be done from distance using the mobile application. Once activated, the system would stream video signals to the cellphone.

4 Hardware and Firmware

The system for measuring the different vital signs includes e-Health sensor shield, connection bridge and arduino or Raspberry Pi board.

E Health sensor shield is designed for biomedical researched by Cooking Hacks. The board is compatible with Intel Galileo Boards, with Raspberry Pi, and Arduino. The vital signs acquired such as body position, pulse oxygen, body temperature, airflow, …. Are obtained using the eHealth board and its associated sensors.

The shield is compatible with many available electronic boards especially with Arduino and Raspberry Pi. However, its connection to a Raspberry Pi requires a special shield: The Raspberry Pi to Arduino shield connection bridge.

Even though the designed E-Health shield is not certified for medical use, it can be used for educational and product development purposes especially in the prototyping phase.

Moreover, a dedicated C++ library has been developed for the E-Health with examples showing the use of each sensor. The library is originally available for Arduino. A special library called "ArduPi" has been developed to allow the use of this library and Arduino codes on Raspberry Pi. All the needed libraries are freely available online from the cooking hacks website [9].

Example of application could be using this shield with arduino or with raspberry Pi.

Raspberry Pi is a very small, low cost computer, which was designed for education in order to help improving the hardware and software understanding. Even though it is slower than a modern computer, it still supports different operating systems such as Raspbian, windows 10 and Arch Linux ARM.

Different models of Raspberry Pi exist. The second generation of Raspberry Pi is the Raspberry Pi 2 model B. This model has a 900 MHz quad-core ARM cortex-A7 CPU and 1 GB RAM. Therefore, it can run full range of ARM GNU/Linux distributions, as well as Microsoft windows 10 [10]. This big upgrade in processor means that the performance is two times better and for really multi-thread-friendly code, the speed can increase up to 7.5 times.

As for the hardware, this mini-computer has four USB ports, 40 GPIO pins, and camera and display interface.

Concerning the power draw, just having the Pi 2 model B running idle, it draws 200 mA. However, doing heavier computational tasks, such as using Bluetooth, having inputs connected to the GPIO and using the USB ports, this will need a 5 V 2000 mA power supply.

5 System Block Diagram

Three different blocs: signal acquisition, signal processing and communication could be identified in any architecture.

In this example, signal acquisition consists of five different sensors. The pulse oximeter sensor, the temperature sensor and the airflow sensor that are connected to the E-Health platform. Whereas the microphone and the USB camera are directly connected to Raspberry Pi.

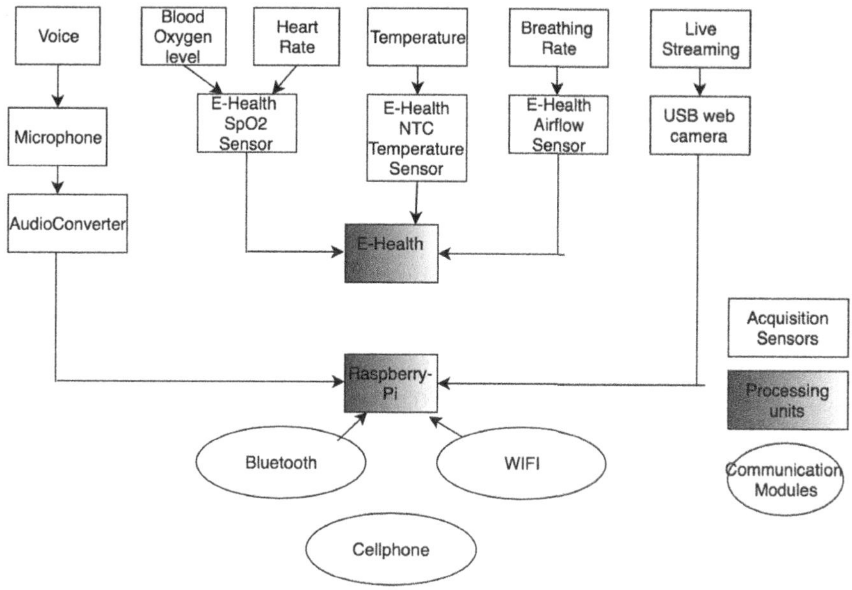

Fig. 3 Vital sign monitoring block diagram

Next, the signal processing bloc is represented by the raspberry Pi, which contains the dedicated libraries and developed codes. This bloc mainly processes all acquired data. It is considered as the main processor.

Finally, the communication layer involves the Bluetooth and the Wi-Fi module which stream the data to a dedicated mobile application. Figure 3 illustrates the detailed block diagram.

6 Wireless Communications Modules

6.1 *Bluetooth Communication*

Bluetooth technology was created in 1994 as a solution permitting data transfer using radio transmissions instead of cables. This technology has many benefits including wide availability in many systems (cellphones, laptops ...), low power consumption, ease of use and low cost.

Since there is no strong evidence that radio frequency waves are dangerous, Bluetooth is considered safe.

Bluetooth has two major types:

Bluetooth Basic Rate/ Enhanced Data Rate (BR/EDR) adopted in versions 2.0 and 2.1 which permits short range continuous wireless connection.

Bluetooth with Low Energy (BLE) adopted in versions 4.0, 4.1 and 4.2 which allows short periods of long range connections. Dual mode chipsets permits the connection to the two types of Bluetooth (Bluetooth) [11].

6.2 Wi-Fi Communication

"Wi-Fi is the name of a popular wireless networking technology that uses radio waves to provide wireless high-speed Internet and network connections" (www.webopedia.com). In fact, Wi-Fi, which is a wireless local area network technology (WLAN), is the trademark name corresponding to the IEEE 802.11x standards which specifies a fixed channel bandwidth of 25 MHz for 802.11b and 20 MHz for either 802.11a or g networks.

Moreover, the 802.11 family includes different versions with different specifications [12]:

- 802.11: Uses 2.4 GHz frequency band and provides 1- or 2-Mbps transmission
- 802.11a: Uses 5 GHz frequency band and provides 54 Mbps transmission
- 802.11b: Uses 2.4 GHz frequency band and provides au to 11 Mbps transmission
- 802.11 g: Uses 2.4 GHz frequency band and provides 20+ Mbps transmission

In addition, Wi-Fi networks are half-duplex, which makes a communication between an access point and a mobile station possible in both ways but not at the same time.

Finally, "IEEE 802.11 wireless LANs use a media access control protocol called Carrier Sense Multiple Access with Collision Avoidance (CSMA/CA)". In order to detect collisions, a Distributed Control Function (DCF) was developed. DCF states that transmission by a Wi-Fi station will occur only when clearness of the channel is ensured. Once a transmission is made, the station waits for an acknowledgement which will insure the absence of collision.

6.3 TCP-IP Protocol

The basic communication language of the internet is the TCP/IP (Transmission Control Protocol/Internet Protocol). It consists of 2 layers:

- TCP layer is responsible of assembling a message into smaller packets that will be sent over the internet. This message will be received and reassembled by the TCP layer.
- IP layer assigns addresses to each packet. Packets from the same message might be routed differently, but they will be reassembled at the destination.

TCP/IP is used by web browsers via different protocols like HTTP (Hyper Text Transfer Protocol), HTTPS (secure http) and FTP (File Transfer Protocol).

TCP/IP is also used by email programs via different protocols like SMTP for sending emails (Simple Mail Transfer Protocol), MIME for multimedia files transmission (Multi-purpose Internet Mail Extensions), POP for email retrieve from the server (Post Office Protocol) and IMAP also for email retrieve from the server (Internet Message Access Protocol) [12].

7 Data Processing and Analysis

Some data processing examples are illustrated in this paragraph.

7.1 Sound Processing

Once the sounds were acquired, data analysis followed the flowchart shown in Fig. 4.

In the features extraction, twelve temporal and frequential features are extracted in order to identify the significant ones for sound detection (Table 2).

The events' types are the following: Cries, Silence, Music (Lullaby) and Talk.

In order to differentiate between the events, different classifiers could be tested: Support Vector Machine (SVM) with different kernels, K-Nearest Neighbor (KNN), ANN-Artificial Neural Network.

Fig. 4 Audio processing flowchart

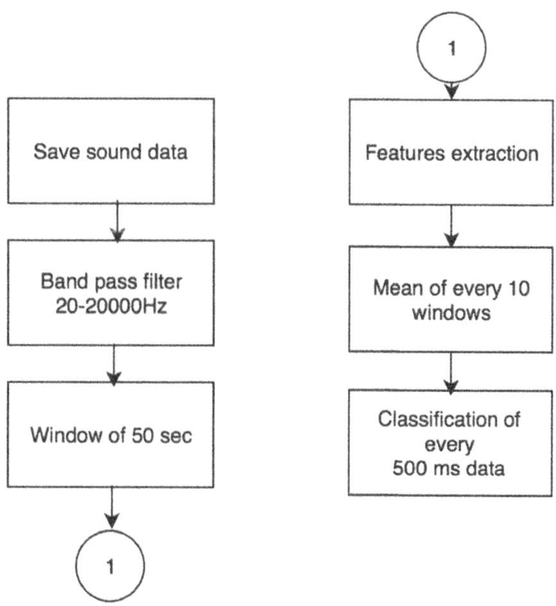

Table 2 Features extracted from audio signals

Feature	Formula
Mean	$\frac{\sum x}{N}$
Standard Deviation	$std = \sqrt{\frac{\sum\left((x-\underline{x})^2\right)}{N-1}}$
Variance	$var = std^2$
Hjorth Parameters: Mobility	$Mobility(x) = \sqrt{\frac{var(x')}{var(x)}}$
Hjorth Parameters: Complexity	$\frac{Mobility(x')}{Mobility(x)}$
Mean Frequency	Frequency corresponding to the mean of the Power Spectrum Density
Maximal Frequency	Frequency corresponding to the maximum of the Power Spectrum Density
Median Frequency	The median of the Power Spectrum Density
Minimal Frequency	Frequency corresponding to the minimum of the Power Spectrum Density
Skewness	$\sum \frac{[(x-\underline{x})]^3}{std(x)^3}$
Energy	$\sum_i x(i)^2$
Power	$\sum fft(x).conj[fft(x)]$

where:
- x is the signal's sample
- N is the number of samples
- $\underline{x}$ is the signal's mean

In order to evaluate the performance of a classifier, accuracy, specificity and sensitivity can be computed.

$$\text{Specificity} = \frac{TN}{TN + FP} \times 100$$
$$\text{Accuracy} = \frac{TP + TN}{P + N} \times 100$$
$$\text{Sensitivity} = \frac{TP}{TP + FN} \times 100$$

where:

- TP: True Positive
- TN: True Negative
- FP: False Positive
- FN: False Negative

7.2 AHI Apnea Event Detection

"AHI" is considered as the number of "Apnea Event" (represented in the Arduino code as apneaCount) detected in 60 min so, after the detection of "Apnea Events" basing on the SpO_2 and "AirFlow" sensor; the total number of events will be calculated in order to have the "apneaCount", than "apneaCount" is compared to the previously mentioned ranges in order to classify the severity of Apnea that the patient is facing. The algorithm of this implementation is as follow:

- If: $5 \leq \text{apneaCount} < 15$:

  In this case, the "Apnea Event" is classified as: Mild Apnea.

- If $15 \leq \text{apneaCount} < 30$:

  In this case, the "Apnea Event" is classified as: Moderate Apnea.

- If $\text{apneaCount} \geq 30$:

  In this case, the "Apnea Event" is classified as: Severe Apnea (Fig. 5).

Fig. 5 Sleep apnea detection

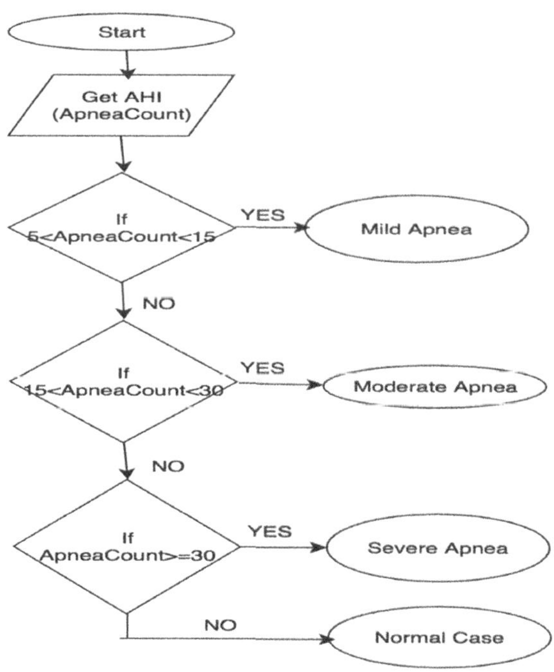

8 Conclusion

In this chapter, we have addressed the basic steps in an innovative design thinking of a vital signs monitoring device. Globally be related to Medical Biometrics, the topics of this chapter are general citing of some examples which are non exhaustive. Moreover, the block diagram and the proposed architecture design are generally applicable to any medical device design. The current trend shows that mobile application, communication modules and advances algorithms software are more integrated and implemented in portable monitoring devices. This chapter is a small introductory to the innovative thinking design to the proof of concept.

References

1. Device Discovery and Concept, The Device Development Process, FDA, US Food and Drug Administration. https://www.fda.gov/
2. V. Kamat, Pulse oximetry. Indian J. Anaesth. **46**, 261–268 (2002)
3. Sleep Apnea Guide, Symptoms, Causes, Cures and Treatment Options [Online]. Available: http://www.sleep-apnea-guide.com/sleep-apnea-oxygen-level.html
4. Ganong's Review of Medical Physiology (24 edn.), p. 619. ISBN 0071780033
5. S.L. DeBoer, *Emergency Newborn Care* (Trafford Publishing, 4 Nov 2004), p. 30. ISBN 978-1-4120-3089-2. [self-published source]
6. W.Q. Lindh, M. Pooler, C. Tamparo, B.M. Dahl, *Delmar's Comprehensive Medical Assisting: Administrative and Clinical Competencies* (Cengage Learning, 9 Mar 2009), p. 573. ISBN 978-1-4354-1914-8
7. A. Rodríguez-Molinero et al., Normal respiratory rate and peripheral blood oxygen saturation in the elderly population. J. Am. Geriatr. Soc. **61**(12), 2238–2240 (2013)
8. "Healthcare 4 home", [Online] Available: https://www.healthcare4home.com/oximeter-sleep-monitor/p.html
9. "EHealth Biometric Sensor platform using arduino and raspberry-pi" [Online] https://www.cooking-hacks.com/documentation/tutorials/ehealth-biometric-sensor-platform-arduino-raspberry-pi-medical
10. "www.raspberrypi.org", [Online] Available: www.raspberrypi.org/products/raspberry-pi-2-model-b/)
11. "Bluetooth", Bluetooth SIG, Inc, [Online] Available: www.bluetooth.com
12. L. Goleniewski, *Telecommunications Essentials: The Complete Global Source for Communications*, 2002

From Medical to Security Biometrics: Some Case Studies

Dalila Cherifi, Kamel Aloui and Amine Nait-Ali

Abstract This chapter focuses on some study cases involving medical image processing. The first case study specifically describes one kind of medical Biometrics applied on mammography images. The same concept is then extended to security Biometrics particularly on hidden Biometrics.

1 Introduction

Through some case studies, this chapter aims to highlight the link between what it is called Medical Biometrics, and its extension to Security Biometrics field. Generally speaking, Medical Biometrics is known to be a field in which collected data (e.g. biosignals, medical images, etc) are processed in order to measure some useful parameters related to pathologies and disorders. Moreover, some measurements can also be used to understand and analyze some biomedical phenomena that can be used to improve some medical devices such as prosthesis. Actually, some scientific communities don't explicitly use the term "Medical Biometrics" and, instead they prefer to use "Biomedical engineering applications". Actually, we would like to say that Medical Biometrics is a restricted field, strongly related to measurements, analysis and identification of abnormal medical cases.

In this chapter, we will focus basically on some study cases involving medical image processing. In particular, the first study case (Sect. 2) describes a kind of Medical Biometrics, applied on mammography images acquired by low-energy X-ray devices. The purpose is not to describe in details all the steps of images pro-

D. Cherifi
Institute of Electrical and Electronic Engineering,
University of Boumerdès, Boumerdès, Algeria

K. Aloui
LTSIRS, Lab., Tunis, Tunisia

A. Nait-Ali (✉)
LiSSi, University-Paris-Est (UPEC), Vitry-sur-Seine, France
e-mail: naitali@u-pec.fr

A. Nait-Ali (ed.), *Biometrics under Biomedical Considerations*,
Series in BioEngineering, https://doi.org/10.1007/978-981-13-1144-4_13

cessing, but the idea here is to highlight the fact that the aim of such processing is to measure and quantify what is called a "mass lesions". As a second Medical Biometrics application (Sect. 3), cardiac Magnetic Resonance Imaging (MRI) is considered to detect eventual myocardial infarction on cardiac cine MRI data by estimating the myocardium thickness. Afterwards, as a third study case (Sect. 4) brain MRI is considered on volumetric images (3D) where the aim is to apply a basic clustering approach, which may be useful to distinguish between normal and abnormal tissues. Finally, in Sect. 5, we will show that the classical Medical Biometrics can be extended to some Security applications, through what it is called Hidden Biometrics. The border between the Medical Biometrics and the Security Biometrics is that in Medical Biometrics, the purpose is to identify pathologies, disorders, etc. However, once extended to Security Biometrics, and more specifically to Hidden Security Biometrics, the aim is to process medical data for individual identification and verification (authentication).

2 Medical Biometrics in Mammography

The majority of early detection in the breast is indicated by presence of calcifications in the breast tissue. In mammogram images, these calcifications appear as mass lesions, macrocalcifications or microcalcifications. The variation of calcifications in size, shape, position and density makes their detection very difficult for the radiologist.

A mass is defined as a space-occupying lesion seen in at least two different projections [1]. Masses are described by their shape and margin characteristics. They are characterized by their type and distribution properties. The detection of masses can be hampered by the wide diversity of their shape, size and subtlety. The tumoral masses present themselves as thickenings, which appear on images as lesions with a size ranging from 3 to 30 mm. Indeed there are also masses with size greater than 30 mm but they are quite obvious. The lesions can vary considerably in optical density, shape, position, size and characteristics at the edge. In addition, the visual manifestation in the mammogram of the shape and edge of a lesion does not only depend on the physical properties of the lesion, but is also affected by the image acquisition technique and by the projection considered. A mass may appear round or oval, according to the projection, because other normal architectural structures of the breast could be superimposed on the lesion.

Mammogram images necessitate that the quality of the image to be interpreted should be free from noise and other aberrations. Thus it is important to perform preprocessing operations on the image so that the resultant preprocessed image is better suited for interpretation. To achieve this goal, image enhancement is most suitable. Enhancement refers to accentuation or sharpening of image features, such as contrast, boundaries, edges, etc. However, the process of image enhancement, in no way increases the information content of the image data.

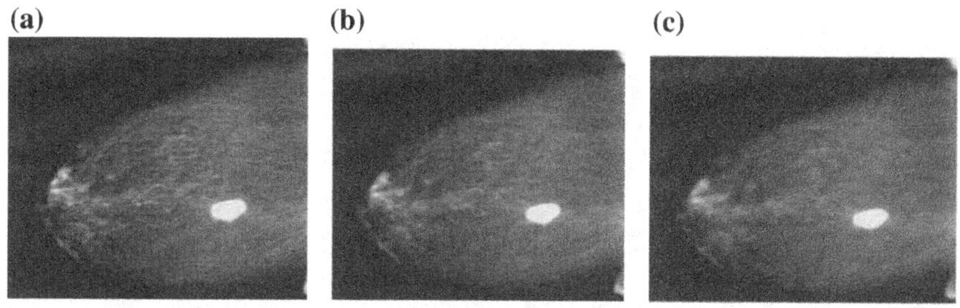

Fig. 1 Image filtered by different median filter windows **a** 3 × 3, **b** 5 × 5, **c** 7 × 7

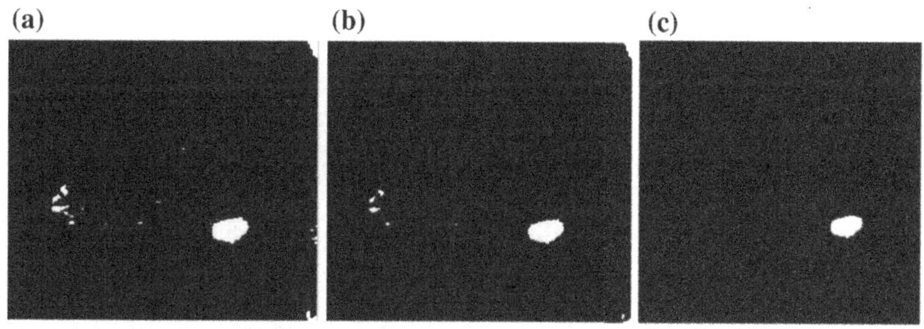

Fig. 2 Thresholded image by three different values using bi-level thresholding

The first step in the analysis of mammogram images involves enhancement or denoising of the image. The denoising technique should not deteriorate or destroy the information content in the image. The objective is to remove background noise while preserving the edge information of suspicious areas in the images.

There are several techniques of mammogram de-noising such as Kuan filter, Mean filter, Lee filter, Median filter etc. In our case we will use median filter.

As we can notice from Fig. 1, a median filtering is achieved in order to smooth the mammorphy image.

This operation is followed by a thresholding in order to identify the mass (i.e. separate the mass from the background). An appropriate threshold value can be calculated from the histogram to highlight the calcifications (masses). Very often, finding the optimal threshold is one of the most difficult tasks (Figs. 2 and 3).

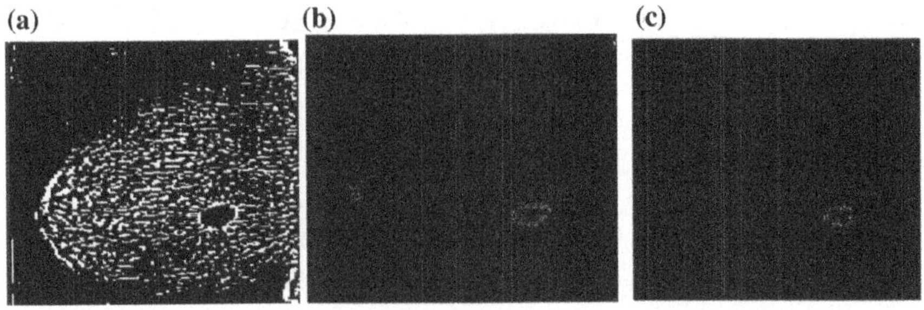

Fig. 3 Thresholded image using entropy based thresholding (three different values)

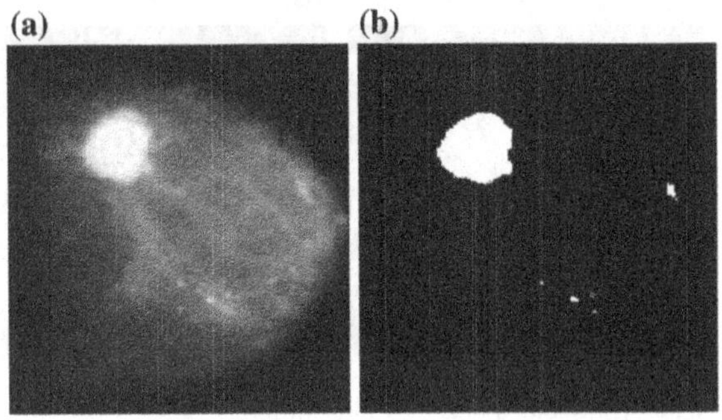

Fig. 4 a Original mammogram image, b thresholded image

In some cases, thresholding is useless if it is used alone for detection because it cannot separate completely the mass from the background (as shown in Fig. 4) especially when the mass is associated with other types of calcifications such as microcalcifications.

3 Medical Biometrics for Cardiac MRI

Medical image analysis is a challenging and important process used for a range of purposes such as segmentation of different organs and tumor quantification [2, 3]. In this part we describe an algorithm for automatic segmentation of the left ventricle in order to detect eventual myocardial infarction on cardiac cine MRI data by estimating the myocardium thickness. We investigated deformable models and their performance in cardiac MRI data to detect myocardial borders as accurately as possible in order to extract some quantitative measurements of myocardial infarction.

Fig. 5 Gradient vector flow of the edge map

(a) (b)

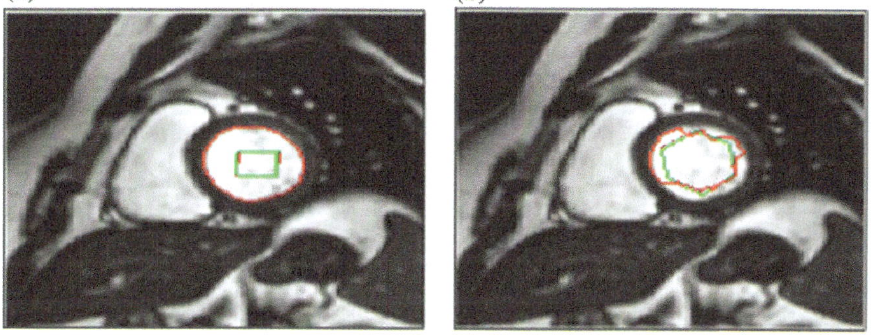

Fig. 6 The convergence of the GVF- snake to the endocardial borders of **a** pre-processed image, **b** initial original image

3.1 Detecting Myocardial Borders

Gradient Vector Flow (GVF) snake (Fig. 5) is often used [4–7] in MRI images in order to detect myocardial borders. In presence of noise, it happens that the snake converges to wrong boundaries. To overcome this problem, Gaussian filter and edge detector filters can be employed to improve the results, as it is illustrated in Fig. 6.

In this Figure, the snake is initialized within the left ventricle to trace the endocardial borders. In particular, it is highlighted two different cases where the snake deformation is applied on both the original image and the pre-processed image.

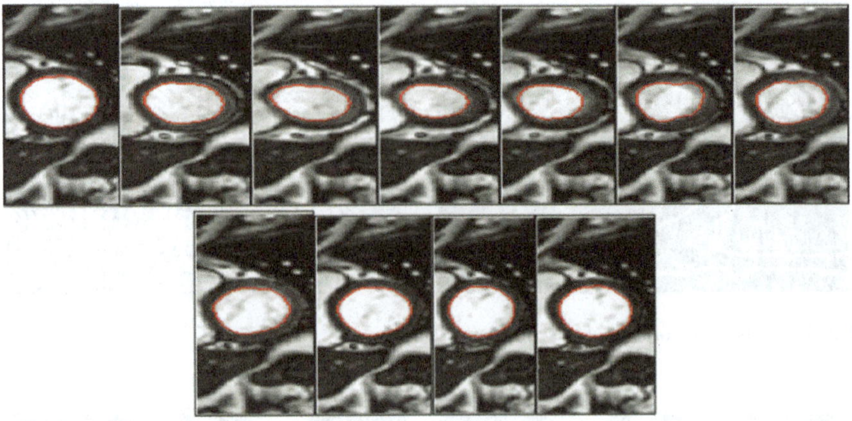

Fig. 7 A multislice segmentation of the endocardial surfaces of the left ventricle

Fig. 8 Variation of the inner wall of the LV area with respect to time

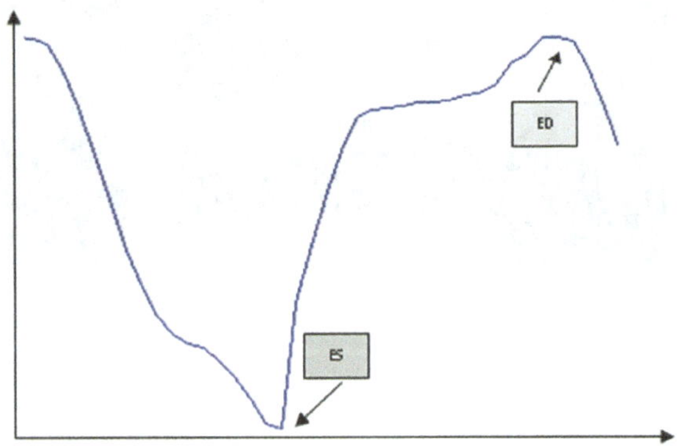

An automatic algorithm independently segment each slice to form a layered collection of contour lines which may be suitably interpolated to approximate the myocardial surface. First, the snake is initialized within the left ventricle in the first slice of the sequence. When the snake converges to the endocardial borders, the surface is computed, stored and became the initial snake of the next slice and so on until sequence is completed (Fig. 7).

At the end, the variation of the LV surface with respect to time (Fig. 8) is obtained. Consequently, the surfaces of the end-systolic and end-diastolic of the LV endocardial can be easily determined. It represents the minimal and maximal area of the LV respectively. In this example, the minimum (systole) corresponds to the 18th frame in the sequence, and the maximum (diastole) to the 35th frame [8].

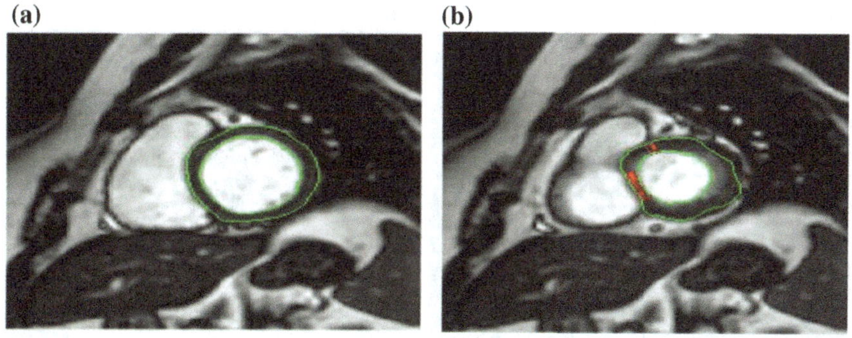

Fig. 9 The contours extracted from the endocardial and epicardial borders of the LV using snakes-GVF, superimposed with the original images: **a** at end diastole, **b** at end systole

3.2 Estimating the Wall Thickness and Motion

To determine the myocardium thickness, one can delineate the outer wall (epicardium). Since the areas of the LV at end diastole and end systole are supposed to be detected, corresponding frames are used to compute the snake and delineate the outer wall. By subtraction, the myocardium thickness is obtained (see Fig. 9).

Note: in some cases, the epicardial surface is not easily distinguished by local edge strength alone, due to the presence of dark regions in the surrounding. These dark regions produce weak edges around the heart. Corrections can be considered by increasing the rigidity, in order to avoid the bending of the curve or convergence to the wrong edges and delineate the borders as accurately as possible. The infarcted tissues are characterized by their inability to contract and dilate. Otherwise the myocardium thickness remains constant during all the cardiac cycle where the tissues are damaged. This is illustrated in Fig. 9, the cross hatched regions determines the infarcted tissues.

3.3 A Case Study

To investigate the approach described above which consists in quantifying and detecting myocardial infarction by estimating the myocardium thickness using snakes-GVF, it is considered a specific pathological case of a 72 years man attained by a myocardial infarction.

By running the algorithm described above on this cine MRI, the variation of the area during the heart cycle, and after determining the end systole and end diastole, are obtained. The infarcted regions can be determined as illustrated in Fig. 10. Using a contrast agent (Gd-DTPA), one can point out that the necrosis is located in the inferior region of the LV. The attained region is pointed to by an arrow, obtained using the snakes-GVF segmentation.

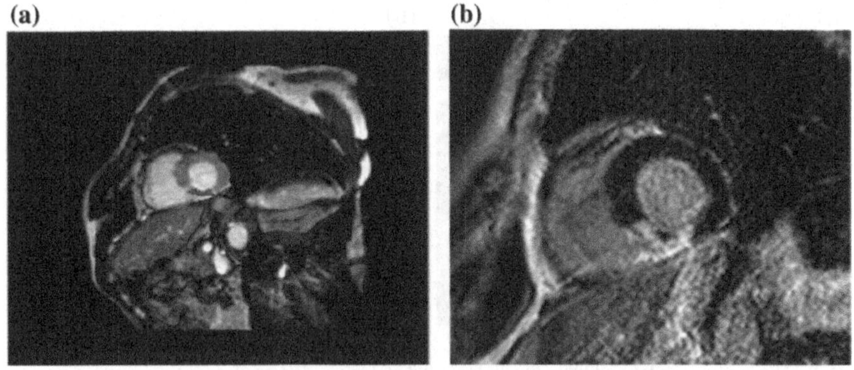

Fig. 10 Detection of myocardial infarction: **a** using snakes-GVF on MRI images of a man, 72 years old, attained by that disease in the inferior part of the left ventricle (shown by the cross hatched region), **b** using a contrast agent (Gd-DTPA) on the same MRI images (it is pointed to by an arrow)

4 Medical Biometrics in 3D Brain MRI

The human brain can be studied using many different medical imaging techniques. Many imaging systems are capable of producing very detailed 3D anatomical information about the subject. One such system is the MRI. This technique is widely used as a noninvasive diagnostic tool, since it provides excellent spatial resolution, and is very efficient in distinguishing between soft tissues. The digital information, thus obtained, can be automatically processed. Nevertheless, the evaluation of this data still requires expert visual inspection evaluation from medical doctors. Segmentation of the medical imagery is a challenging problem due to the complexity of the images. The brain has particularly a complex structure and its segmentation is very important step for tissues recognition, change detection of the morphology and visualization for surgical planning.

In numerous applications, the purpose is to detect, segment, extract, classify and measure properties of normal brain tissues and abnormal cases (i.e. tumor tissues). Therefore, several clustering methods are commonly implemented for tumor extraction. The aim of clustering is to reduce the amount of data by categorizing similar data items together. Such grouping is pervasive in the way humans process information. Clustering is hoped to provide an automatic tool for constructing categories in data feature space. For instance, one can cite the well known K-means algorithm, or Fuzzy K-Means algorithm or classification approaches based on EM segmentation [9].

Let's consider here, K-means clustering. This a method of cluster analysis which aims to partition n observations into k clusters in which each observation belongs to the cluster with the nearest mean. It is similar to the expectation-maximization algorithm for mixtures of Gaussians in that they both attempt to find the centers of natural clusters in the data [10]. The K-means algorithm is a partitional clustering algorithm, which is used to distribute points in feature space among a predefined

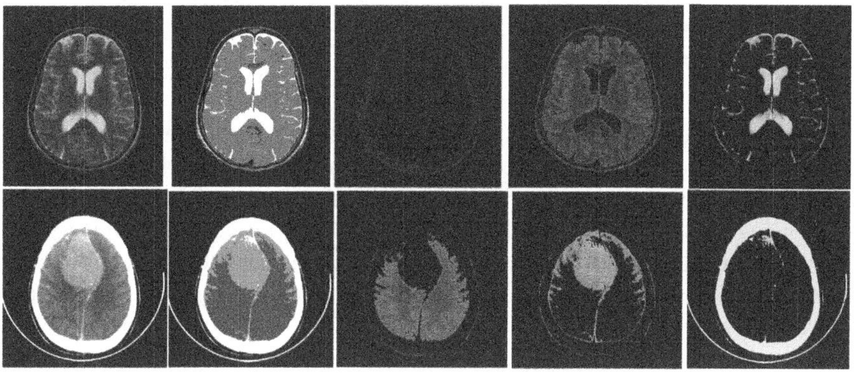

Fig. 11 K-means algorithm applied on a set of CT and MRI slices

number of classes [11]. Clustering each point is assigned to the cluster whose center (also called centroid) is nearest. The center is the average of all the points in the cluster that is, its coordinates are the arithmetic mean for each dimension separately over all the points in the cluster. An implementation of the algorithm applied in pseudocode can be summarized as follows [12–14]:

(1) Choose the number of clusters, k.
(2) Randomly generate k clusters and determine the cluster centroids m_k(centers of each class), or directly generate k random points as cluster centers.
(3) Assign each point S to the nearest cluster centroids.

$$C(s) = agmin\|y_s - m_k\|^2, S = 1, \ldots, N \tag{9}$$

where

C(s) is the class to which the pixel is assigned
Y_s grey level of the pixel s

(4) Recompute the new cluster centroids.

$$m_i = \frac{\sum_{j:N_i} x_j}{N_i}, \quad i = 1, \ldots, K \tag{10}$$

(5) Repeat the two previous steps until some convergence criterion is met (usually that the assignment hasn't changed).

As an illustration, K-means algorithm applied on a set of CT and MRI slices are shown in Fig. 11.

By processing the different slices MRI slices, tumor can be detected, characterized, and measured in terms of surface, volume or evolution over time. This measurement is a part of Medical Biometrics which allows to identify accurately the pathology.

5 Extension of Brain MR Image Processing to Security Biometrics

As it will be seen in this section, processing Brain Magnetic Resonance Images is not restricted exclusively to medical applications, or more specifically to Medical Biometrics. Actually, recent researches show that Brain MRI can be used as a challenging modality in Security Biometrics. In other words, the purpose is to identify/verify the identity of individuals; instead identifying pathologies as it is the case in Medical Biometrics. Regarded as "Hidden Biometrics", this approach can be used to prevent forgeries, for both identification of verification. It is considered as a very robust anti-spoofing modality. This approach differs from the common modalities (e. g, fingerprint recognition, iris, face, etc) by the fact that it is not subject to external or internal modifications, except in pathological cases that are not considered here. Moreover, only adult and healthy individuals are retained. Actually, even if nowadays, this technique seems to be somewhat complex to use in identification routines; our aim consists basically in validating the feasibility of this new biometric tool though a certain number of experiences.

Within this context, the extracted features from the brain MR images constitute what is called "Brainprint", or "Braincode". Brainprint can be used as a signature for recognition purposes. Actually, by analyzing human brain shapes, one can notice that brain shape and folds are discernibly different among individuals. Therefore, Brainprint (Braincode) can be extracted using geometric and textural features.

To understand this context, let's consider the brain from the macroscopic point of view, where it has been mentioned in [16, 17] that it different stages of development to pass from a completely smooth brain to a strongly wrinkled brain. According to the appearance date of the folds, their shapes and their variabilities: three types of cortical folds can be distinguished, namely:

(1) Primary folds: characterized by low inter-individual variability, visible from the 16th week of gestation [16],
Secondary folds: characterized by intermediate level variability occurring towards the thirty-second week of gestation, establishing the degree of gyrification of the cerebral cortex
(3) Tertiary folds: characterized by great inter-individual variability, develop towards the thirty-sixth week of gestation.

Thus, two brains will never have the same shape and same fold patterns.

Moreover, numerous anatomical studies of the brain and some works on the cerebral asymmetry consider a brain partitioning into different areas which are often delineated by folds and sulci, showing a significant inter-individual variability of the brain morphology (Fig. 12). From an individual to another, brain sulci differ in their positions, shapes and numbers of components [15, 18]. In addition, studies have shown that the two hemispheres of a same individual aren't necessarily symmetrical. This inter-hemispheric variability allows that the shape and Sulco-gyral patterns of the brain is unique for each individual [19–21]. Several researches have also proved

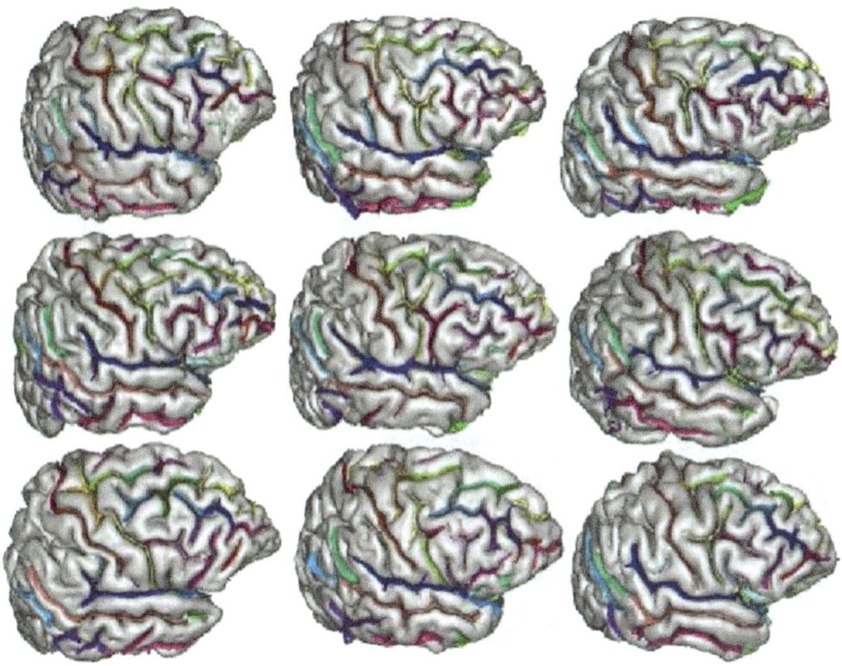

Fig. 12 Anatomical variability of Sulco-gyral patterns for 9 different brains. Each color corresponding to the same sulci [15]

that macroscopic structures of the brain are relatively stables for a period of life [22–24].

Technically speaking, the biometric system can fellow the generic scheme, as described in Fig. 13. By considering the simplest scenario, one brain MR image is considered. From this slice, a Region of Interest is defined (Fig. 14). Using the same algorithm (Daughman's algorithm [25]) as the most common one used for Iris recognition, Braincode is extracted. Consequently, this signature can be used for both identification or verification [26–30]. Within the same context, the reader can refer to [31] in which an advanced technique is proposed by considering the extraction of the Braincode from a 3D volumetric image.

6 Conclusion

In this chapter, the aim was to highlight the border between the Medical Biometrics, and Security Biometrics (through Hidden Biometrics) when it comes to process Medical data. Basically, we pointed out the fact that medical data can be used for security purpose. The most important difference between both applications is the final

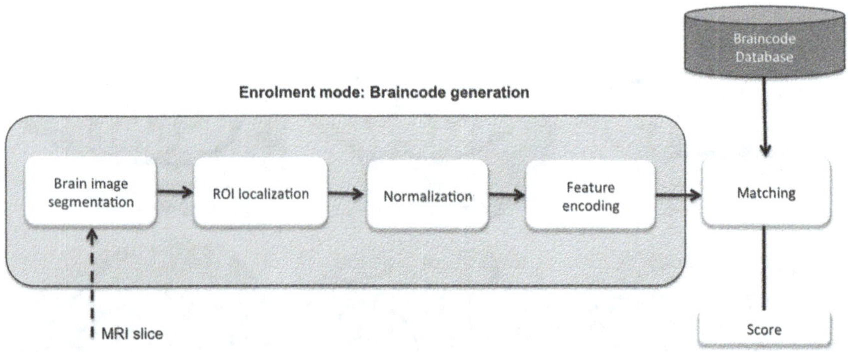

Fig. 13 Operation principle of brain biometrics system

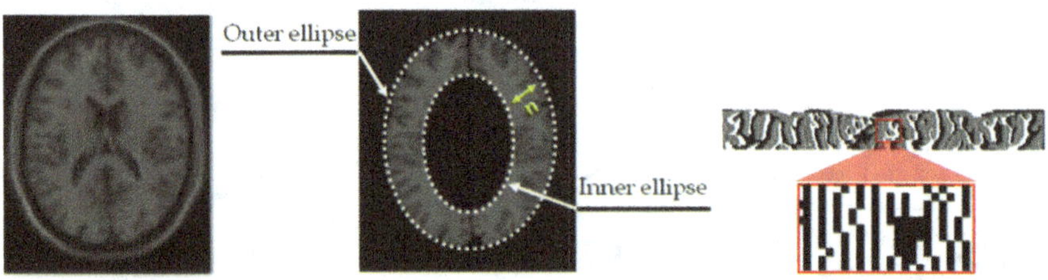

Fig. 14 (Left) Original Brain MRI slice, (middle) region of Interest (ROI) extraction, (right) Braincode (Brainprint) extraction using Daugman's algorithm

processing step. Obviously, both applications share, almost the same generic scheme. In Medical Biometrics, the measurements are quantified to identify pathologies (we saw an example of mass lesions identification in mammography), however in Security Biometrics, only normal subjects are considered. Therefore, the purpose is to identify or to verify the identity of individuals, excluding any potentials pathology. Finally, we would like to mention that within this topics, biosignals can also be considered for both applications. For instance, this may concern, electrocardiogram (ECG), Electroencephalogram (EEG), etc.

References

1. M.P. Sampat et al., Computer-aided detection and diagnosis in mammography. Handbook of image and video processing **2**, 1195–1217 (2005)
2. T. Acharya, A.K. Ray, *Image Processing: Principles and Applications* (Wiley, New York, 2005)
3. B.R. Wilcox et al. *Surgical Anatomy of the Heart*: *The Postnatal Development of the Human Cerebral Cortex*. J Am Med Assoc, vol. 113 (Cambridge University Press, 2005), p. 1156
4. M. Kass et al., Snakes: active contour models. Int. J. Comput. Vis. **1**, 321–331 (1988)
5. L.D. Cohen, On active contour models and balloons. CVGIP: Image Underst. **53**, 211–218 (1991)

6. L.D. Cohen, I. Cohen, Finite-element methods for active contour models and balloons for 2-D and 3-D images. IEEE Trans. Pattern Anal. Mach. Intell. **15**, 1131–1147 (1993)
7. Xu Chenyang et al., Snakes, shapes, and gradient vector flow. IEEE Trans. Image Process. **7**, 359–369 (1998)
8. D. Cherifi, M. Djohor, N. Benamer, A. Debiane, Myocardial infarction detection by estimating the wall thickness and motion using snakes-gvf, in *10th Algerian-French Radiology Days and National Congress*, 4–5 June 2010 in Algiers
9. D. Cherifi, M. Zinelabidine, A. Nait-Ali, Z. Aici, Abnormal Tissue Extraction in MRI Brain Medical Images, in *7th International Workshop on Systems, Signal Processing and Their Applications* (WOSSPA), Tipaza, Algeria, 9–11 May 2011
10. http://en.wikipedia.org/wiki/K-means_algorithm
11. M. Ramze, C. Rezee, P.M. Nyqvist, J. Van Der Zwet et al. Segmentation of MR images by a fuzzy c-mean algorithm, in *Computers in Cardiology* (IEEE, 1995), pp. 21–24
12. J.C. Bezdek, *Pattern Recognition with Fuzzy Objective Function Algorithms* (Plenum Press, New York, 1981). ISBN 0306406713
13. J.C. Bezdek, *Partition Structures: A Tutorial, Analysis of Fuzzy Information*, vol. 3 (CRC Press, Boca Raton, 1987), pp. 18–108
14. J. Shi, J. Malik, Normalized Cuts and Image Segmentation, in *Conference on Computer Vision and Pattern Recognition* (IEEE, 1997), pp. 731–737
15. Y. Cointepas et al., BrainVISA: Software Platform for Visualization and Analysis of Multi-Modality Brain Data. Neuroimage **13**, 98 (2001)
16. J. Le, The Postnatal Development of the Human Cerebral Cortex. The cortex of the newborn, vol. 1 (Harvard University Press, Oxford, England, 1939)
17. W. Welker, Why Does Cerebral Cortex Fissure and Fold?, in *Cerebral Cortex*, ed. by E.G. Jones, A. Peters (Springer, US, 1990), pp. 3–136
18. C. Renault, Courbures et lignes de crête sur des images en niveaux de gris : Etude comparative et application aux sillons corticaux (Caen) (2001)
19. J. Dubois et al., Mapping the early cortical folding process in the preterm newborn brain. Cereb. Cortex **18**, 1444–1454 (2008)
20. J.J. Hutsler et al., Individual variation of cortical surface area asymmetries. Cereb. Cortex **8**, 11–17 (1998)
21. P.M. Thompson, C. Schwartz, R.T. Lin, A.A. Khan, A.W. Toga, Three-dimensional statistical analysis of sulcal variability in the human brain. J. Neurosci. **16**, 4261–4274 (1996)
22. F.H. Gage, A.R. Muotri, What makes each brain unique. Sci. Am. **306**, 26–31 (2012)
23. J. Régis et al., 'Sulcal root' generic model: a hypothesis to overcome the variability of the human cortex folding patterns. Neurol. Med. Chir. **45**, 1–17 (2005)
24. D.C. Van Essen, A tension-based theory of morphogenesis and compact wiring in the central nervous system. Nature **385**, 313–318 (1997)
25. J. Daugman, How iris recognition works. IEEE Trans. Circuits Syst. Video. Technol. **14**, 21–30 (2004)
26. A. Nait-Ali, Beyond Classical Biometrics: When Using Hidden Biometrics to Identify Individuals, in *3rd European Workshop on Visual Information Processing* (2011a)
27. A. Nait-Ali, Hidden Biometrics: Towards Using Biosignals and Biomedical Images for Security Applications, in *International Workshop on Systems, Signal Processing and Their Applications* (WOSSPA), Tipaza, Algeria, 9–11 May 2011
28. K. Aloui et al., New Biometric Approach Based on Geometrical Human Brain Patterns Recognition: Some Preliminary Results, in *3rd European Workshop on Visual Information Processing* (2011a)
29. K. Aloui et al. A Novel Approach Based Brain Biometrics: Some Preliminary Results for Individual Identification, in *Workshop on Computational Intelligence in Biometrics and Identity Management* (CIBIM) (IEEE, 2011)

30. K. Aloui et al., A new useful biometrics tool based on 3D brain human geometrical characterizations. J. Sig. Inf. Process. **03**, 198–207 (2012)
31. Aloui K et al., Using Brain Prints as New Biometric Feature for Human Recognition. Pattern Recogn. Lett. (2017) https://doi.org/10.1016/j.patrec.2017.10.001

The manufacturer's authorised representative in the EU is Springer Nature Customer Service Centre GmbH, Europaplatz 3, 69115 Heidelberg, Germany. If you have any concerns regarding our products, please contact ProductSafety@springernature.com

Printed and bound by CPI Group (UK) Ltd, Croydon, CR0 4YY
28/09/2026
02224086-0007